Two Manuals Combined into One Book

Cavalry Troop

And

MK 19, 40-MM Grenade Machine Gun

Department of Defense

Pentagon
Publishing

Field Manual
No 17-97

HEADQUARTERS
DEPARTMENT OF THE ARMY
Washington, DC 3 October 1995

CAVALRY TROOP

TABLE OF CONTENTS

*This publication supersedes FM 17-97, September 1988.

Preface

This publication serves as a doctrinal guide for commanders and subordinate leaders of both light (HMMWV equipped) and heavy (M1 and M3 equipped) cavalry troops. Heavy cavalry troops are found in Armored Cavalry Regiments (ACR), in Armor/Mechanized Divisions, and in Separate Armor/Mechanized Brigades. Light cavalry troops are found in Light Armored Cavalry Regiments (LACR), in Light Infantry Divisions, and in Separate Light Infantry Brigades.

Although FM 17-97 does not specifically address the TOE of the Light Division Cavalry Troop or the troops of a Separate Heavy or Light Brigade, the tactics, techniques, and procedures outlined in this publication still apply.

This publication lays out the organization, command and control, tactical employment, and service support of the cavalry troop in combat. It establishes the responsibilities and duties of key personnel in the troop during combat.

FM 17-97 is one of three publications in the cavalry troop family. It provides troop leaders with a doctrinal reference to link training and fighting. ARTEP 17-487-30-MTP, *Mission Training Plan for the Regimental Armored Cavalry Troop*, outlines how to train the troop. FKSM 17-97-3, *Cavalry Troop Common SOP*, describes troop operating procedures. This family of publications supports FM 17-15, *Tank Platoon*; FM 7-91, *Tactical Employment of Antiarmor Platoons, Companies, and Battalions*; and FM 17-98, *Scout Platoon*.

The information in this manual is based on the objective table of organization and equipment (TOE) for the ACR and LACR troops. Differences between actual unit organizations and equipment and those described in this manual are caused by differences in the modified table of equipment and organization (MTOE).

The proponent for this publication is HQ TRADOC. Submit changes for improving this publication on DA Form 2028 (Recommended Changes to Publications and Blank Forms), and forward it to Commandant, USAARMS, ATTN: ATSB-SBA-F, Fort Knox, KY 40121-5211.

Although this manual does not implement any particular international agreement, the material presented herein is in accordance with related international agreements. A list of related international agreements and other references can be found in the References section. The paragraph in Chapter 2, Section I, entitled Issue a Warning Order is in compliance with STANAG 2014. Chapter 7, Section III, Relief in Place, and Section IV, Battle Handover and Passage of Lines, are in compliance with STANAG 2082. Tables 7-2 and 7-4 are in compliance with STANAG 2083.

Unless this publication states otherwise, masculine nouns and pronouns do not refer exclusively to men.

Article I. Chapter 1

Section 1.01 Introduction

The role of the cavalry troop in Army operations remains unchanged from the traditional role of cavalry throughout the history of warfare. The troop is organized, equipped, and trained to protect and preserve the fighting ability of other combined arms forces. While its primary missions are reconnaissance and security, the cavalry troop may be called upon to execute attack, defend, and delay missions as part of squadron and regimental missions. *The troop accomplishes its missions by communicating, moving, and shooting in that order.*

The purpose of this chapter is—

- To depict organizations of the cavalry troop in—

  Heavy Troop:

  - The Armored Cavalry Regiment (ACR) (M1 and M3 equipped).
  - The Heavy Division Cavalry Squadron (M1 and M3 equipped).

  Light Troop:

  - The Light Armored Cavalry Regiment (LACR) (HMMWV equipped).

Contents

- To outline missions each troop performs and to highlight each troop's capabilities and limitations.
- To establish responsibilities of key personnel in combat.

(a) Section I. Organizations

(I) THE HEAVY CAVALRY TROOP

The heavy cavalry troop consists of 6 officers and 126 enlisted soldiers. The troop is organized into a headquarters section, two scout platoons, two tank platoons, a mortar section, and a maintenance section (see Figure 1-1).

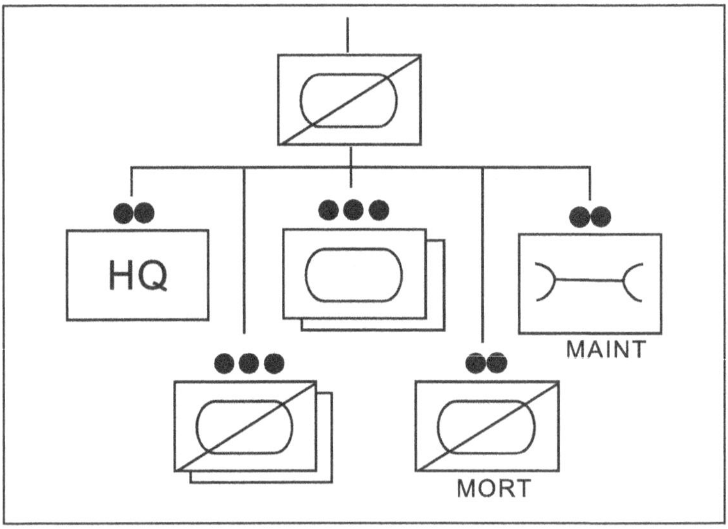

a) Figure 1-1. Heavy cavalry troop organization.

2) The Heavy Troop Scout Platoon

The scout platoon is organized and equipped to conduct reconnaissance and screening in support of its parent troop. However, when the parent unit is performing missions within an economy-of-force role, the scout platoon may conduct offensive, defensive, and retrograde operations in support of the troop mission. The platoon consists of 1 officer and 29 enlisted soldiers, and is equipped with six M3 cavalry fighting vehicles (CFV) organized into three scout sections (see Figure 1-2).

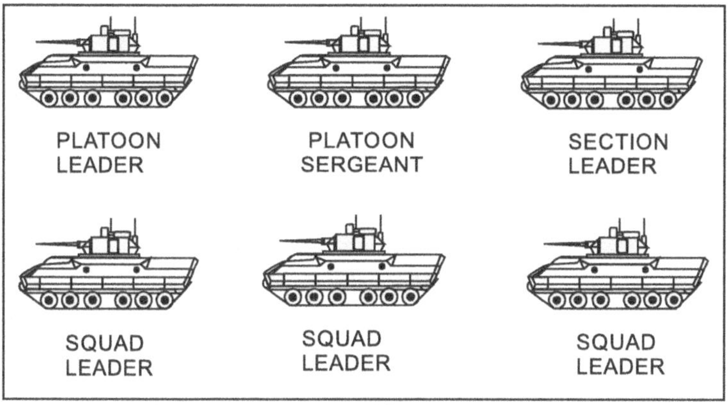

3)

a) Figure 1-2. Scout platoon organization.

4)

5) The Heavy Troop Tank Platoon

The tank platoon is organized and equipped to perform its three primary missions—attack, defend, and move. The platoon consists of 1 officer and 15 enlisted soldiers manning four M1A1 Abrams main battle tanks organized into two sections (see Figure 1-3).

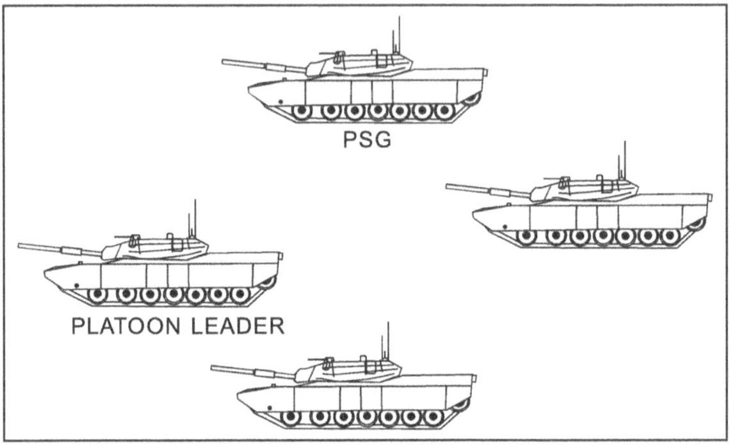

6)

a) Figure 1-3. Tank platoon organization.

7) The Heavy Troop Mortar Section

The mortar section is organized and equipped to provide immediate indirect fires in support of troop operations. Such supporting fires are usually suppression, screening, obscuration, or illumination. The section consists of nine enlisted soldiers. It is equipped with two 107-mm mortars mounted in two self-propelled mortar carriers (see Figure 1-4).

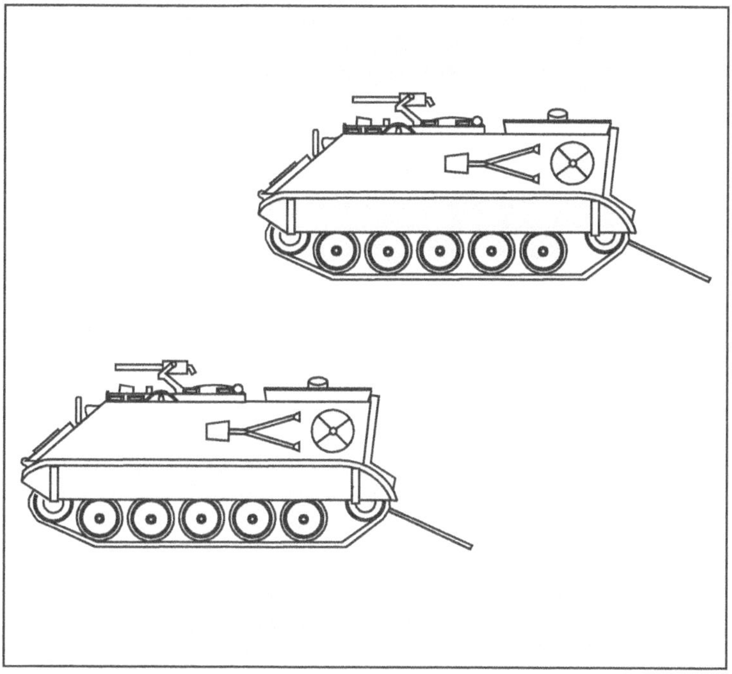

Figure 1-4. Mortar section organization.

8) The Heavy Troop Maintenance Section

This section is organized and equipped to diagnose and repair most equipment faults at troop level. It has the capability to recover all troop vehicles and to maintain the troop's equipment records. The section consists of 18 enlisted soldiers equipped with one armored personnel carrier (APC), one heavy recovery vehicle, one utility truck with cargo trailer, and two cargo trucks with cargo trailers (see Figure 1-5).

a) Figure 1-5. Maintenance section organization.

9) The Heavy Troop Headquarters Section

a)

The troop headquarters section is organized and equipped to perform command and control and logistical support functions for the troop. The section consists of 2 officers and 11 enlisted soldiers. It is equipped with one main battle tank, one command post (CP) carrier, one APC, one M3 CFV, one cargo truck with a 400-gallon water trailer, and two utility trucks with one cargo trailer (see Figure 1-6).

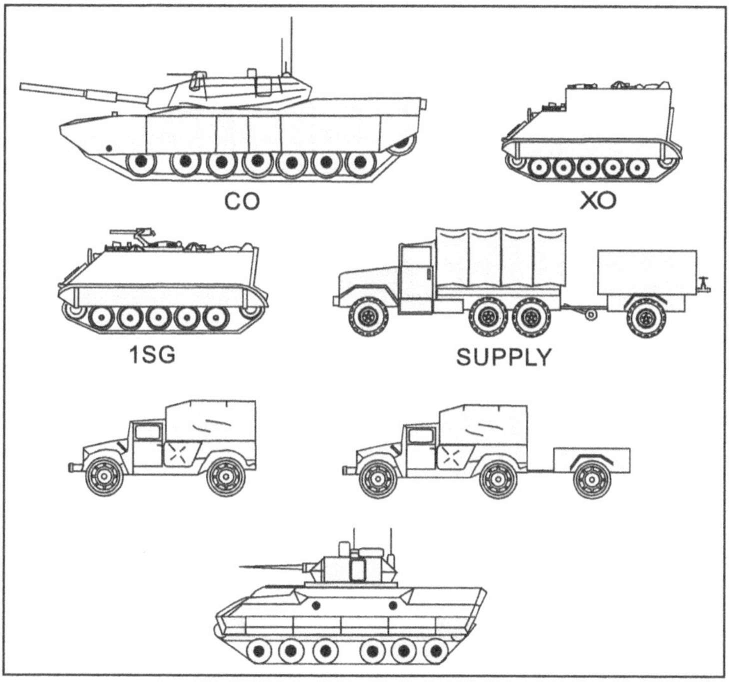

b) Figure 1-6. Headquarters section organization.

10) The Fire Support Team (FIST)

The FIST comes from the squadron howitzer battery in the heavy cavalry regiment and the division artillery in the division cavalry squadron. The FIST consists of one fire support officer (FSO), one NCO (team chief), one enlisted fire support specialist, and one enlisted radio operator. The FIST is responsible for coordinating indirect fires for the troop. In the heavy troop the team is mounted on an M981 FIST vehicle (see Figure 1-7).

11)

a) Figure 1-7. FIST organization.

(II) THE LIGHT CAVALRY TROOP

The light cavalry troop consists of 6 officers and 107 enlisted soldiers. The troop is organized into a headquarters section, two scout platoons, two antitank (AT) platoons, a mortar section, and a maintenance section (see Figure 1-8).

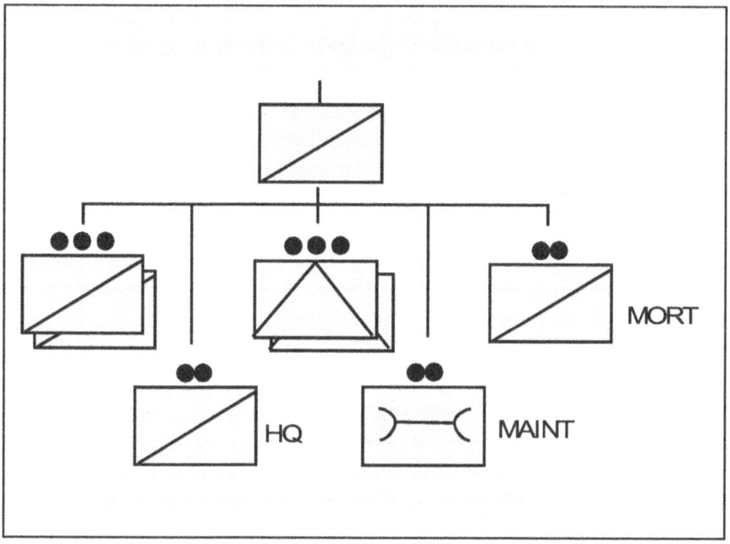

a) Figure 1-8. Light cavalry troop organization.

2) The Light Troop Scout Platoon

The scout platoon is organized and equipped to conduct reconnaissance and screening in support of its parent troop. However, when the parent unit is performing missions within an economy-of-force role, the scout platoon may conduct offensive, defensive, and retrograde operations in support of the troop mission. The platoon consists of 1 officer and 29 enlisted soldiers. It is equipped with ten M1025/M1026 HMMWVs (five MK-19 equipped and five cal .50 equipped). The platoon can organize into various configurations, usually from two to five sections, depending on factors of METT-T (see Figure 1-9). FM 17-98 details the tactics, techniques, and procedures for employing the HMMWV scout platoon.

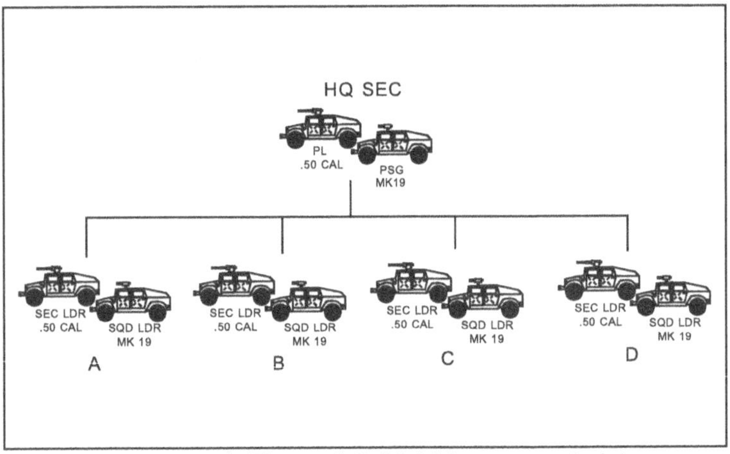

a) Figure 1-9. Light troop scout platoon organization.

3) The Light Troop Antitank Platoon

The AT platoon is organized and equipped to perform its three primary missions—attack, defend, and move. The platoon consists of 1 officer and 11 enlisted soldiers manning four M996 HMMWVs (TOW carriers) organized into two sections (see Figure 1-10).

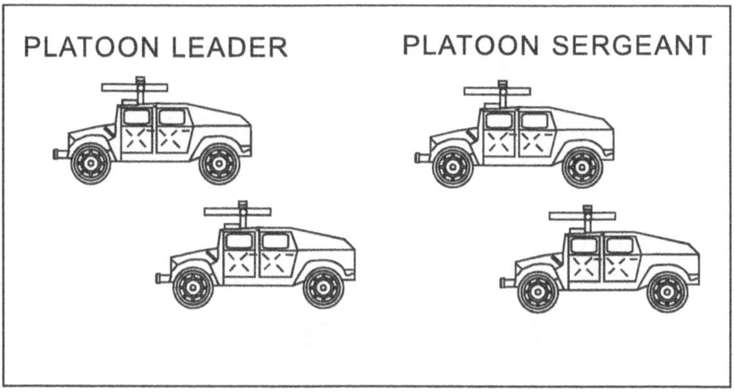

a) Figure 1-10. Light troop AT platoon organization.

4) The Light Troop Mortar Section

The mortar section is organized and equipped to provide immediate indirect fires to support troop operations. Such supporting fires are usually suppression, screening, obscuration, or illumination. The section consists of nine enlisted soldiers. It is equipped with two 120-mm mortars trailered by three M1025/M1026 HMMWVs (see Figure 1-11).

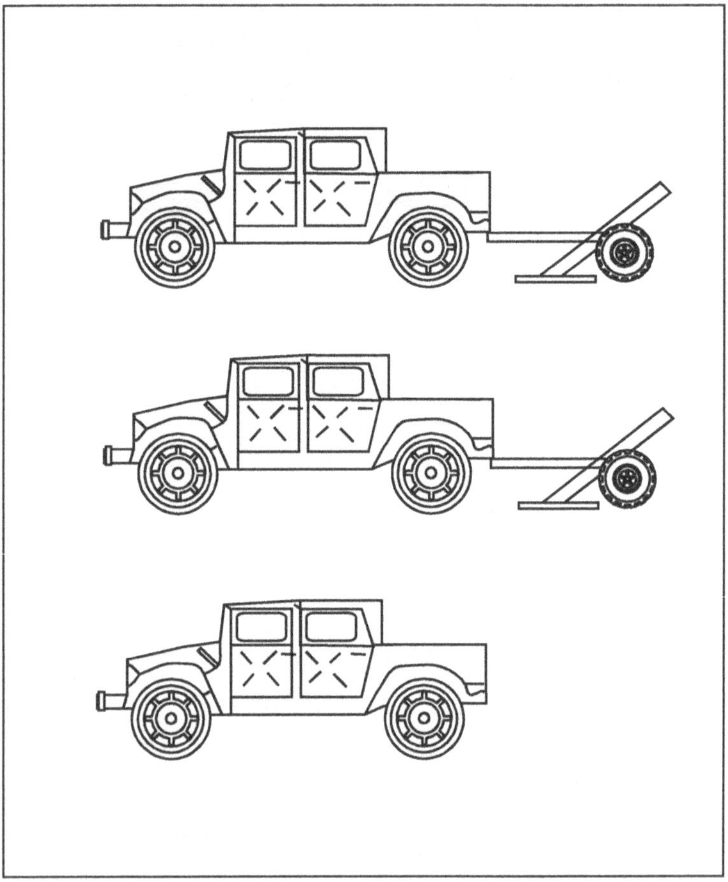

a) Figure 1-11. Light troop mortar section organization.

5) The Light Troop Maintenance Section

This section is organized and equipped to diagnose and repair most equipment faults at troop level. It has the capability to recover all troop vehicles and to maintain the troop's equipment records. The section consists of eight enlisted soldiers equipped with one M998 cargo/troop carrier HMMWV, one cargo truck with cargo trailer, and one 5-ton wrecker (see Figure 1-12).

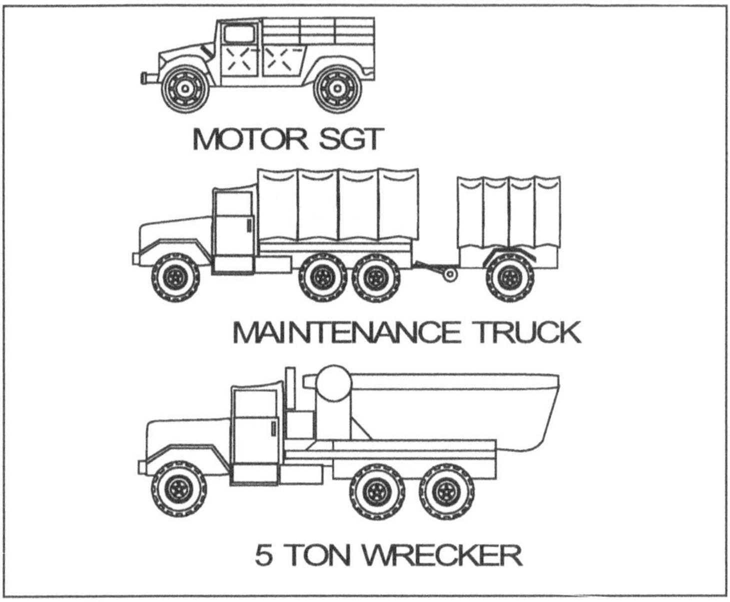

6)

a) Figure 1-12. Light troop maintenance section organization.

7) The Light Troop Headquarters Section

The troop headquarters section is organized and equipped to perform command and control and logistical support functions for the troop. The section consists of two officers and ten enlisted soldiers. It is equipped with one M1025 HMMWV, one M1037 HMMWV shelter carrier with S250 shelter, one M998 HMMWV with cargo trailer, and one 2 1/2-ton cargo truck with 400-gallon water trailer (see Figure 1-13).

a) Figure 1-13. Light troop headquarters section organization.

8) The Fire Support Team (FIST)

The FIST comes from the squadron howitzer battery in the light cavalry regiment. The FIST consists of one officer (FSO), one NCO (team chief), one enlisted fire support specialist, and one enlisted radio operator. The FIST is responsible for coordinating indirect fires for the troop. In the light troop the team is mounted on a HMMWV (see Figure 1-14).

(b)

a) Figure 1-14. Light troop FIST organization.

(c) Section II. Missions, Capabilities, and Limitations

(I) MISSIONS

Heavy and light cavalry troops perform reconnaissance and security missions to protect and preserve the fighting ability of the units to which they are assigned or attached. Both troops also conduct offensive, defensive, and retrograde operations in an economy-of-force role. Cavalry troops in general have limitations and capabilities associated with their TOEs and METT-T that must be considered when employing them in a specific mission role (see Figure 1-15).

	Hvy Trp	Lt Trp
Reconnaissance Missions		
Route Recon	✓	✓
Zone Recon	✓	✓
Area Recon	✓	✓
Security Missions		
Screen	✓	✓
Area Security	✸	✸
Convoy Security	x	x
Economy-of-Force Role		
Offensive Missions		
Hasty Attack	✓	✸
Attack	✓	✸
Movement to Contact	✓	✸
Defensive Missions		
Defend a Battle Position	✓	✸
Defend in Sector	✓	✸
Retrograde Missions		
Delay	✓	✸

✓ = fully capable
x = capable when reinforced
✸ = capable under permissive METT-T

a) Figure 1-15. Cavalry troop mission profiles.

(II) CAPABILITIES

a) The heavy cavalry troop is a unit that can be deployed by rail, sea, or both into a theater of operations. It can perform its missions under all visibility conditions and in any terrain that supports heavy armor movement. The integrated thermal sights on the M1A1 tank and M3 CFV along with ancillary passive night sights and night vision devices authorized to the troop provide an outstanding nighttime/reduced visibility acquisition and fighting capability. Additionally, the firepower and survivability organic to the heavy cavalry troop allow it to execute missions aggressively across the spectrum of warfare.

The light cavalry troop is a unit that can be rapidly deployed by air, rail, or sea into a theater of operations. It can perform its missions under all visibility conditions and in any terrain that supports wheeled vehicle movement. The ancillary thermal and night vision devices authorized to the troop provide an excellent reduced visibility acquisition capability. However, the light cavalry troop's ability to fight under reduced visibility conditions is limited because of its lack of integrated passive and thermal sights on stabilized weapons platforms.

(III)

(IV) LIMITATIONS

a) Given suitable terrain conditions, the ability of heavy and light cavalry troops to accomplish their assigned missions is mainly limited by the size and strength of threat forces encountered. Both troops are limited in close terrain, especially urban environments, due to their limited number of dismounted scouts. Also, the troops are dependent on their parent squadrons and regiments for additional combat support (CS) and combat service support (CSS) assets.

(d)
(e)

(f) Section III. Responsibilities

(I) THE TROOP COMMANDER

The troop commander is responsible to the squadron commander (SCO) for the discipline, combat readiness, and training of the troop, and for the maintenance of its equipment. He must be proficient in the tactical employment of the troop and its assigned and attached CS elements. He must also know the capabilities and limitations of the troop's personnel and equipment as well as those of CS elements attached to him.

a) The troop commander's responsibility in combat is twofold. He will—

- Accomplish all missions assigned to the troop in accordance with the SCO's intent.
- Preserve the fighting capability of the troop.

(II) THE EXECUTIVE OFFICER

In combat, the troop executive officer (XO) is second in command. He supervises the troop tactical operations center (TOC), where he stays abreast of the tactical situation in the troop's area of operations (AO). He manages the flow of combat information between the troop and the squadron from the troop TOC. With the assistance of the troop first sergeant (1SG), he plans and coordinates CSS for the troop.

(III) THE FIRST SERGEANT

The primary responsibility of the troop 1SG is sustaining the troop's ability to fight. He supervises the procurement and distribution of fuel, ammunition, food, water, clothing, equipment, replacements, and repair parts. He receives incoming personnel and assigns them to subordinate elements as needed. He is responsible for the medical evacuation of sick, injured, and wounded soldiers to the supporting medical treatment facility. He is also responsible for the evacuation of soldiers killed in action to the supporting graves registration collection point. He is also responsible for the recovery and evacuation of damaged combat equipment.

(IV) THE TROOP FIRE SUPPORT OFFICER

The troop FSO is responsible for the planning and coordination of the troop fire support plan. He relays troop requests for indirect fire and advises the commander on the employment of indirect-fire weapons in support of the troop's maneuver. The troop FSO may control and position the mortars during combat operations.

(V) THE PLATOON LEADER

The platoon leader is responsible to the troop commander for the discipline, combat readiness, and training of the platoon, and for the maintenance of its equipment. He must be proficient in the tactical employment of the platoon and know the capabilities and limitations of the platoon's personnel and equipment.

The platoon leader's responsibility in combat is twofold. He must—

- Accomplish all missions assigned to the platoon in accordance with the troop commander's intent.

- Preserve the fighting capability of the platoon.

(VI) THE PLATOON SERGEANT

The platoon sergeant (PSG) leads elements of the platoon as directed by the platoon leader, and assumes command of the platoon in his absence. The PSG assists the platoon leader in maintaining discipline, conducting training, and exercising control. He supervises platoon CSS, which includes supply and equipment maintenance.

(VII) THE MORTAR SECTION SERGEANT

The mortar section sergeant is responsible for providing responsive indirect fires to support the troop commander's concept of the operation.

(VIII) THE SUPPLY SERGEANT

The supply sergeant picks up, transports, and issues supplies and equipment to the troop. He works closely with the 1SG to accomplish these tasks. He also evacuates enemy prisoners of war and assists in the evacuation of soldiers who are killed in action to the graves registration collection point.

(IX) THE MAINTENANCE SERGEANT

The maintenance sergeant is responsible for the prompt battlefield repair or recovery of damaged or inoperable equipment. He works closely with the 1SG to accomplish these tasks.

(X) THE COMMUNICATIONS SERGEANT

The communications sergeant ensures the troop TOC and its crew are prepared for combat operations, and assists the XO in the TOC during combat operations. Within his capability, he repairs the communications equipment of subordinate elements.

(XI) THE NUCLEAR, BIOLOGICAL, CHEMICAL (NBC) NCO

The troop NBC NCO is responsible for troop NBC defense activities. He supervises radiological monitoring, chemical detection, and decontamination operations. He assists in maintaining NBC equipment and training NBC equipment operators and decontamination teams.

Chapter 2

Battle Command

Battle command is the art of battle decision making, leading and motivating soldiers and their organizations into action to accomplish missions. Battle command includes visualizing the current state and the future state, then formulating concepts of operations to get from one state to the other at least cost. Other functions of battle command include assigning missions; prioritizing and allocating resources; selecting the critical time and place to act; and knowing how and when to make adjustments during the fight.

Instantaneous response to orders is the hallmark of effective battle command in any cavalry operation. Leaders must quickly analyze information, make tactical decisions, and turn those decisions into successful battlefield actions.

Battle command of cavalry units is typically decentralized due to the size of the area of operations, vagueness of the enemy situation, and terrain unknowns. This places the burden of sound, timely decision making at the lowest levels. Leaders must develop a keen sense of situational awareness and constantly track the actions of subordinate units as well as those to the front, flank, and rear.

Contents

Effective battle command begins in the planning phase and continues through the execution phase of each mission. This chapter outlines the tools and techniques a troop commander needs to effectively command and control a cavalry troop in combat.

Section I. Troop Leading Procedures

The nine troop leading procedures are—

- Receive and analyze the mission.
- Issue a warning order.
- Make a tentative plan.
- Start necessary movement.
- Conduct a reconnaissance.
- Make final decisions and complete the plan.
- Issue an operations order.
- Rehearse.
- Supervise and refine.

RECEIVE AND ANALYZE THE MISSION

The troop commander will receive missions from squadron in the form of written, briefed, or radioed operation orders (OPORD) and fragmentary orders (FRAGO). Upon receipt of a mission, he will conduct a mission analysis to determine the *who*, *what*, *when*, *where*, and *why* elements of the mission and how much time is available until mission execution.

Mission. The mission analysis must identify the following:

- Specified tasks.
- Implied tasks.
- Essential tasks.
- Intent of the higher commander.
- Any constraints or limitations.

The items for analysis are derived from the squadron order as noted below.

- Paragraph 2, MISSION.
  - Mission Statement.
- Paragraph 3, EXECUTION.
  - Concept of the Operation.
  - Commander's Intent.
  - Maneuver.
  - Specific Instructions.
  - Coordinating Instructions.
- Execution matrixes.
- Thorough map reconnaissance.

List all specified and implied tasks not covered by unit SOP in a mission list. Identify those tasks that are vital to the squadron accomplishing its mission. These are essential tasks. Put the essential tasks in the form of a restated mission for the troop. The restated mission statement should be short and simple, and should cover *who, what, when, where,* and *why.*

One final note on this—keep the mission list at hand. It makes the job easier when preparing a course of action and assigning missions or tasks to platoons and sections.

Time. Because time is usually the most limited resource available, it must be used as efficiently as possible. Figure out what has to be done and what can be done in the time available. Use the backwards planning process to determine when critical events in the planning process must occur and stick to the timeline developed. Get the timeline out to subordinates in the warning order. Strive to use no more than one-third of the time available in the planning process at troop level. The remaining two-thirds is for subordinate leaders to plan and prepare.

Consider the following in terms of time when conducting mission analysis and preparing the troop warning order:

- Combat service support.

- Precombat inspections.

- Route reconnaissance (time/distance factors):
  - Assembly area to route start point.
  - Start point to command posts.
  - Command posts to release point.
  - Release point to line of departure.

- Subordinate's planning time.

- Platoon rehearsals:
  - Battle drill rehearsals.
  - Actions on contact.
  - Obstacle breaching/emplacement.
  - Movement formations.

Note. The preparatory actions above are triggered by receipt of a warning order at platoon level.

- Time to issue troop OPORD.

Note. One-third of the available time includes the time it takes to issue the troop OPORD.

- Troop rehearsals:
  - Backbriefs.
  - Rock drills.
  - Radio rehearsals.

- Time to move from—
  - Assembly area to line of departure.
  - Phase line to phase line.
  - Line of departure to objective.

- Effect of weather on movement.

Once the mission and time available have been analyzed, move to the next step in the trooping leading procedures.

ISSUE A WARNING ORDER

Issue the warning order promptly. Elements of the warning order include—

- Situation.
- Mission.
- Earliest time of movement.
- Time and place of OPORD.
- Attachments and detachments.
- Actions triggered:
  - Security.
  - Reconnaissance.
  - Combat service support actions.
  - Precombat inspections.
  - Sleep plans.
  - Rehearsals.

MAKE A TENTATIVE PLAN

Develop a course of action (COA). Consider the factors of METT-T when developing two, or better yet, three separate COAs.

- **Mission.** See *Receive and Analyze the Mission* paragraph on page 2-2.

- **Time.** In addition to the factors mentioned in the *Receive and Analyze the Mission* paragraph on page 2-2, the commander should consider the following details in relation to time when developing his tentative plan.

  - Time to move from—
    Line of departure to phase lines.
    Phase line to phase line.
    Line of departure to objective.

- Time for enemy to—
  - Attack.
  - Counterattack.
  - Reinforce.
- Effect of weather on movement and identification of friend or foe.
- Effect of NBC operations on movement.

• **Enemy.** Conduct intelligence preparation of the battlefield (IPB). IPB integrates enemy doctrine with weather and terrain to determine and evaluate enemy capabilities, vulnerabilities, and probable courses of action. IPB is a key part of preparing for battle. Much of the IPB workup can be found in Paragraph 1 of the squadron OPORD.

• **Terrain.** Analyze terrain in the area of operations. Consider the factors of OCOKA. (Arranged here in a more logical, systematic approach, based on IPB procedures and preparation of the modified combined obstacle overlay [MCOO].)

■ ***Obstacles.*** Identify all obstacles and restricted and severely restricted terrain throughout the sector or zone (a terrain sketch may be beneficial). Consider the following as obstacles:

- Water.
- Power lines.
- Bridges.
- Defiles.
- Slopes.
- Towns.
- Embankments.
- Railroads.
- Existing engineer obstacles.
- Wooded areas.

Terrain that flows around, or in some instances through or over, the obstacles identified is the primary maneuver corridor, or avenue of approach.

■ ***Avenues of Approach.*** Consider the following for each avenue of approach:

- What size force does it support? In what formation?
- Does it support long range observation and fields of fire? From where?
- Does it provide any cover and concealment? Where?
- Where does it lead?
  Through sector/zone?
  Enters/exits a flank?

The answers to these questions will help determine what terrain, if any, is key terrain.

■ ***Key Terrain.*** Key terrain is that terrain which control over would provide a clear advantage over another force. Seizing, securing, or even avoiding key terrain in sector/zone will figure prominently in the final plan. Identify what terrain is key and evaluate it in terms of observation and fields of fire along each avenue of approach.

■ ***Observation/Fields of Fire.*** Evaluate all terrain along the avenues of approach in terms of observation and fields of fire. Determine the following:

- What terrain provides the best observation of critical areas or along avenues of approach?
  At night?
  When visibility is obscured by smoke or fog?
- What terrain offers effective range of friendly and enemy weapon systems?
- What terrain provides the best fields of fire? For what size force?

- ***Cover and Concealment.*** Evaluate all terrain along the avenues of approach in terms of cover and concealment. Determine the following:

  - What terrain affords the best cover and concealment to the enemy and to the friendly forces?
  - Where are likely hide positions that support potential enemy and friendly battle positions?
  - What routes would a force use to move from hide positions to battle positions?

- **Troops Available.** Analyze the troops available to execute the mission. Determine the following:

  - ***Soldiers.***

    - What is the strength of the platoons in the troop?
    - Can all weapon systems be manned?
    - How is the morale of the troop?
    - What is the experience level of the troop?
    - What training and skills are required? Are specialists such as engineers needed?
    - What is the general health status and fatigue level of the soldiers?

  - ***Equipment.***

    - How many weapon systems are operational and employable?
    - Is special equipment or additional equipment needed to accomplish the mission?

START NECESSARY MOVEMENT

While developing his tentative plan, the commander may decide to reposition some of his forces before the operation starts so that he is prepared to meet mission requirements. If the entire troop needs to move, he will often execute the movement in accordance with a squadron plan. The amount of time he has to reposition his forces is determined by the readiness condition (REDCON) in effect.

Note. See Chapter 7 for a discussion of REDCON.

CONDUCT A RECONNAISSANCE

If sufficient time is available after determining his tentative course of action, the commander should make every effort to get out into his AO and look at the terrain. Ground reconnaissance is the norm; however, air transportation may be available. Staying on the ground gives a better feel for the terrain than does reconnoitering by helicopter, but it is slower. Although faster, aerial reconnaissance does not provide an appreciation of the terrain from a vehicle commander's point of view in terms of fields of fire and cover and concealment. If circumstances permit, subordinate leaders should accompany the commander to help perform the reconnaissance. It gives them an opportunity to become familiar with the area.

MAKE FINAL DECISIONS AND COMPLETE THE PLAN

Based on the new information gathered by his personal reconnaissance, the commander makes final adjustments to his plan and nails down the details. Once the plan is complete, he prepares the order. Troop orders need not be written and handed to subordinate leaders. Speak face to face with them if possible. Put the plan in the form of a standard five-paragraph field order. Jot it down in a notebook so every detail can be recalled.

ISSUE THE OPERATIONS ORDER

Assemble the orders group. Make sure everyone in the orders group is present. Issue necessary graphics to subordinate leaders and give them time to prepare their maps. Ensure the overlays are neat and accurate. Messy overlays with broad pen strokes cause confusion and waste valuable time. Once the subordinate's overlays are prepared, check them for accuracy. (A technique for ensuring accuracy is to include in the order a list of grid coordinates for key locations that might be ambiguous in the operational graphics. Even neat graphics, once copied from squadron down to platoon, may often deviate by as much as 500 meters.) Arrange the orders group in a semicircle from left to right in the order in which they will be addressed when giving specific instructions.

Issue the order. Use notes when issuing the order; do not rely on memory. If possible, issue the order while overlooking the AO. Keep in mind that the order will be translated a couple of times before the privates receive it, and it must make sense to them. Talk in their language. The order should be short and simple, and must be logical and easy to follow; be clear, organized, and concise so that no one becomes confused. Misunderstanding the commander's intent most often occurs in the transmission of orders to the platoon leader and his subordinates.

<u>Note</u>. See Appendix C for the operations order format and sample OPORD.

Conduct the confirmation brief. Immediately after the order is issued, call for questions from the subordinate leaders. After all questions concerning the order have been answered, begin the confirmation brief by each of the troop's subordinate leaders.

The confirmation brief is used by the commander to confirm that his intent and guidance for the conduct of the operation are clearly understood by everyone in the orders group before they are dismissed to begin their planning. The confirmation brief adjourns only when the commander is confident his subordinates understand their mission, his and the higher commander's intent, his concept of the operation, the scheme of maneuver, the time plan, and the type and location of the rehearsal.

REHEARSE

Rehearsals are of paramount importance before executing any plan. Rehearsals help in the following ways.

- Clarify the commander's intent.
- Expose combat, combat support, combat service support, or disconnected activities in the plan.
- Reinforce the scheme of maneuver and fire support plan.
- Focus on actions and decision points critical to mission accomplishment.
- Ensure subordinates explicitly understand their missions, how their missions relate to each other, and how each mission relates to the commander's plan.
- Outline conditions that, when present, would necessitate execution of branch plans.
- Provide feedback to the commander.

Rehearsals instill confidence in participants by—

- Giving participants faith in the success of their own plan as well as in their commander's plan.
- Providing subordinates with purpose, direction, and motivation.
- Enabling leaders to execute missions with speed, flexibility, and audacity.

There are generally seven types of rehearsals—full, key leader, terrain model, sketch map, map, radio, and backbrief. Preparation time and resources for each range from extensive to minimal. At troop level, the most common rehearsal types are backbrief, map, sketch map, terrain model, and radio.

Backbrief. During the backbrief, each subordinate leader briefs the commander of how he intends to accomplish his mission before he issues his OPORD to his respective unit. By having subordinates explain their intent and concept of operation, the commander can ensure their plans support his own. Flaws or potential problems with the plan may be revealed at this time.

Do not confuse the backbrief rehearsal with the confirmation brief the commander uses immediately after he issues an OPORD to determine how well his subordinate leaders understand the order.

Map/Sketch Map/Terrain Model. Using either a map and overlay of the same scale as used to plan and execute the operation, a large scale sketch map, or a terrain model, the commander and subordinate leaders move unit markers in a sequential (either by phase, event, or time), interactive, verbal execution of the operation. These type rehearsals are essentially conducted like a war game to show the planned sequence of action-reaction-counteraction to critical events or phases of the operation. This is a good opportunity to coordinate not only actions of the unit but also critical locations such as contact points, checkpoints, boundaries, battle positions, and hide positions among subordinate leaders.

Radio. Using existing communications networks (either FM or wire), the commander and subordinate leaders verbally and interactively execute critical portions of the operation. This technique can have obvious communications security disadvantages; if so, then only the essential, most critical portions of the operation are rehearsed. Radio rehearsals require few resources and little time and are best used in conjunction with other methods or to further refine the plan.

SUPERVISE AND REFINE

The last step in troop leading procedures must not be neglected. Failure to supervise the execution of the order, or to refine the plan as the situation changes, is the road to ruin. The importance of this step cannot be overemphasized. Actions during the final step may include, but are not limited to, higher commander and his staff being present in the subordinate unit staff planning process.

Section II. Situational Awareness

Situational awareness is always keeping a clear picture of the tactical situation, both mentally and graphically. This picture includes both the friendly and enemy situation and an understanding of the relevant terrain. Since the troop normally operates dispersed over wide frontages, it is essential that all leaders maintain situational awareness so they can make sound, quick tactical decisions. Situational awareness also permits the leaders to anticipate events and relate separate pieces of information to form logical conclusions. One of the critical outcomes of situational awareness observed by all leaders is a reduction of fratricide incidents.

BATTLEFIELD FRAMEWORK

How the commander has structured the battlefield impacts the troop commander's ability to maintain good situational awareness. A commander will structure the battlefield based on the conditions of METT-T. The framework of the battlefield can vary from a very rigid extreme with an obvious front and rear boundary and closely tied adjacent units, to a very dispersed and decentralized structure with no secure areas, unit boundaries, or definable front. Between these extremes is an unlimited number of possible variations. Maintaining situational awareness will become more difficult the more unstructured the battlefield is. Modern, highly mobile operations with small forces lend themselves to a less rigid framework which will challenge the ability to maintain a good picture of the battlefield.

PICTURING THE BATTLEFIELD

To have a good picture of the battlefield, all leaders must have virtually perfect knowledge of the friendly situation one level higher than their own. This means that all principal subordinate leaders must know the troop situation and the troop commander must know the squadron situation. All leaders must have accurate knowledge of the terrain, and they must know as much as possible about the enemy. The requirement to maintain a real-time picture of the battlefield one level higher does not relieve leaders of the requirement to understand the situation two levels higher. The difference is that a leader's understanding of the situation two levels higher than his own does not have to be as specific or in real time.

Most of the information passed between elements is in the form of reports over FM radio. The troop commander receives many reports as a result of his troop graphics. Good graphics require that the subordinate elements report periodically as they accomplish tasks. The troop commander must be aware of when subordinates report so he will know how current his visualization of the situation is. If an element does not report in a timely manner, the commander must quickly determine the situation of the overdue element.

Although many reports may not be addressed specifically to him, particularly on the squadron net, the commander must monitor them by eavesdropping on the nets as traffic is sent. How effectively he can accomplish this is, to some degree, experience dependent; however, there are techniques he can use to relate the information to his map and thereby track the tactical situation. The troop XO plays an important role in assisting the troop commander and platoon leaders in maintaining situational awareness. The troop XO provides leaders with periodic updates of friendly units to the front, flanks, and rear, based on traffic from the squadron operations and intelligence (OI) net.

The commander's map is the key to maintaining situational awareness (see Figure 2-1). He should plot all friendly position reports up to one level higher than his own. Information from spot reports should also be plotted. Using different colors for friendly and enemy elements allows quick distinction. To avoid cluttering the map, he should place a dot or symbol with a number on his map where the element is located. The same number is then written in the map margin (or beyond the area of operations) with the complete spot report or unit identification next to it. This notation should also include the time. As positions or reports are updated, the old symbol is crossed off and a new one with a corresponding notation is added. This system allows the commander to easily track and monitor the tactical situation. This system is augmented by a formal operations log kept in the troop TOC by the XO or NCOIC.

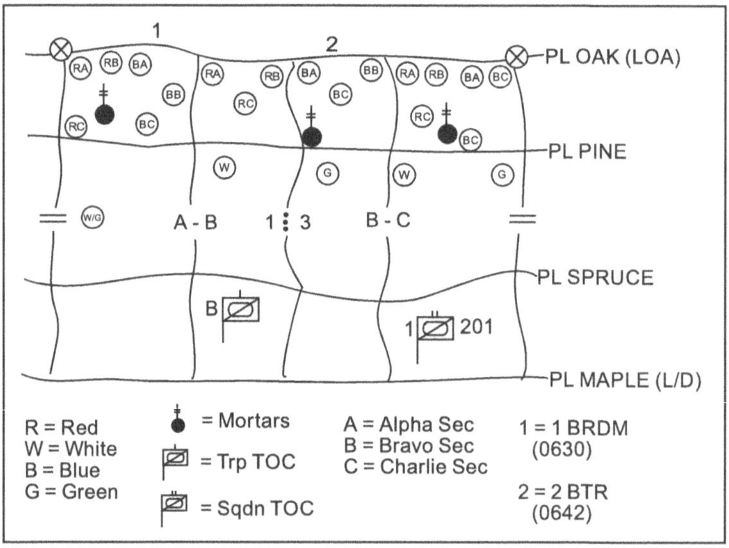

Figure 2-1. Sample situation map (troop commander).

BATTLE SPACE

Situational awareness, as previously discussed, is critical to successful reconnaissance and security missions; however, the troop commander's primary attention must be on his battle space. Battle space for the cavalry troop is determined by the location of individual platoons and sections, the range of direct-fire weapons, observation, sensors, and the terrain on which they are applied. This space is the immediate area over which the troop has influence. The troop commander must be aware of the general situation, but he must actively direct and manage all activity within his battle space. Most of the troop commander's command and control efforts focus on what is happening within his battle space and shaping his battle space to make it more efficient.

FRATRICIDE

Fratricide is a significant danger to all forces operating on a mobile battlefield where weapon system lethality is significantly greater than friend or foe identification ability. For this reason, situational awareness on the part of all leaders, particularly the commander, is critical not only to mission success but also to survival.

Under these types of conditions, it is critical that the commander know where other friendly elements are operating. With this knowledge of the situation, he must anticipate dangerous situations and take steps to either avoid or mitigate them. The commander must constantly be vigilant to changes and developments in the situation that may place his elements in danger. He must also ensure all subordinate unit positions are constantly sent to higher headquarters so all other friendly elements are aware of where they are and what they are doing. When the commander perceives a potentially dangerous situation, he must personally use the squadron command net to coordinate directly with the friendly element involved.

Section III. Communications

NETS

Several troop radio nets are used to circulate the volume of information in any combat operation. A troop command net links the troop commander with his subordinate units (see Figure 2-2). A troop fire support net links subordinate units with the fire support team (FIST) and mortar section (see Figure 2-3). Each net has an assigned radio frequency, but not necessarily a dedicated radio.

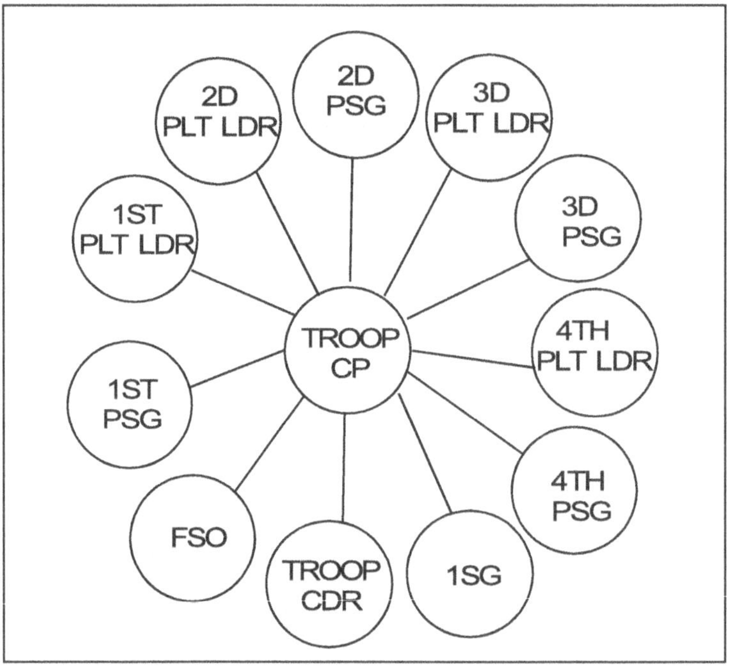

Figure 2-2. Troop command net.

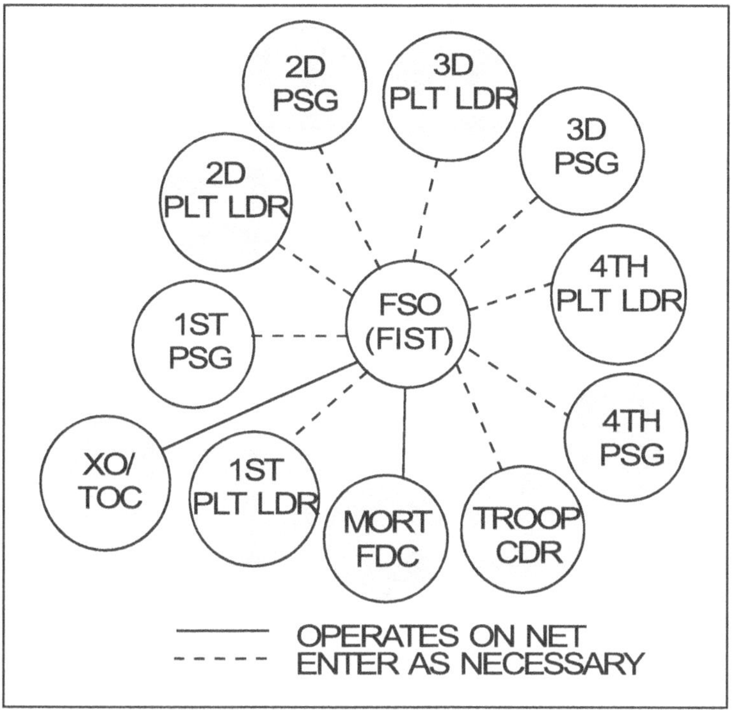

Figure 2-3. Troop fire support net.

In the troop CP, three additional radio nets are established to link the troop with the SCO and his headquarters. The squadron operations and intelligence (OI) net is used primarily to transfer enemy and friendly information by FM radio. All routine and recurring reports from the troop to squadron are transmitted on this net. Squadron headquarters uses this net to transmit information about current enemy and friendly situations within the squadron's AO. It is often used to issue FRAGOs. The troop commander and the XO are linked to the SCO by an FM squadron command net. This net is reserved strictly for use by the SCO, XO, S3, and CPs; and by subordinate troop, company, and battery commanders. In addition, an AM squadron command net allows the SCO to command and control the squadron when it is spread over distances that exceed the range of FM radios (see Figure 2-4).

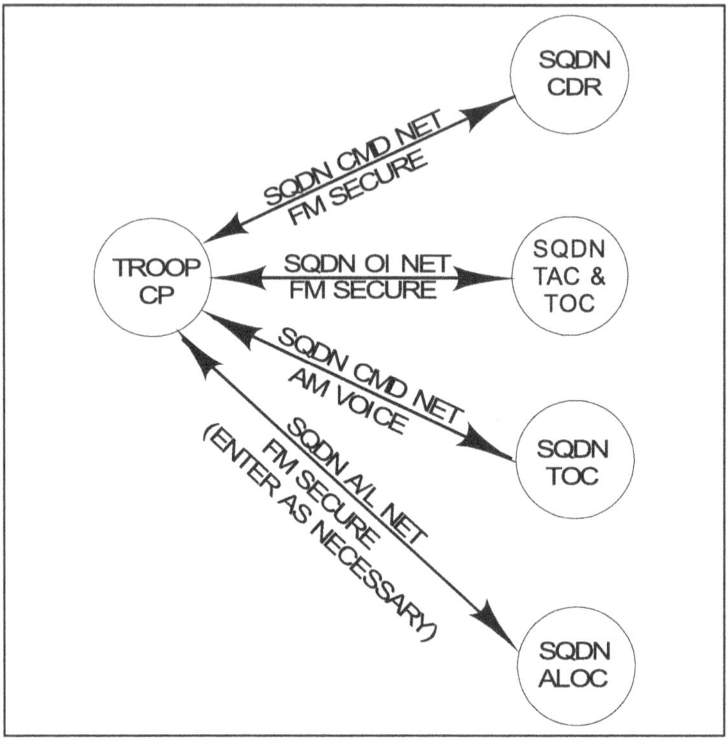

Figure 2-4. Troop command post external nets.

EAVESDROP

This tool is used by all cavalrymen. It is designed to give leaders time to think before they act. Most leaders have a hard time making decisions while they are talking on the radio. Decisions are usually better if the commander listens closely to what is said, and if he has uninterrupted time to consider information received. Eavesdropping is a method of communicating based on that principle. Everyone simply listens to the information sent from one leader to another, and unless necessary, stays off the radio net. In the troop, for example, reports from scout platoons are transmitted to the troop CP, where the XO reviews and records them. All others listen in and record the information themselves. Everyone stays off the net yet remains informed. If all or part of a transmission is missed, enter the net and quickly get the needed information from the troop CP.

FIXED CALL SIGNS

Fixed call signs are tools that establish immediate identity of subunits and leaders, thereby reducing the length of transmissions on the radio. These call signs eliminate the confusion often caused by the ever-changing signal operation instructions (SOI). Each platoon is referred to by a color, and leaders are referred to by numbers. These call signs do not change. An example of fixed call signs for a cavalry troop is outlined below.

Fixed Call Signs

Element	**Call Sign**
TROOP HEADQUARTERS	BLACK
FIST	Black 1
Medics	Black 2
TOC	Black 3
Supply	Black 4
XO	Black 5
Commander	Black 6
1SG	Black 7
Maintenance	Black 8
Mortars	Black 9
NBC	Black 10

1ST PLATOON (Scout)	RED
Plt Ldr 11	Red 1
Sec Ldr 12	Red 2
Sqd Ldr 13	Red 3
PSG 14	Red 4
Sec Ldr 15	Red 5
Sqd Ldr 16	Red 6
*Sec Ldr 17	Red 7
*Sqd Ldr 18	Red 8
*Sec Ldr 19	Red 9
*Sqd Ldr 10	Red 0
2D PLATOON (Tank or AT)	WHITE
Plt Ldr 21	White 1
TC 22	White 2
TC 23	White 3
PSG 24	White 4
3D PLATOON (Scout)	BLUE
Plt Ldr 31	Blue 1
Sec Ldr 32	Blue 2
Sqd Ldr 33	Blue 3
PSG 34	Blue 4
Sec Ldr 35	Blue 5
Sqd Ldr 36	Blue 6
*Sec Ldr 37	Blue 7
*Sqd Ldr 38	Blue 8
*Sec Ldr 39	Blue 9
*Sqd Ldr 30	Blue 0
4TH PLATOON (Tank or AT)	GREEN
Plt Ldr 41	Green 1
TC 42	Green 2
TC 43	Green 3
PSG 44	Green 4

*Light troop only

Section IV. Techniques of Tactical Control

CHAIN OF COMMAND

The troop commander cannot personally supervise the execution of all combat tasks he assigns. He must depend on a trusted team of subordinate leaders. The chain of command in the troop extends from the commander to the platoon leaders and section leaders. Within the platoons, it extends from platoon leaders and their platoon sergeants to section sergeants to vehicle/crew commanders. The troop XO and the first sergeant serve as the commander's principal assistants. They help coordinate the actions of subordinate units to ensure unity of effort. They also ensure the troop has what it needs to accomplish its assigned tasks.

ORDERS GROUP

The troop orders group consists of all subordinate leaders required to receive and execute the troop commander's orders. As a minimum, the orders group should include the XO, the first sergeant, the platoon leaders, the mortar section sergeant, the fire support officer (FSO), and the leaders of any units that are attached or under the troop's operational control (OPCON).

STANDING OPERATING PROCEDURES

A troop SOP is a written collection of standing orders. It establishes procedures for performing routine combat tasks. For example, the SOP tells subordinates what procedures to follow when the troop occupies an assembly area, breaches an obstacle, conducts a passage of lines, or reacts to air attack. It lists reports that are routinely sent to troop and squadron headquarters, and shows their correct formats. It also states procedures for communicating within the troop, obtaining indirect fire, or getting additional fuel or ammunition. The SOP is a valuable time saver for troop leaders. Most important, the SOP keeps everyone in step, reduces confusion, and provides the backbone of combat proficiency. Disciplined execution of the troop SOP is the hallmark of outstanding units.

TERRAIN INDEX REFERENCE SYSTEM

Terrain index reference system (TIRS) is used to maneuver the troop, rapidly report friendly locations without fear of compromise, and pass out control measures. TIRS is not used to report enemy locations.

Description

Each TIRS point is shown by a tick mark located on a grid line intersection. Each point is given a designator of one letter and two numbers, such as X56, placed in the upper right quadrant of the tick mark. TIRS point designation is SOP, and units determine which letters they will use. They may designate specific letters for specific unit sectors or AO. The TIRS point is then recorded as X56, located at PA2450 (four-digit grid).

TIRS point-to-point references designate kilometers. For example, 500 meters is given as “POINT FIVE,” 1,000 meters as “ONE,” and 3,500 meters as “THREE POINT FIVE.”

For shifts from the TIRS point, use cardinal directions instead of left, right, up, or down. Using TIRS, “FROM X-RAY FIVE SIX—EAST ONE POINT SEVEN—NORTH POINT SEVEN” translates to “From the tick mark for TIRS point X56, shift east 1,700 meters and north 700 meters.” When each TIRS point is placed on a four-digit grid intersection, the use of shifts makes the TIRS as accurate as the six-digit grid system. For an example of a TIRS point in use, refer to Figure 2-5.

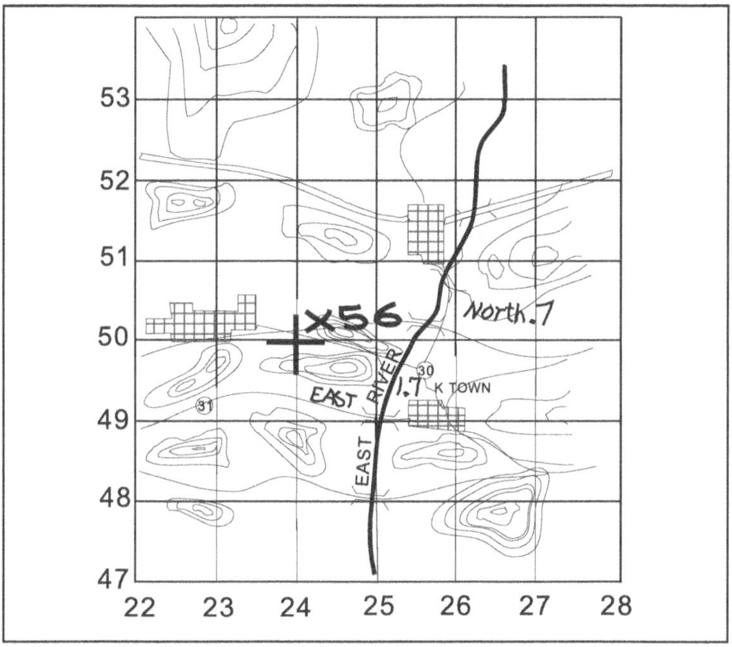

Figure 2-5. Placing TIRS on the map.

Application

The squadron normally issues the TIRS to be used for the operation as early as possible, perhaps with the warning order. The TIRS list should be issued to elements as an annex to a written OPORD.

The squadron should designate four to six TIRS points in each 10-kilometer square. The squadron TIRS should normally be sufficient for the troop to operate; however, if the troop designates additional TIRS, it should always ensure only squadron TIRS are referenced in its communications with higher headquarters.

TIRS should be used routinely to control combat operations. Use them—

- To identify BPs and to pass out control measures (such as LDs, PLs, and boundaries) quickly.
- To report friendly unit locations.

Passing control measures, such as BPs, sector boundaries, and PLs, are quick and accurate using TIRS.

Figure 2-6 shows BP B22. A troop commander could report "TROOP ESTABLISHED VICINITY ALFA ZERO SEVEN." To be more precise, he could report "TROOP BRAVO ESTABLISHED ALFA ZERO SEVEN WITH RED AT FROM ALFA ZERO SEVEN EAST POINT THREE—NORTH POINT FIVE—WHITE AT FROM ALFA ZERO SEVEN WEST POINT THREE—NORTH POINT EIGHT—BLUE AT FROM ALFA ZERO SEVEN EAST ONE POINT NINE—NORTH ONE POINT NINE—TO FROM ALFA ZERO SEVEN WEST POINT NINE—NORTH ONE POINT SEVEN—AND GREEN AT FROM ALFA ZERO SEVEN EAST ONE POINT ZERO—NORTH POINT FIVE." In the second transmission, the troop commander gives the precise location of the center of mass for all his platoons: red, white, blue, and green represent the 1st, 2d, 3d, and 4th platoons.

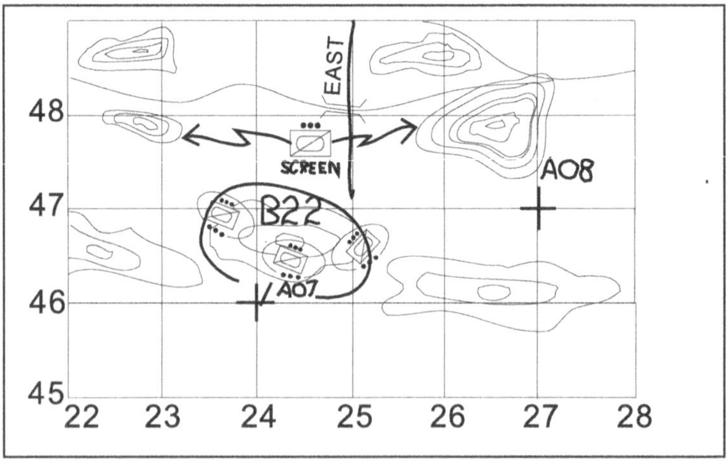

Figure 2-6. Reporting location using TIRS.

Using this transmission, the SCO orders Troop B to occupy a BP that is not on the troop commander's map: "BRAVO SIX SIX OCCUPY A BP EXTENDING FROM ALFA ZERO EIGHT WEST POINT SEVEN—SOUTH ONE POINT ZERO—TO FROM ALFA ZERO EIGHT EAST POINT SIX—SOUTH POINT FIVE—ORIENT NORTHEAST."

Figure 2-7 shows how the SCO could quickly establish new phase lines and boundaries between troops.

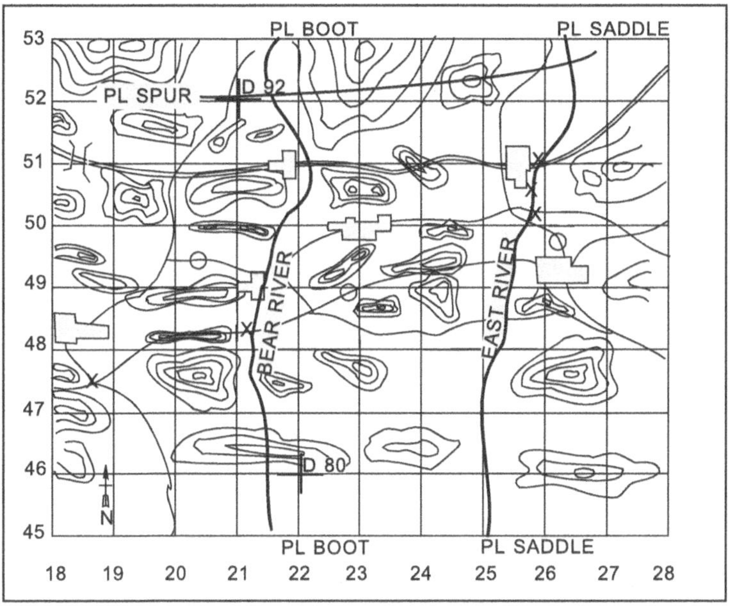

Figure 2-7. Establish graphic control measures using TIRS.

The SCO could transmit a new boundary between troops B and C by radio: "BRAVO SIX SIX AND CHARLIE SIX SIX BOUNDARY IS FROM DELTA EIGHT ZERO EAST POINT NINE—NORTH POINT EIGHT—TO FROM DELTA NINE TWO EAST TWO—SOUTH THREE—TO FROM DELTA NINE TWO EAST ONE POINT NINE—SOUTH POINT ONE."

PL SPUR could be sent as "PL SPUR IS FROM DELTA NINE TWO WEST POINT NINE—TO FROM DELTA NINE TWO EAST FIVE POINT THREE—NORTH POINT SEVEN."

Section V. Command Guidance and Organizational Control

TROOP COMMANDER

The troop commander is mounted in an M1A1 tank in the heavy troop or a HMMWV in the light troop. His first priority is the command and control of the troop.

The commander must provide the troop with clear intent and guidance for the mission. Intent and guidance facilitate positive command and control over subordinate platoons yet provide flexibility in execution for subordinate leaders.

The commander positions himself where he can best command and control the troop. He may maneuver along with one of the tank/AT platoons or between the tank/AT platoons and scout platoons. It is imperative that the troop commander not become decisively engaged with fighting his vehicle. He must retain the ability to control the actions of his troop. The troop commander operates on both troop and squadron command radio nets. He must have the capability to communicate to subordinates as well as to the squadron commander.

TROOP TOC

The purpose of the troop tactical operations center (TOC) is to report information to the higher headquarters. The troop TOC is the critical link between squadron and troop. One of the primary purposes of the cavalry troop is to report, and the troop TOC is central to that function. The troop TOC operates under the direction of the troop XO who is responsible for assisting the commander in the command

and control of the troop. The troop TOC takes reports from subordinate units in the troop, processes the information, and reports the information to the higher headquarters. The TOC tracks the battle at the troop and squadron level and relays information to the commander and subordinate platoons pertaining to the friendly/enemy situation. The TOC is the net control station for the troop command net and operates on the squadron command and OI net. The troop TOC also monitors and operates as necessary on the squadron A/L and the troop fire direction net.

In the heavy troop the TOC vehicle is an M577. An M3 CFV moves with the heavy troop TOC to provide local security for the M577 and to serve as a command and control vehicle for the XO if the commander is out of action. The XO should not conduct TOC operations out of the CFV because his experience is needed in the TOC vehicle to handle the information flow between troop and squadron over the OI and command nets. If the commander is out of action, the XO should avoid, if possible, rushing forward to assume command of the troop because of the importance of the activities in the TOC. The XO should assume command of the troop from the TOC until the battle is over. However, if the XO does need to move forward to command and control the troop, a qualified replacement must be placed in the TOC to assume his duties. This individual may be the FIST chief, one of the platoon leaders, or the first sergeant. The correct replacement for the XO in the TOC should be based on the depth available at any one of the above mentioned positions.

The light troop TOC is a cargo HMMWV with a shelter.

The primary concern when positioning the troop TOC is based on the TOC's ability to communicate with the squadron and the subordinate elements of the troop. The TOC position should also be an area that provides good cover and concealment from enemy and limited access to civilian population. During reconnaissance or offensive operations, maneuver the TOC at least one terrain feature behind troop combat elements. During security or defensive operations, the TOC should be positioned in sufficient depth to avoid contact with the enemy yet maintain communications with the forward scout platoons. The TOC and the troop trains may collocate for increased local security.

FIST

The FIST is the critical link with the supporting artillery and responsible for coordinating indirect fires (both mortar and FA) for the troop. The team processes calls for fire from the platoons and allocates the appropriate indirect-fire system based on the commander's guidance for fire support. The FIST can also assist the squadron ALO with the employment of close air support.

The FIST operates on three radio nets: troop command, troop fire direction, and squadron FSE digital/voice. The FIST monitors at least one of the following nets: squadron command, squadron OI, and howitzer battery (supporting artillery headquarters in the heavy and light division).

The FIST vehicle also may serve as the alternate troop TOC. The FSO has ready access to the higher level situation and the radio systems to replicate the troop TOC if it becomes damaged or destroyed.

Command guidance to the FIST should include the following:

- **Purpose of indirect fires.** How does the commander intend to use field artillery and mortar fires to support his maneuver?
  - Screening.
  - Suppression.
  - Development of the situation.
  - Disengagement.

- **Engagement/attack criteria.** How many rounds and of what type and mix will be fired at a particular target? Which targets will be engaged with artillery and which with mortars?

- **Control of troop mortars.** If the FIST controls movement of troop mortars, how far forward of the scouts will the mortars be able to range? Where are the mortars going to move? When are the mortars going to move?

- **FIST movement.** The primary considerations when positioning the FIST are security of the team and the ability to communicate with the squadron FSE, howitzer battery, or the DS artillery. The FIST is not the forward observer team for the troop; the troop has 19Ds who act as forward observers. The five techniques to maneuvering the FIST are—

  - Maneuvers with the commander.

  - Maneuvers with or near the mortar section.

  - Maneuvers with the TOC.

  - Maneuvers alone to maintain communications.

  - Maneuvers with the scouts or tanks to directly control fires or to use the ground laser designator (GLD).

Note. See Chapter 8 for a more in-depth discussion of troop fire support command and control techniques.

TROOP MORTARS

The troop mortars provide responsive indirect-fire support to the troop. In the light troop they are equipped with three cargo HMMWVs and two 120-mm towed mortars. The heavy troop is equipped with two M106 mortar carriers with 107-mm mortars. The mortars operate on the troop command and fire direction nets.

Command guidance for the mortars should include the following:

- **Purpose of mortar fires.** How does the commander intend to use mortar fires to support his maneuver?
  - Screening.
  - Suppression.
  - Development of the situation.
  - Disengagement.

- **Engagement/attack criteria.** How many rounds and of what type and mix will be fired at a particular target?

- **Positioning.** How far forward of the scouts will the mortars be able to range?

- **Movement.** Where will the mortars move and what will the mortars orient their movement on, for example, checkpoints, a route, mortar firing positions? Who will control the movement of mortars—the mortar section sergeant, the FIST, or the troop XO?

SCOUT PLATOONS

The scout platoons, HMMWV or Bradley equipped, perform reconnaissance and security missions for the troop. They may also perform missions within an offensive or defensive role in certain METT-T situations. To enhance command and control and to provide scout platoons the freedom of action required in cavalry operations, the troop commander should give the following guidance to scouts, regardless of what missions they are performing.

- Engagement, disengagement, destruction, and bypass criteria.

- Tempo of the operation. Tempo is dictated by the higher commander's intent for the operation, and is a function of the enemy situation as it relates to the survivability and firepower of the systems conducting the operations (METT-T). Tempo for the scout platoon determines the rate at which they conduct their mission and develop the situation. High tempo operations, for example, could be characterized by very aggressive reconnaissance focusing on mounted avenues of approach and likely enemy defensive positions. Engagement criteria in high tempo operations would be unrestricted, allowing the scouts to rapidly develop the situation by fire and maneuver. Low tempo operations could be characterized by scouts conducting a very detailed reconnaissance and maximizing the use of stealth to develop the situation.

The scout platoon leader and platoon sergeant operate on both the troop command and platoon command nets. As required, they will operate on the troop fire direction net. For platoon organization for combat, see FM 17-98, Chapter 2, Section VI.

TANK/AT PLATOONS

The tank/AT platoons are the principal killers in the troop. They assist the troop commander in the rapid development of the situation during operations by fire and maneuver. The troop commander should give the platoons the following guidance to facilitate their command and control during battle.

- Location in the troop formation and distance behind the scout platoon.

- Engagement, disengagement, and destruction criteria. When and where will they be employed and against what size force? Will they be used to suppress in support of another maneuver force, or destroy by fire and maneuver?

- Tempo of the operation. (Same as for the scout platoons above.)

The platoon leader and platoon sergeant operate on troop command and platoon command nets. As required, they will operate on the troop fire direction net.

TROOP TRAINS

The troop trains provide logistical support and casualty evacuation to the troop. The trains are under the direction of the first sergeant and consist of the troop combat and field trains. The first sergeant monitors the troop command and squadron A/L nets. He directs resupply, treatment and casualty evacuation, and maintenance/recovery operations for the troop.

The troop trains may be consolidated or echeloned. Either way, the trains should move one terrain feature behind the trail combat elements of the troop. If echeloned, the field trains move with the squadron field trains under the direction of the squadron HHT commander. The troop combat trains consist of the first sergeant in his vehicle, a maintenance contact team with recovery vehicle, and a medical evacuation vehicle. The field trains are made up of the supply sergeant and the remaining members of the maintenance team.

Command guidance to the troop trains should include the following:

- Trains organization. Should they move consolidated or echeloned?

- Movement guidance. Where do the trains move in the troop formation and what is their orientation for movement? Use checkpoints, routes, or trains locations to orient the movement of the trains.

Note. See Chapter 9 for a more in-depth discussion of the TTP for combat service support.

Section VI. Tactical Movement Formations

During tactical operations, the troop normally moves in one of four movement formations: troop column, troop line, troop vee, or troop split-vee.

TROOP COLUMN

Use the troop column when moving on a designated route, when speed is essential, and when contact with the enemy is not expected. This formation moves the troop quickly and efficiently from one place to another. Control of the troop while moving in column is eased, but the troop is vulnerable to enemy air or ground attack.

The troop usually moves at a designated speed and with a set distance between vehicles when moving in column formation. The march speed and distance between vehicles are designated by the troop SOP, based on the situation.

Note. See Chapter 7 for a discussion of column formation.

TROOP LINE

The troop line formation can be used when conducting defensive or retrograde operations in the heavy and light troops (see Figure 2-8). The formation may also be used by the light troop during reconnaissance operations when the troop is scrambled.

Figure 2-8. Troop line formation.

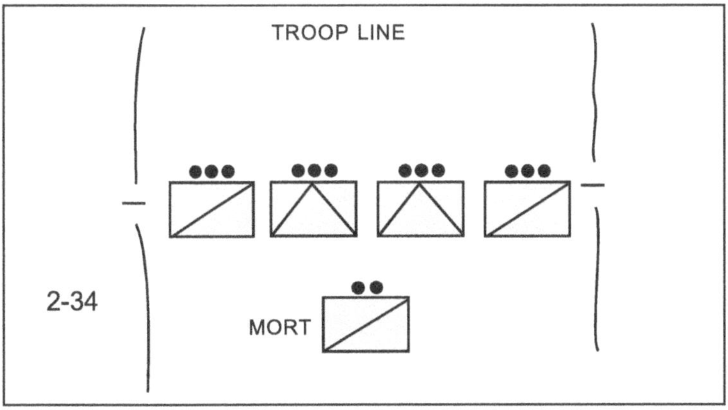

TROOP VEE

The troop vee formation is used to conduct reconnaissance or security missions (see Figure 2-9). The purpose of the vee formation is to keep maximum reconnaissance forward; the two tank or AT platoons are positioned in depth, ready to attack or defend any where in the troop zone/sector. This formation allows the troop to mass combat power at the decisive point (see Figure 2-10).

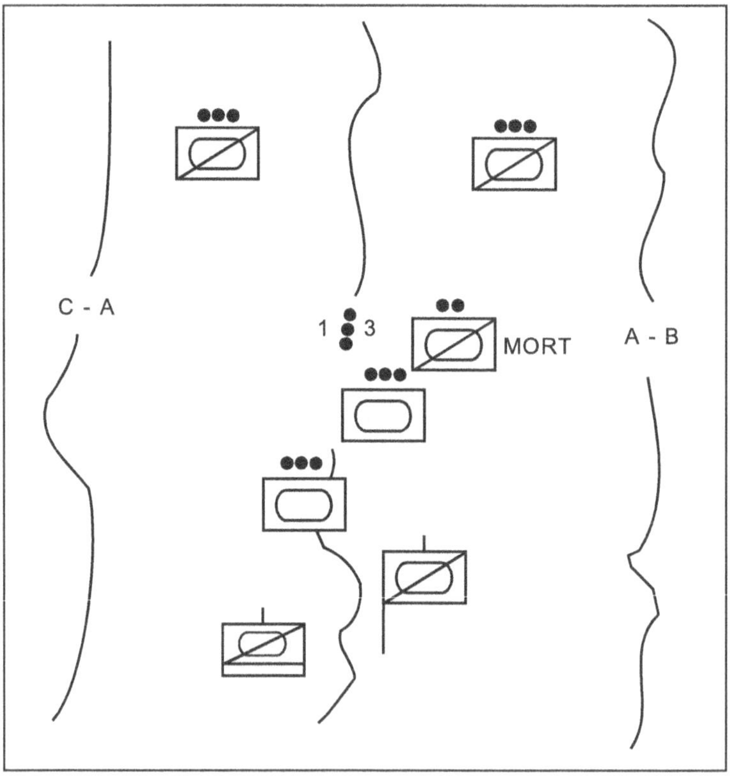

Figure 2-9. Troop vee formation.

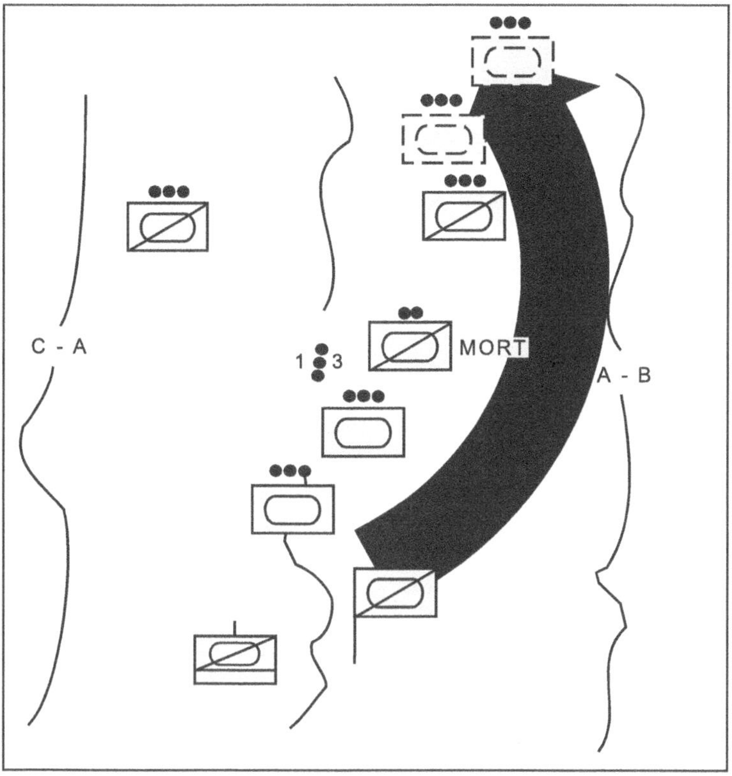

Figure 2-10. Troop commander maneuvers his tank platoons (troop vee).

In the troop vee formation, the two scout platoons work abreast in their respective zone/sector. The two tank/AT platoons are collocated in depth of the scouts, and they maneuver in the troop zone/sector.

Heavy Troop. Normally the distance between the scouts and tanks varies from 1 kilometer to 4 kilometers. The tanks should be close enough to respond quickly, but far enough back to respond to enemy contact based on the commander's guidance for engagement. Generally, do not position the tank platoons to overwatch the scout platoons. When the scouts make contact with the enemy, the tanks should be far enough in depth to maintain their ability to maneuver. The scouts are usually responsible for providing their own security as they move.

Light Troop. Normally the distance between the scouts and TOWs varies between 500 meters to 1,500 meters. TOW vehicles should respond quickly to enemy contact in order to assist in the scout platoon's development of the situation. Because of the reduced fire power in the HMMWV scout platoon, the TOW systems may provide the scouts with additional security in the form of overwatch during reconnaissance operations.

Position the mortars in or near the center of the troop zone/sector to provide indirect-fire support across the troop front. Keep them positioned to fire about one third to two-thirds of their range (about 3 kilometers to 3.5 kilometers, terrain dependent) beyond the scouts, so the scouts can engage enemy forces at long range with indirect fires. (Line of sight is the determining factor in range forward of the scouts. In densely wooded terrain, the scout's visibility may be only 100 to 200 meters forward of their front line trace, so mortar range need not always be 3,000 meters forward of the scouts.)

TROOP SPLIT-VEE

The troop split-vee is a variation of the troop vee, and can be used in reconnaissance and security missions (see Figure 2-11). In the split-vee formation, the scout platoons work abreast and forward of the troop. The two tank/AT platoons work in depth of the scout platoons, but each tank/AT platoon initially follows a single scout platoon in its zone/sector. The scout platoons conduct their reconnaissance or screening mission within their zone/sector, and the tank/AT platoons key their movement on the progress of the scout platoons. Give the tank/AT platoon leaders command guidance on where they should move in relation to the scout platoons. Tank/AT platoon leaders eavesdrop on the troop command net to monitor the progress of the scouts and to move their platoons to maintain their orientation with the scouts based on the troop commander's guidance.

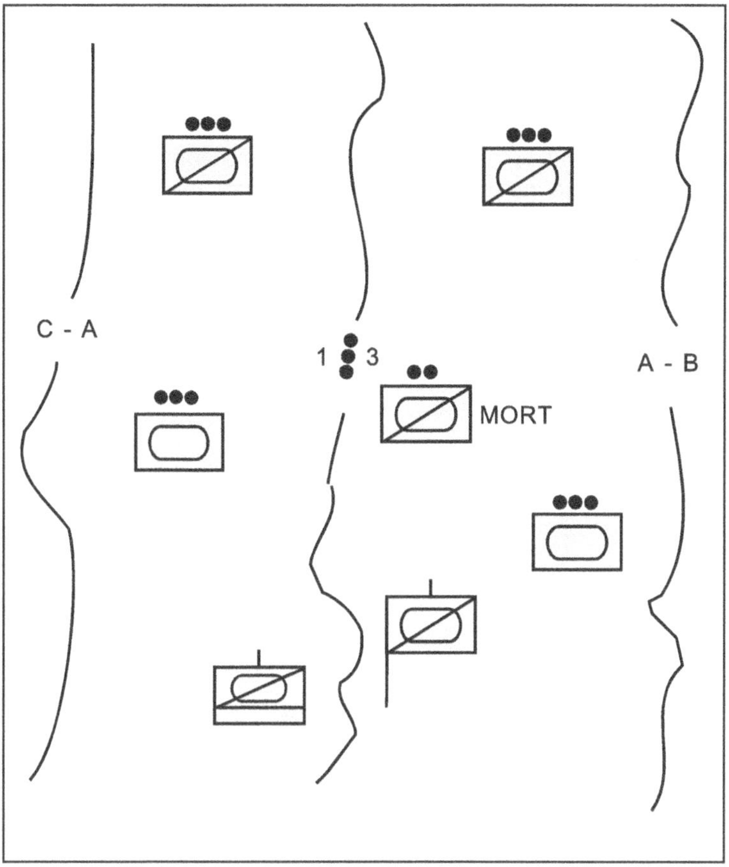

Figure 2-11. Troop split-vee formation.

Develop a habitual relationship between a scout platoon and a tank/AT platoon. First and second platoons work together, and third and fourth platoons work together. This builds teamwork between the scout and tank/AT units, and gives them confidence in each other.

All four platoons still work for the troop commander in this formation. The purposes of having a tank/AT platoon work with a scout platoon are to have combat power spread across the entire troop front, to keep the tanks/TOWs in a position to respond rapidly in support of a scout platoon, and to have scouts available to reconnoiter routes and positions for use by the tanks/TOWs. The troop split-vee is often used in close or restrictive terrain that reduces the tank/AT platoon's mobility so that it cannot quickly maneuver across the troop zone/sector. Generally the troop commander is the only one who commits the tank/AT platoons to conduct a hasty attack, to counterattack, or to move into overwatch positions. If he must mass the combat power of both tank/AT platoons, the troop commander can still shift the tank/AT platoons across the troop zone/sector during the mission (see Figure 2-12).

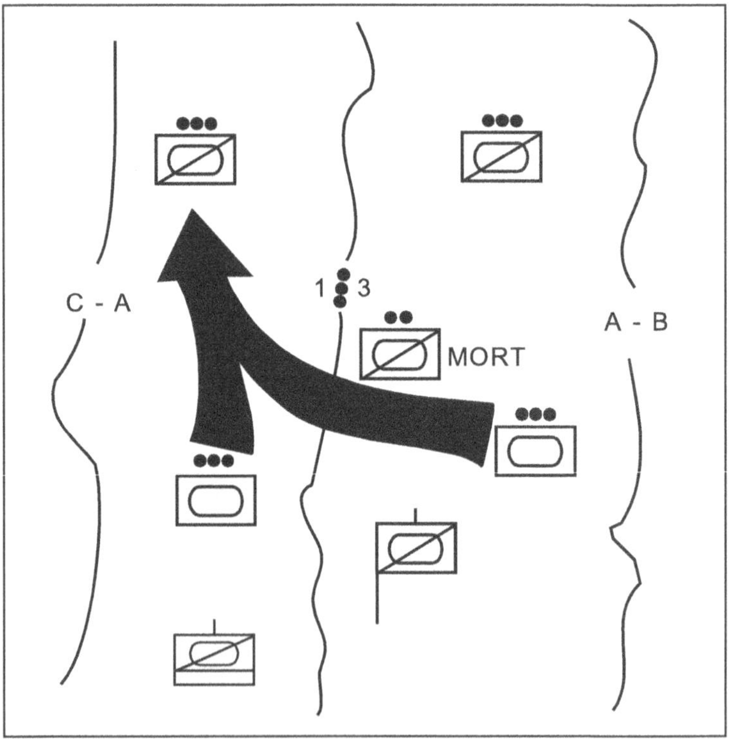

Figure 2-12. Troop commander maneuvers his tank platoons (troop split-vee).

LIGHT TROOP SCRAMBLE

The light troop scramble is a task organization that allows the troop commander to use a mix of weapon systems to enhance acquisition and destruction of the enemy under reduced visibility conditions and in restrictive terrain, where massed TOW fires are not possible. This task organization can also be used to respond to multiple but separate taskings of a troop usually associated with operations other than war, for example, manning checkpoints and area security missions. The troop commander may maneuver his troop in a vee, split-vee, column, or line formation when task organized into a scramble.

The troop is task organized into four like platoons of seven vehicles each. Five scouts and two AT vehicles are in each platoon (see Figure 2-13). The platoon may task organize and operate two ways:

- TOWs integrated with the scout sections (see Figure 2-14).
- TOWs maneuvering as a section to overwatch the movement of the scouts (see Figure 2-15).

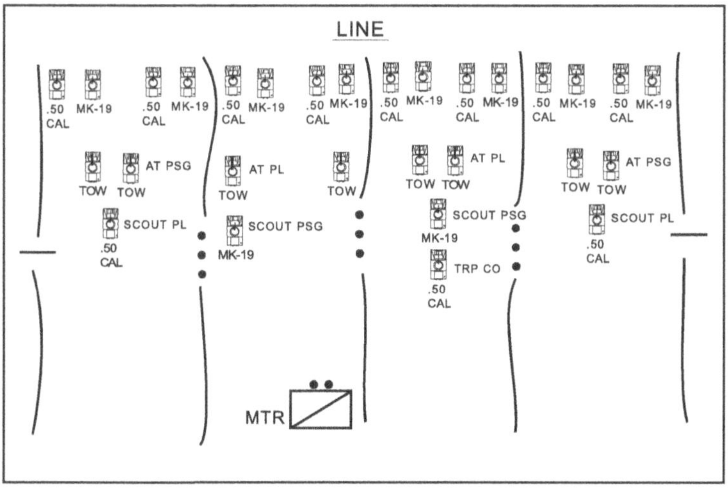

Figure 2-13. Light troop scramble.

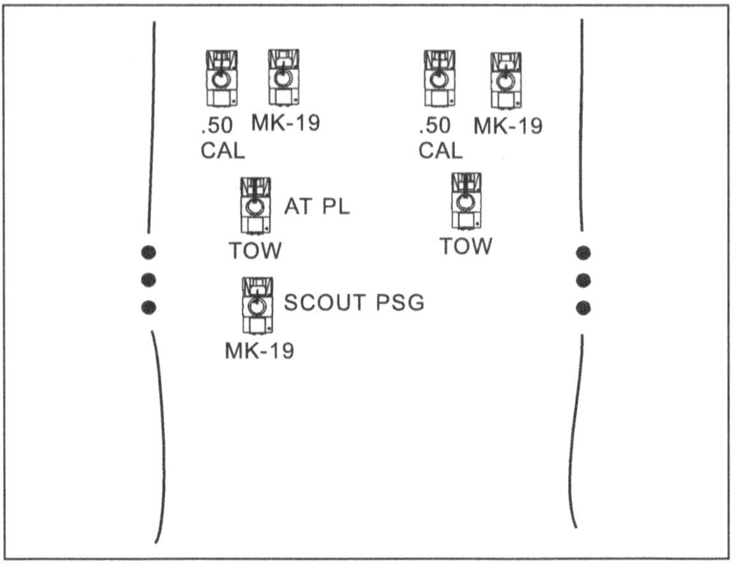

Figure 2-14. TOWs integrated with scout section.

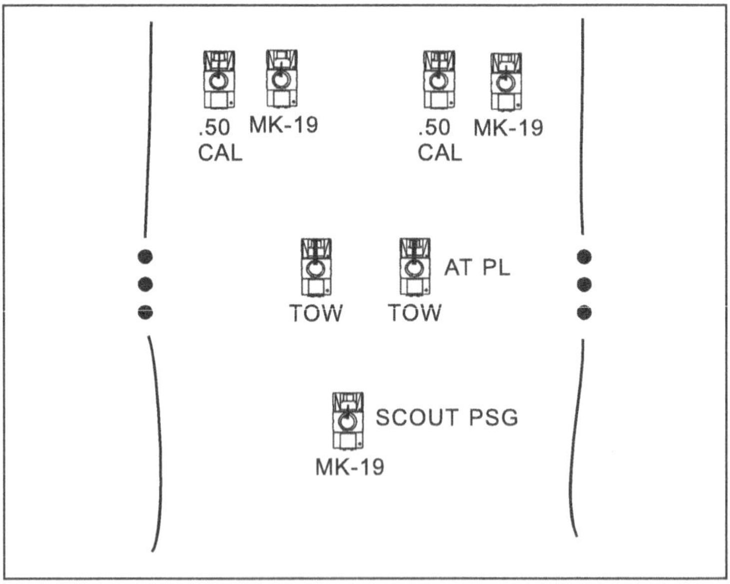

Figure 2-15. TOWs maneuvering as a section to provide overwatch.

Chapter 3

Reconnaissance

Reconnaissance is a mission undertaken to obtain information about the activities and resources of an enemy or about the meteorologic, hydrographic, or geographic characteristics of a particular area. Reconnaissance produces combat information. Combat information is a by-product of all operations acquired as they are in progress. Reconnaissance, however, is a focused collection effort. It is performed before and during other combat operations to provide information used by the commander to confirm or modify his concept.

Cavalry is the corps or division commander's principal reconnaissance organization. Cavalry troops perform reconnaissance using a combination of mounted and dismounted techniques. The cavalry troop is uniquely organized, trained, and equipped to perform the crucial task of reconnaissance for other combined arms forces. Reconnaissance is the cavalryman's specialty.

CONTENTS

Section I. Purpose, Fundamentals, and Capabilities

PURPOSE

Cavalry troops conduct reconnaissance forward of another friendly force to provide information about the terrain and enemy within the area. The reconnaissance allows the follow-on force an opportunity to maneuver freely and rapidly to its objective. Reconnaissance keeps the follow-on force from being surprised or interrupted, and protects it against losing soldiers and equipment on the way to the objective. The cavalry troop performs three types of reconnaissance—route, zone, and area.

FUNDAMENTALS

Successful reconnaissance operations are planned and performed with six fundamentals in mind.

- Maximum reconnaissance force forward.
- Orient on the location or movement of the reconnaissance objective.
- Report all information rapidly and accurately.
- Retain freedom to maneuver.
- Gain and maintain enemy contact.
- Develop the situation rapidly.

Maximum Reconnaissance Force Forward. In reconnaissance, every pair of eyes makes a difference. Do not keep scouts in reserve. This does not mean that every scout should be forward in a strictly linear sense, but actively employed in the conduct of the reconnaissance.

Orient on the Location or Movement of the Reconnaissance Objective. From the IPB (intelligence preparation of the battlefield) the S2 will identify gaps in the squadron's knowledge of the enemy and terrain it will be operating in. These gaps together with the commander's guidance form the PIR (priority intelligence requirements) and direct the reconnaissance efforts of the troop. Focus the efforts of the troop in selecting a course of action and a scheme of maneuver that continuously orients the troop on its reconnaissance objective based on the commander's guidance. The troop's focus may be a terrain feature, a specific area, or an enemy force.

Report All Information Rapidly and Accurately. Higher commanders base their decisions and plans on the battlefield information cavalry troops gather during reconnaissance. Combat information loses value as it ages, so the fresher the better. Scouts report exactly what they see. Troop commanders and executive officers may summarize or clarify reports to the squadron, but must never delete raw information reported by the scouts. Information that seems unimportant to the troop may be extremely valuable at higher echelons.

Retain Freedom to Maneuver. Cavalry troops must be able to maneuver on the battlefield to continue the reconnaissance mission. If the troop becomes decisively engaged, the reconnaissance stops. Use of proper movement techniques and overwatch helps prevent decisive engagement. IPB provides the troop commander information that allows him to anticipate battlefield events and to retain freedom of maneuver.

Gain and Maintain Enemy Contact. Contact is any condition ranging from a surveillance sighting to engaging in close combat. Surveillance is often sufficient and is the preferred method of maintaining contact. When necessary or required, troops will use fire and maneuver to maintain contact with an enemy force. Once contact is gained, however, it is not lost unless ordered by higher headquarters.

Develop the Situation Rapidly. During reconnaissance operations cavalry troops frequently and repeatedly encounter situations that require action to determine what the troop must face. These situations may be terrain oriented, obstacles, or enemy. Terrain or obstacle situations require close reconnaissance, bypass, hasty breach, if necessary, and marking. If an enemy force is encountered, the troop will determine the size, composition (What is the enemy force made up of—tanks, personnel carriers?), disposition (Is the enemy force in an offensive or defensive posture? Is he dug in? What is his orientation?), and activity. IPB provides the threat situational information that guides the effort. Reconnaissance techniques, often in the form of drills, are used while developing the situation.

Troops must develop the situation rapidly to get inside the enemy's decision cycle and force him to react to the actions of the troop rather than the troop reacting to the enemy. This requires the troop to rapidly execute development of the situation at the platoon level through the execution of rehearsed battle drills.

CAPABILITIES

The ability of a cavalry troop to conduct reconnaissance is a function of the enemy situation, the terrain the troop is operating in, and the type of troop conducting the reconnaissance.

- Heavy Troop
  - Can reconnoiter up to a 10-kilometer wide zone.
  - Can reconnoiter up to two routes simultaneously.
  - Can conduct reconnaissance at the rate of about 1 kilometer per hour, depending on the terrain.
  - When faced with a heavy-equipped threat, will conduct either aggressive or stealthy reconnaissance, depending on the higher commander's guidance.
- Light Troop
  - Can reconnoiter up to a 10-kilometer wide zone.
  - Can reconnoiter up to two routes simultaneously.
  - Can conduct reconnaissance at the rate of about 1 kilometer per hour, depending on the terrain.

- When faced with a heavy-equipped threat, will conduct stealthy reconnaissance and aggressive reconnaissance when reinforced with heavy forces.
- When faced with a light-equipped threat, will conduct either aggressive or stealthy reconnaissance, dependent on the commander's guidance.

Section II. Planning Considerations, Methods, and Techniques

The purpose of this section is to outline the planning considerations, methods, and procedures a cavalry troop uses to execute reconnaissance missions.

PLANNING CONSIDERATIONS

When planning a reconnaissance mission, the troop commander must take into account his unit's capabilities and limitations and consider the following:

- Time available from mission receipt to completion.
- Threat size, composition, disposition, and will to fight.
- Terrain and weather effects on the troop's ability to maneuver.
- Tempo of the operation.
- The squadron commander's (SCO) intent and guidance, as follows:
  - What are the SCO's focus and desired endstate of the reconnaissance?
  - What triggers the squadron's employment of the tank/AT company?
  - What does the SCO want the troop to destroy, fix, and bypass?
- Task organization or reinforcements.
- Critical tasks to be accomplished by the troop. Specifically, identify which critical tasks may be deleted during the reconnaissance.

Based on the considerations above, the troop commander determines the following:

- What is the focus of the reconnaissance? What critical tasks must be accomplished within the constraints of time and terrain?
- What specified or implied missions are associated with the squadron endstate?

Note. If given a limit of advance (LOA) for the reconnaissance mission, the troop commander should plan to screen along the LOA (see Chapter 4).

- How will the troop deal with enemy contact? What are the troop's criteria for engagement, destruction, and bypass?
- Under what situations does the commander see employment of his tank (heavy troop) or antitank (AT) (light troop) platoons?
- How will the commander use indirect fires from mortars and artillery to support his maneuver?
- Who controls the troop's attachments and how are they integrated into the reconnaissance?

The troop may receive attachments from higher headquarters. These assets may be maintained under troop control or tasked down to platoons for their use in the execution of the platoon's specified tasks. Examples include attached engineers, ground surveillance radar (GSR), or a chemical reconnaissance element.

- **Engineers.** If an engineer platoon is attached to the troop, the commander may elect to keep them under his control and treat them as a maneuver platoon. He may assign them to conduct the reconnaissance of the route, while the scout platoons move just ahead and reconnoiter terrain on either side. If the troop receives a squad or section, the troop commander may elect to task it to the scout platoon conducting the reconnaissance of the route.

- **GSR**. If a GSR squad or section is attached, the troop commander may elect to task organize them to a platoon or keep them under troop control. During reconnaissance operations, GSRs may be focused on flank avenues of approach into the troop zone with their movement controlled by the scout platoon leader or commander. In either case the commander should give clear guidance for positioning and orientation of the reconnaissance.
- **Chemical reconnaissance element.** If a chemical reconnaissance element (squad, section, or platoon) is task organized to the troop, the commander may maintain the element in reserve to reinforce a scout platoon that comes in contact with a contaminated area, or he may task organize the element down to one of the scout platoons.

METHODS OF RECONNAISSANCE

There are three methods of conducting reconnaissance at the cavalry troop level: *dismounted, mounted, and reconnaissance by fire.* The troop commander may use any method or combination of methods to accomplish the reconnaissance mission under the restrictions placed on him by METT-T (mission, enemy, terrain, troops, and time available) and the higher commander's intent and guidance. A fourth method of reconnaissance, *aerial reconnaissance*, may be executed by an air cavalry troop conducting a coordinated reconnaissance forward of the ground troop.

Dismounted Reconnaissance

The troop commander may direct scouts to conduct dismounted reconnaissance when—

- Time is not a limiting factor.
- Detailed information is required.
- Stealth is required.
- Enemy contact is expected or has been achieved through visual means.

- Scout vehicles cannot move through an area because of terrain.
- Security is the primary concern.

Dismounted reconnaissance permits the cavalry troop to collect the most detailed information about the terrain and enemy within a given zone, area, or route. However, dismounted reconnaissance is also the most time-consuming of all the reconnaissance methods. The cavalry troop, heavy or light, is limited in the number of dismounted scouts it can employ at any time.

Mounted Reconnaissance

The troop commander directs scouts to conduct mounted reconnaissance when—

- Time is limited.
- Detailed reconnaissance is not required, or mounted method affords the same opportunity to collect information as the dismounted method.
- Enemy locations are known.
- Enemy contact is not likely.
- An air cavalry troop is conducting a coordinated air reconnaissance.

Reconnaissance by Fire

When conducting reconnaissance by fire, the troop places direct and/or indirect fire on positions the enemy is suspected of occupying. This action causes the enemy to disclose his presence by moving or by returning fire. The troop commander may use reconnaissance by fire when—

- Time is critical.
- Natural or man-made obstacles that could be overwatched by an enemy force are encountered.
- A suspected enemy position fits the situational template.
- Bunker complexes that may or may not be occupied are encountered.
- Enemy locations are known.

The disadvantage of the reconnaissance-by-fire method is that the troop will lose any element of surprise it may have had. However, reconnaissance by fire may reduce the chance of some portion of the troop being caught in an enemy kill zone. Reconnaissance by fire may not always provide the desired effect. A well-disciplined force will resist the inclination to move when probed by weapon fires.

When indirect-fire situations exist, the troop commander ensures scouts are in a position to observe the target area. Once the decision is made to use reconnaissance by fire, weapons should be used in the following priority.

- Indirect-fire systems.
- Machine gun.
- 25-mm chain gun or MK-19.
- TOW or tank cannon fire.

Reconnaissance by fire does not mean the indiscriminate use of direct and indirect fires at all woodlines and hilltops in the hopes of causing the enemy to react. Not only will the enemy recognize this ploy for what it is and not react to it, but also it wastes valuable ammunition.

Aerial Reconnaissance

Air cavalry troops perform aerial reconnaissance when time is critical. Aerial reconnaissance is often coordinated closely with a ground reconnaissance troop. The air and ground forces complement each other. The air element can move forward of the ground unit and reconnoiter key pieces of terrain or restrictive terrain, allowing the ground troop to concentrate its efforts in other areas or to increase the tempo of its reconnaissance. The air troop provides the ground troop with added security by clearing the ground forward of the ground unit, thereby facilitating movement of the ground force and quickening the pace of the operation. The ground troop can move rapidly mounted to the areas of interest within its area or zone of operations and have the time to dismount and collect detailed information (see Figure 3-1).

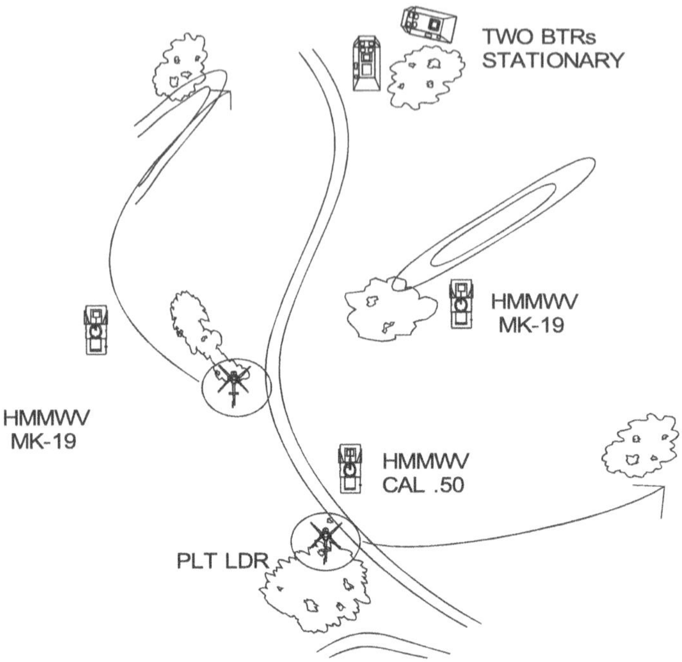

- Air cavalry platoon identifies two stationary BTRs along the woodline.
- Air cavalry platoon leader sends spot report to air cavalry troop commander. Troop commander tells platoon leader to develop situation and get in contact with ground platoon leader in zone.
- Air cavalry platoon leader sends information to ground platoon leader and assists ground scout section in developing situation.
- Air cavalry troop commander and ground troop commander choose a course of action.
- This coordination of air-ground supports squadron commander's intent, which increases troop survivability and squadron operating tempo of reconnaissance.

Figure 3-1. Air scouts conduct coordinated reconnaissance of zone with ground troop.

Aggressive Versus Stealthy Reconnaissance

The cavalry troop uses various combinations of reconnaissance methods. The method or combination of methods chosen by the troop and its scout platoons is based on the higher commander's guidance and METT-T. The method or methods chosen will characterize the reconnaissance as either aggressive or stealthy.

The stealthy approach to reconnaissance is more time-consuming. It emphasizes avoiding detection and engagement by the enemy. To be effective, a stealthy approach must rely primarily on dismounted reconnaissance and maximum use of covered and concealed terrain.

Aggressive reconnaissance is characterized by the speed and manner in which the reconnaissance force develops the situation once contact is made with an enemy force. A troop conducting aggressive reconnaissance uses firepower from direct- or indirect-fire systems and maneuvers to rapidly develop the situation. The troop will primarily use mounted reconnaissance and reconnaissance by fire when conducting aggressive reconnaissance.

Cavalry troops must be trained to conduct both aggressive and stealthy reconnaissance. The troop commander, scout platoon leader, and squad leaders will most likely use many different methods over the course of a reconnaissance operation. The constraints of commander's guidance, the opposing threat, the capabilities of the troop, the terrain the troop is operating in, and the time available to complete the mission (METT-T) all have an impact on the method or methods chosen by the troop leadership.

TECHNIQUES

Actions on Contact

Actions on enemy contact are a series of steps the troop takes when it encounters an enemy force or situation that warrants/demands action. Actions on contact are important because they allow the troop to maintain its tempo of operation by rapidly developing the situation and taking action before the enemy can gain the initiative and force the troop to react. At platoon level, actions on enemy contact consist of four steps.

- Deploy and report.
- Develop the situation.
- Choose a course of action.
- Recommend or execute a course of action.

Note. See FM 17-98 for an in-depth discussion of platoon actions on contact during reconnaissance.

While the platoon that makes contact executes actions on contact, the commander must continue to maneuver the remainder of the troop to ensure a clear picture of the enemy situation across the entire troop front. The following steps demonstrate the actions taken by the platoon in contact and the corresponding actions at the troop level.

Deploy and Report

Platoon Action. The elements of the scout platoon that make initial contact with the enemy immediately deploy to terrain that affords them both cover and good observation. If necessary, scouts return fire to suppress the enemy, and then deploy to their positions. The scout making contact sends a contact report to his platoon leader. The platoon leader forwards the report to the troop commander. Once the scout in contact is in a good covered and concealed position, he sends his initial spot report, which is forwarded through the platoon leader to the troop commander (see Figure 3-2).

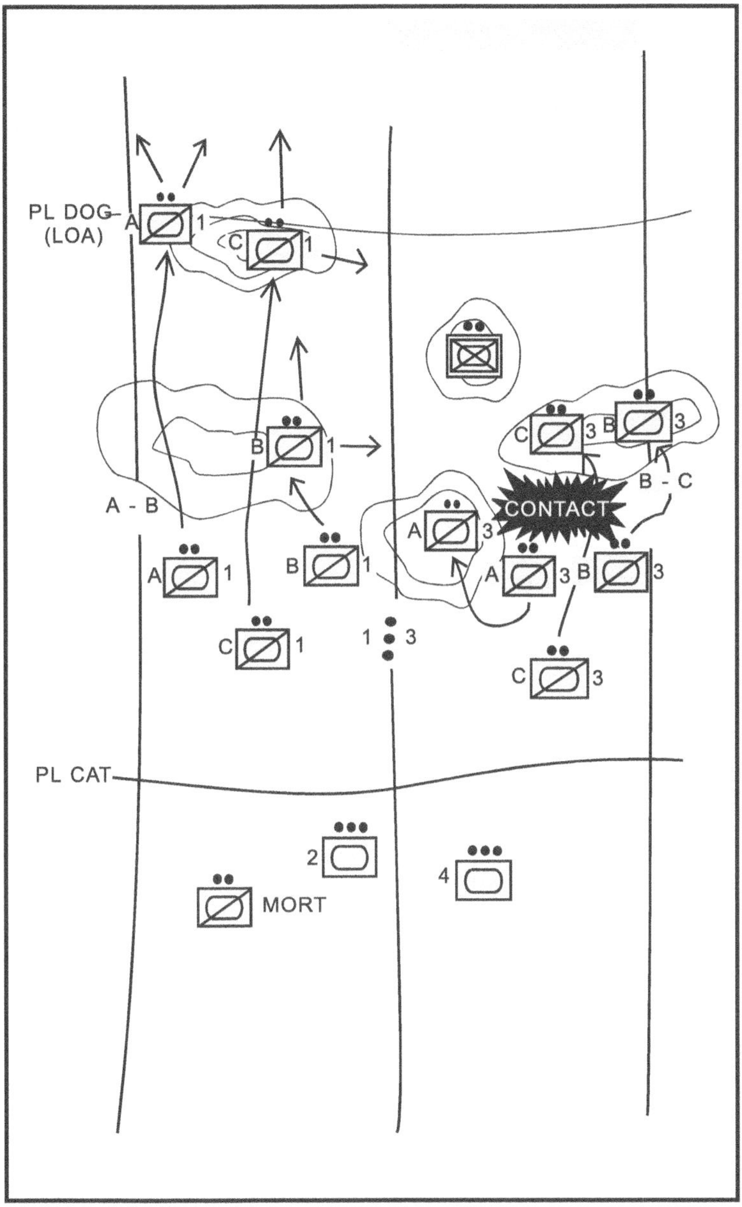

Figure 3-2. Scouts deploy and report then begin to develop the situation.

Troop Action. All other platoon leaders and platoon sergeants monitor the contact report. The troop commander assesses the information and moves, if necessary, to a position where he can observe the action. However, the troop must not lose focus of the reconnaissance mission.

Develop the Situation

Platoon Action. Next, the platoon in contact defines what it is up against. Scouts use dismounted and mounted reconnaissance to determine the enemy's size, composition, and orientation, and the exact location of weapon systems. The platoon may also use reconnaissance by fire to determine the enemy's tactical intentions. The reconnaissance-by-fire technique should, however, be conducted with indirect-fire assets when possible to avoid revealing the scouts' position. The scout platoon will search for minefields, wire, antitank ditches, and other obstacles that could force a friendly unit into a fire sack or cause it to flank itself to the enemy. To determine if the enemy can be supported by any other forces, the scouts search for enemy flanks and scour all adjacent terrain. They identify good counterattack routes into the flanks or rear of the enemy. Once the platoon leader determines the extent of the situation, he forwards a follow-up spot report.

Troop Action. The troop commander will most likely tell the scout platoon not in contact to continue its reconnaissance to a designated LOA to develop the situation across the entire troop front. By doing this, the troop can determine if there are any other enemy forces with which the troop must be concerned. The scout platoon not in contact will establish hasty observation posts along the LOA oriented on likely enemy locations or avenues of approach. The tank platoons will continue to monitor the troop command net and prepare to act based on the commander's intent/guidance and the enemy situation.

Choose a Course of Action

Platoon Action. Now that the platoon leader knows what he is up against, he considers two or three possible courses of action and selects the one that best meets the commander's intent/concept of the operation, is within his capabilities, and allows the troop to resume its reconnaissance mission as soon as possible. The possible courses of action open to him, based on the commander's intent/concept, might be hasty attack, bypass, hasty defense/screen, or support of a hasty attack by another platoon(s). Some courses of action will quickly be ruled out because they do not meet the commander's intent and guidance for the operation.

- **Hasty Attack.** The platoon leader can conduct a hasty attack if he has enough combat power to defeat the enemy quickly. In most cases, the scout platoon does not have the capability to mass enough combat power to defeat an enemy in prepared positions. In addition, the scout platoon leader may not want to risk battle losses that would reduce his effectiveness and his ability to complete the mission.

- **Bypass.** If the platoon does not have enough combat power to conduct a hasty attack, or if it wants to remain undetected and continue the reconnaissance mission, the scout platoon can bypass the enemy. The platoon leader must receive the troop commander's permission to bypass. If he has permission to bypass, the platoon leader must leave scouts in contact with the enemy force unless ordered to break contact by the troop commander. The platoon leader will know from the troop commander's guidance in the operations order if this is a viable course of action.

- **Hasty Defense/Screen.** If the platoon cannot conduct a hasty attack and cannot bypass, it establishes a hasty defense or screen. The platoon will conduct a hasty defense if it can defend against an enemy force. If the enemy contact exceeds the platoon's capability to conduct a hasty defense, it may elect to establish a screen and maintain contact through observation. The platoon concentrates on maintaining contact with the enemy and fixing it in place with indirect or possibly direct fire until additional support comes from the troop.

- **Support a Hasty Attack by Another Platoon(s).** The platoon in contact may become the support element for a hasty attack by a tank/AT platoon(s).

Troop Action. The troop commander continues to maneuver the troop and assists the platoon leader as required. He maintains situational awareness.

Recommend/Execute a Course of Action

Platoon Action. If the course of action the platoon leader chooses meets the commander's guidance for actions on contact, the platoon leader executes the course of action. If the situation dictates that the platoon execute a hasty defense/screen, or support a hasty attack by another platoon, the platoon leader reports this information to the troop. The report is brief. He updates his spot report (Blue 1) with any additional information, tells the troop commander what he is doing about the situation, and recommends the course of action that he thinks best suits the situation. The platoon in contact has then completed its actions on contact.

Troop Action. The commander must approve or disapprove the recommended course of action, based on its effect on the troop and squadron mission.

If the scout platoon is required to establish a hasty defense/screen, the troop commander assumes responsibility for continuing to develop the situation. The commander can do this best in a position from which he can influence troop actions. Based on what the scouts have found and reported and what the commander can see, he must weigh the advantages and disadvantages of each course of action, and choose the one that best meets the troop's mission requirements.

- **Hasty Attack**. The troop can conduct a hasty attack, which throws the weight of one or both tank/AT platoons against the enemy and destroys him quickly, then continue its reconnaissance. The commander may coordinate the efforts of the scout platoon and tank/AT platoon(s) in the hasty attack or delegate the responsibility to the scout platoon in contact. The hasty attack at troop level should have two elements: the *assault/attack-by-fire* element and the *support* element.

  Steps to the execution of a hasty attack are as follows:

  - The scouts determine that the enemy force encountered cannot be bypassed based on the enemy disposition and composition. The enemy encountered meets the commander's intent for destruction.

  - The scout platoon leader recommends hasty attack to the troop commander and identifies a good attack-by-fire or assault position for the tank/AT platoons.

  - The troop commander approves the scout platoon leader's recommended course of action and issues orders to execute a hasty attack.

  - The troop commander decides to use either attack by fire or assault as his method of destruction. Heavy troops equipped with tanks can perform attack by fire or assault. Because of the defensive nature of the TOW, light troops tend to use attack by fire when conducting hasty attack.

  - The scout squad or section moves to link up with the tank/AT platoons and guides them to an assault/attack position. The assault element reports when set in assault positions, and if time is available, reconnoiters the attack-by-fire position or the attack axis.

- The troop commander moves to the scouts who are positioned in overwatch of the enemy contact. He also positions the FIST in overwatch to assist in controlling indirect fires.

- The troop commander designates the scouts in overwatch as the support element. He establishes direction of fire and methods of control for direct-fire weapon systems; for example, left and right limits for fires, and signals for the initiation of direct fires and the shifting of fires as the assault element begins its attack.

- The troop commander directs the FIST to fire for effect on enemy positions when given the signal (AT MY COMMAND). The FIST reports when guns are ready.

- The troop commander issues the fire command for the support force and indirect-fire systems and directs the attack-by-fire or the assault element to execute its movement. The troop commander lifts and shifts direct and indirect fires as necessary to cover the movement of the assault element and to seal off the enemy withdrawal (see Figure 3-3).

- **Bypass**. The troop may bypass the enemy force and continue the reconnaissance to further develop the enemy situation throughout the depth of the zone while maintaining an element of surprise. The decision to bypass is based on the higher commander's intent for the operation. The troop should leave an element in contact with the enemy force; however, the more elements in contact with the bypassed enemy mean the less reconnaissance will be conducted.

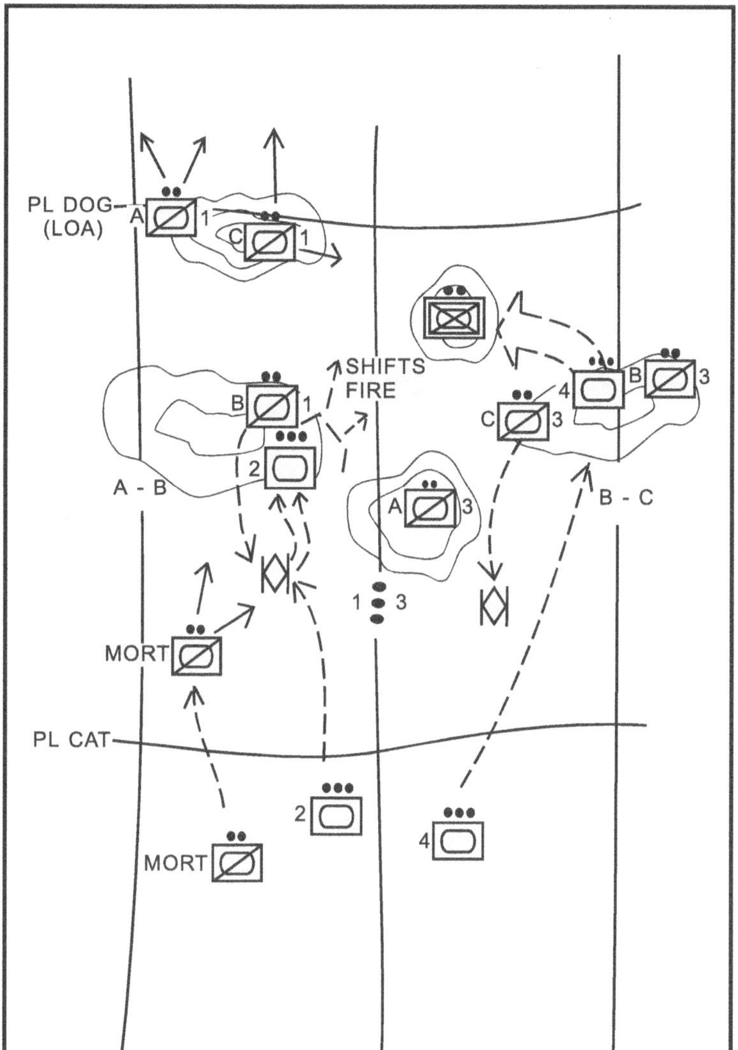

Figure 3-3. Troop positions for hasty attack.

- **Hasty Defense**. If the troop cannot conduct a hasty attack or bypass the enemy, it may establish a hasty defense to fix the enemy while waiting for further orders. If this is the troop's course of action, the squadron commander has the responsibility to develop the situation further.

The overriding considerations in selecting a course of action are the intent of the squadron commander and the troop's ability to complete the mission with minimum losses.

DANGER AREAS

Obstacles

Obstacles encountered during reconnaissance are treated as enemy contact, because all obstacles are assumed to be covered by enemy fire. Upon encountering an obstacle, scouts deploy to covered positions, report, and then begin to develop the situation. The troop commander may move his tank platoon(s)/AT platoon(s) forward to overwatch the obstacle while the scouts conduct dismounted and mounted reconnaissance to determine the following:

- Is the obstacle defended by the enemy? If so, how many enemy soldiers are there and where are they?
- What are the extent and composition of this obstacle? How deep, wide, steep, or long is it?
- Can the obstacle be bypassed, or will breaching be necessary?

Once the scouts have completed the reconnaissance of the obstacle and recommended a course of action (either bypass, hasty attack/breach, or hasty defense), the troop commander must choose the course of action that best meets the commander's intent and guidance for the mission. If the commander's decision is to bypass or breach the obstacle, the troop must mark the obstacle and any lane created through the obstacle (see Figure 3-4).

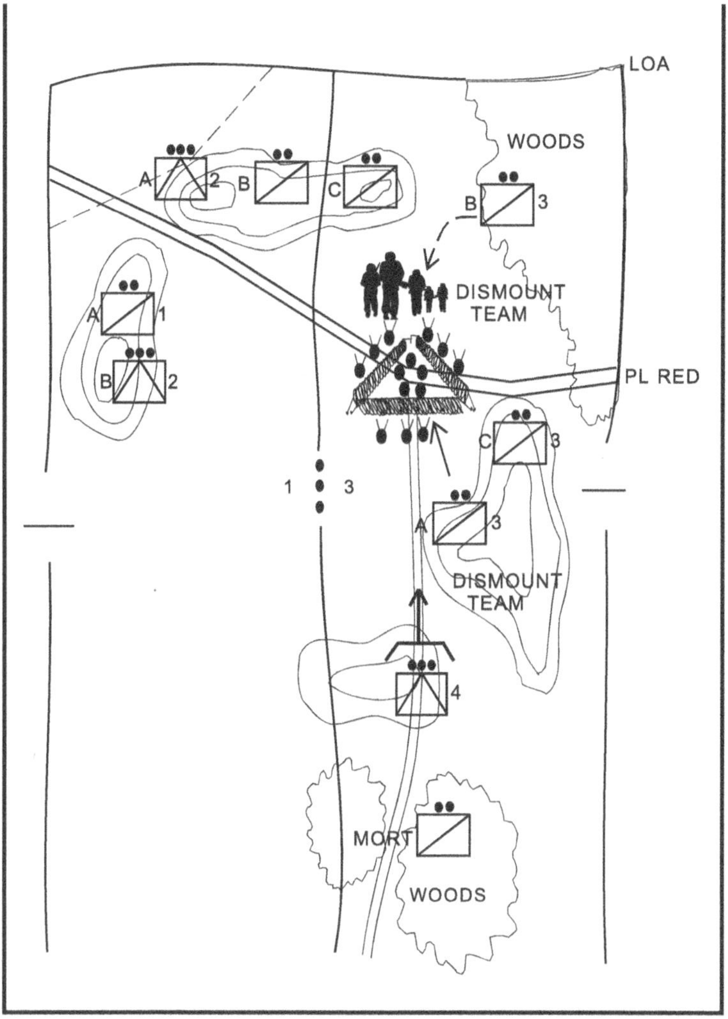

Figure 3-4. The troop develops the situation at the obstacle.

Open Areas

Open areas are dangerous for cavalry troops because they permit the enemy to observe or engage units as they move through them. When reconnoitering a route, zone, or area that includes an open area, do not try to force the reconnaissance through it. Treat it like a danger area. Reconnoiter the flanks using good covered and concealed routes. The troop should try to clear the far side of the open area before attempting to reconnoiter the open area itself. Open areas within the troop's zone or area of operations may force the troop commander to abandon the use of platoon boundaries to facilitate the reconnaissance to the flanks and rear of the open area (see Figure 3-5).

Built-up Areas

In most areas of the world the troop can expect to conduct reconnaissance of built-up areas (BUA). BUAs range from small hamlets to large towns, even portions of cities. BUAs are very dangerous for mounted units and should be avoided when possible. If the situation permits, the troop should conduct a reconnaissance of BUAs from a distance, and then bypass them. However, it may be necessary to move through a town as part of the troop's reconnaissance mission.

The troop's ability to conduct reconnaissance of a BUA is constrained by the lack of dismounted scouts. The troop is capable of finding the following in BUAs:

- Enemy vehicles.
- Enemy command and control facilities.
- Obstacles such as rubble, blown bridges, and craters.
- Logistics elements.
- Bypasses within the BUA.

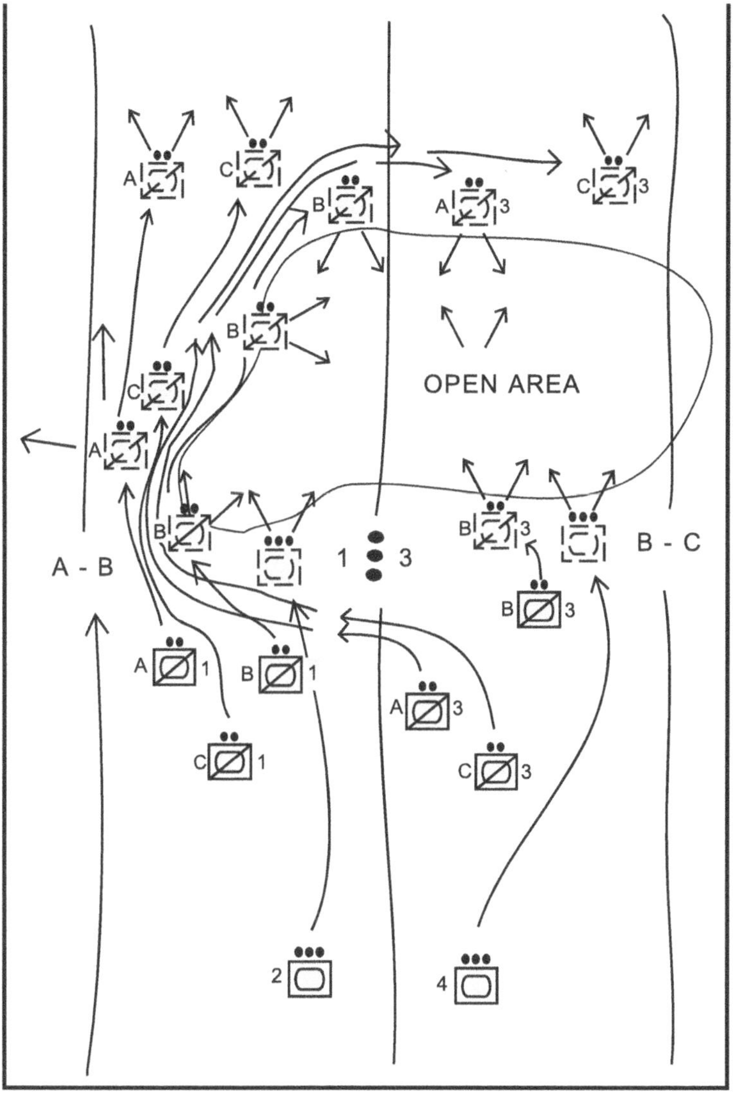

Figure 3-5. The troop reconnoiters to the flanks and rear of an open area.

If the troop is not augmented, it is not capable of performing the following in BUAs:

- Reconnoitering each building and street.
- Reconnoitering underground structures (basements and sewers).
- Determining the strength of dug-in enemy units.
- Determining the detailed disposition of dismounted enemy units.
- Conducting hasty attacks against dismounted enemy units of larger than platoon size.

The following are techniques and considerations for performing reconnaissance in BUAs:

- Dirt roads, alleys, and curves in paved roads are excellent places to employ mines. Structures in BUAs are excellent places to employ booby traps.
- Ask the S2 for detailed street maps of all major BUAs the troop might encounter during an operation. These are more useful than standard 1:50,000 military maps.
- Clearly define platoon zones. Do not divide responsibility for a street between platoons.
- Observe BUAs from outside for signs of enemy activity prior to entering them.
- Main roads through towns are normally accompanied by open areas such as parks, traffic circles, and medians. The enemy may use these same open areas for logistics elements, concentrations of armored vehicles, artillery positions, and combat support vehicle locations.
- Collateral damage constraints may limit the ability to employ some types of weapon systems.

LIGHT TROOP EXAMPLE

The squadron is conducting a zone reconnaissance. The troop zone contains a large town, 2 to 3 kilometers square, and the squadron main supply route, ROUTE GOLD, passes through the town. The troop will have to reconnoiter enough of the town to ensure there are no enemy who can disrupt movement along ROUTE GOLD.

On approaching the town, the troop has options on how to enter it. The techniques used will largely depend on the amount of time available for the mission. If time is limited, the scouts will halt short of the BUA, just off ROUTE GOLD, and observe the town from multiple vantage points before entering. If more time is available, the scouts will move well off the route and observe the BUA before entering. In any case, remember if enemy is present, he will probably cover the main road with fires. Once the platoon is inside the BUA, the main road may provide the best axis for the reconnaissance.

As the scout platoon observes the town, they look for any signs of enemy presence. The antitank platoons move into overwatch positions. The mortars establish a firing position in support of the scouts. Scouts reconnoiter to the flanks and rear of the BUA to find a bypass and establish flank and far side security (see Figure 3-6).

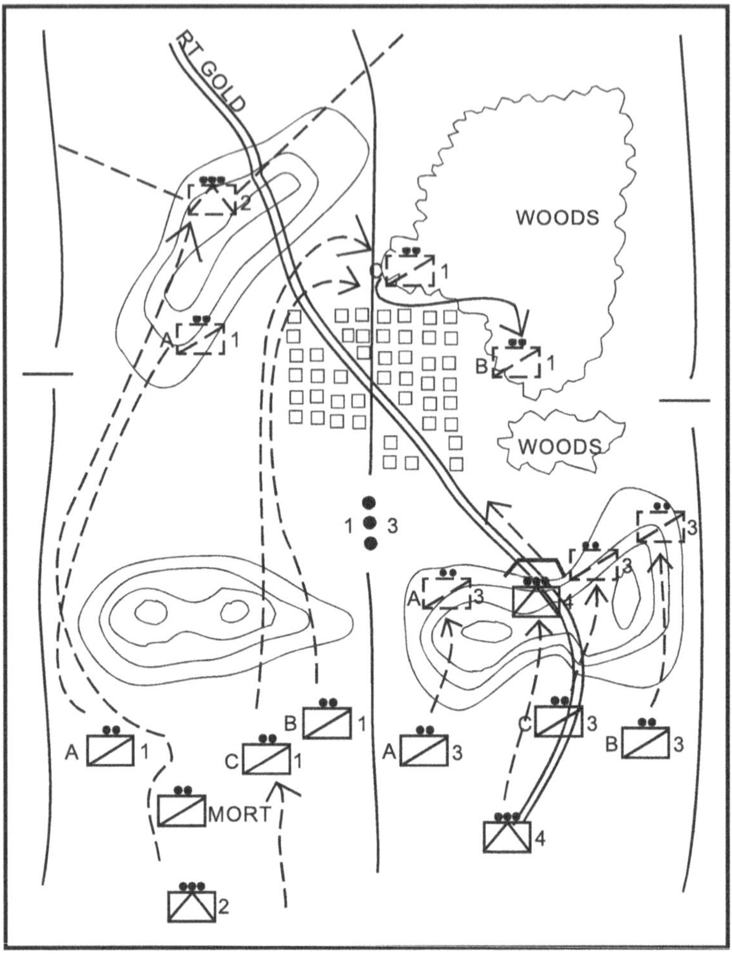

Figure 3-6. The troop securing the BUA with one scout platoon, ATs in overwatch, and one scout platoon preparing to reconnoiter the BUA itself.

Once the area surrounding the BUA is secured, the scouts move mounted to the last concealed position and dismount, searching for suitable entry points into the BUA. *If there is enemy contact*, the overwatching vehicles suppress with direct and indirect fires while the scouts move to a covered and concealed position. At this point the troop commander must decide to maintain contact and bypass, or continue to probe for entry points into the BUA. The determining factors for the commander's decision are the intent of the higher commander and the level of threat in the BUA. In most instances, cavalry troops maintain contact and bypass BUAs occupied by enemy forces. The enemy forces are handed off to follow-on forces for destruction.

If the troop does not encounter enemy contact on its initial reconnaissance of the outskirts of the BUA, the scout platoons continue their search for suitable entry points into the town. Once the scouts determine those entry points into the town are clear, they move into the BUA and reconnoiter nearby buildings (moving through side and back yards, alleys, and other off-street paths) and surrounding areas. The dismounted teams from the scout platoon do not enter or clear each building, but look for obvious signs of enemy presence. HMMWVs may be brought forward to overwatch the dismounted team's movement with caliber .50 machine guns or MK-19s (see Figure 3-7). The AT platoons should continue to overwatch from outside the BUA. After the scouts complete the reconnaissance of the BUA, the troop continues its reconnaissance in zone, ensuring that bypasses around the BUA are marked.

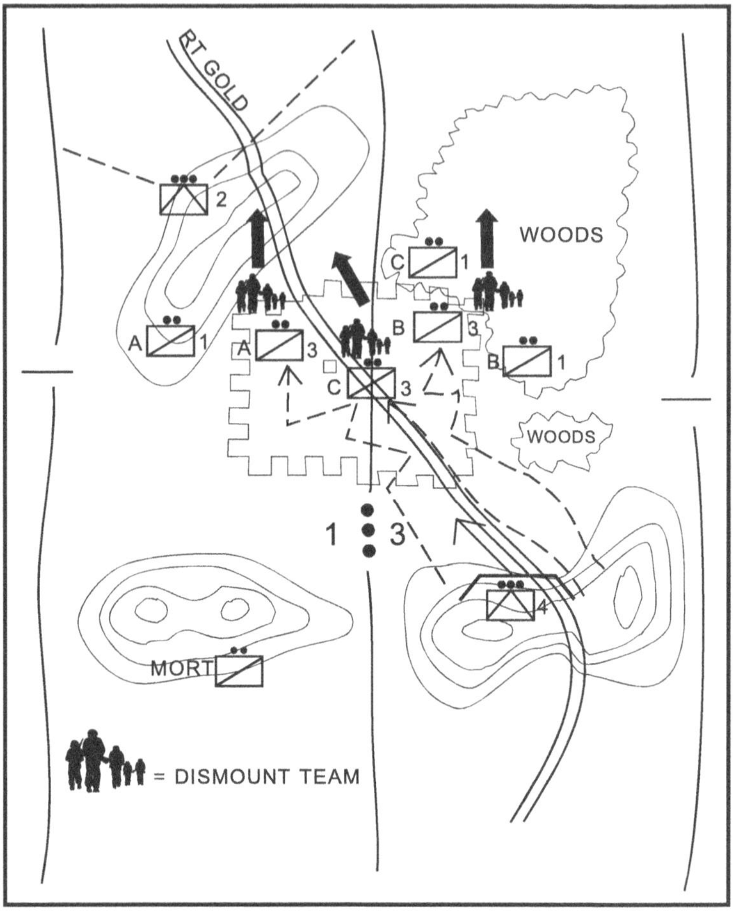

Figure 3-7. Scout section reconnoitering in a built-up area.

Section III. Route Reconnaissance

Route reconnaissance is a directed effort to gain detailed information about a specific route and the terrain on either side of the route that the enemy could use to influence movement along the route. Depending on the terrain surrounding the route, a route reconnaissance mission may be accomplished by a single scout platoon, or it may require the entire troop to cover the terrain and accomplish the required tasks.

A cavalry troop may be assigned the mission to conduct a zone reconnaissance (see Section IV of this chapter), but also have the task of reconnoitering a route that is within its zone. In this case, the troop mission would be zone reconnaissance. However, one of the subordinate units in the troop, scout platoon, or attached engineers may be given the route reconnaissance mission by the troop commander.

CRITICAL TASKS

During a route reconnaissance, the following critical tasks must be accomplished unless the SCO directs the troop to do otherwise:

- Reconnoiter and determine trafficability of the route.
- Reconnoiter all terrain the enemy can use to dominate movement along the route.
- Reconnoiter all BUAs along the route.
- Reconnoiter all lateral routes in the area of responsibility.
- Inspect and classify all bridges along the route.
- Locate fords or crossing sites near all bridges along the route.
- Inspect and classify all overpasses, underpasses, and culverts.

- Reconnoiter all defiles along the route; clear all defiles of enemy and obstacles within its capability, or locate a bypass.

- Locate mines, obstacles, and barriers, and within its capability, clear the route.

- Locate a bypass around BUAs, obstacles, and contaminated areas.

- Report route information.

- Find and report all enemy that can influence movement along the route.

TECHNIQUES

The SCO directs the troop to conduct a route reconnaissance as a mission or as a specific task in another mission. This section discusses route reconnaissance in the context of an assigned troop mission.

The troop commander considers several factors in formulating his concept.

- Start point, release point, and designation of the route.

- Mission to be performed at the start point and after reaching the release point.

- Time the mission is to start, and if required, be completed.

- Critical points along the route identified as checkpoints.

- Any constraints or restrictions.

- IPB information on the route, to include current enemy situation.

IPB provides critical information on the enemy and terrain. Enemy threats may be encountered in two basic forms.

- Ambushes along the route in close or restricted terrain or even tied to obstacles along the route.
- Attack by long-range direct or indirect fires from dominating terrain along the route.

Analysis of the terrain provides an indication of danger areas and the nature of the potential threat. From this information, the commander determines how much terrain on each flank of the route must be reconnoitered and the organization for combat. Any constraints or restrictions may also influence how much terrain is reconnoitered.

The troop commander adds control measures to provide an adequate framework for the mission. He places a boundary on both sides of the route far enough out to provide reconnaissance of the dominating terrain. A line of departure is placed perpendicular to the route short of the start point, allowing sufficient space to deploy into a tactical formation. An LOA is placed far enough beyond the release point to enclose dominating terrain that overwatches the release point. The commander uses additional phase lines to maintain a coordinated effort throughout the troop. Boundaries and phase lines are drawn along recognizable terrain features. The terrain features should be visible from both ground and air to facilitate air-ground integration. The commander uses other control measures as necessary for flexibility in the maneuver and execution of the mission (see Figure 3-8).

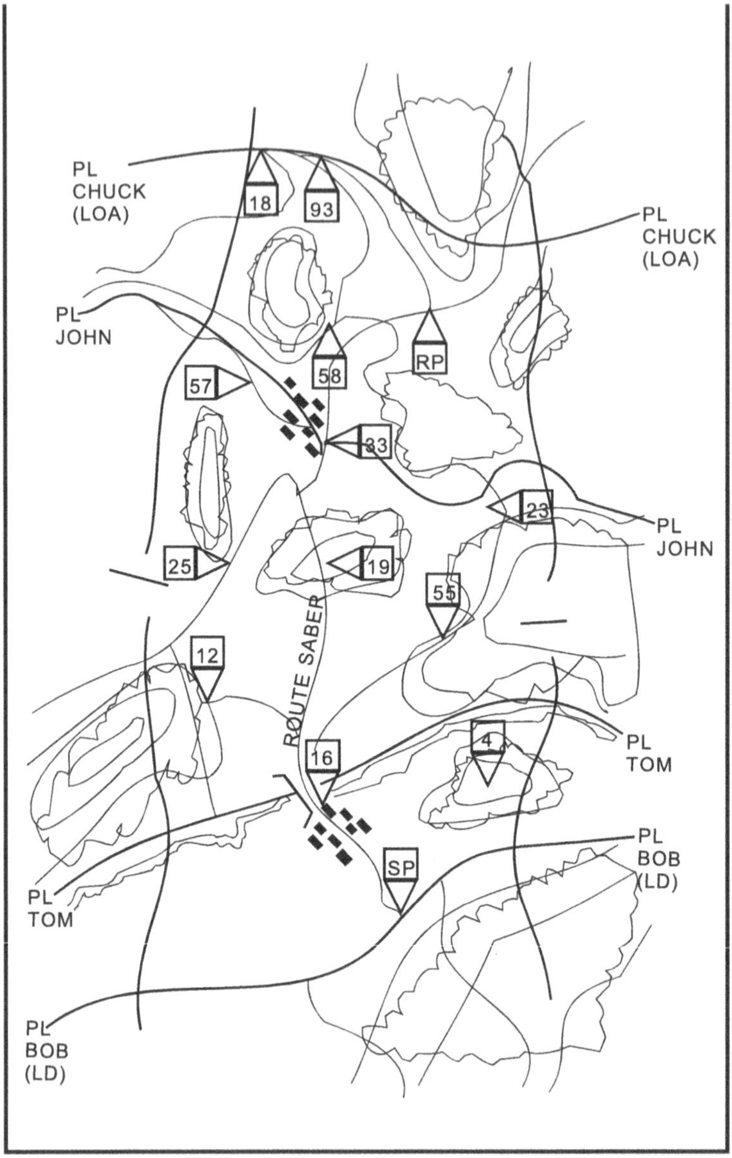

Figure 3-8. Troop graphic control measures for a route reconnaissance.

The cavalry troop normally performs a tactical road march to the line of departure and deploys to execute the reconnaissance of the route. Based on the amount of intelligence known about the enemy, the commander will determine how much security is required for the move forward to the line of departure. Also the commander should consider the effect his final disposition of forces will have on the troop's follow-on mission.

One technique for accomplishing all tasks as rapidly and safely as possible is as follows: Draw a 2.5-kilometer to 3-kilometer boundary on either side of the route. Near the start point, draw a line of departure from one boundary to the other. Next, draw an LOA about 3 kilometers forward of the release point from one boundary to the other. Add phase lines between the line of departure and LOA, as needed, to help control the progress of the troop. To control movement on the route, establish checkpoints. Use checkpoints to identify specific areas or locations to be reconnoitered. These checkpoints will serve as references to orient reconnaissance efforts.

One scout platoon moves out early across the line of departure to reconnoiter the terrain or BUAs on either side of the route. The other scout platoon reconnoiters the route itself, trailing 1 kilometer to 2 kilometers behind the lead scout platoon. This provides a good measure of security for the troop and the platoon working to reconnoiter and classify the route. The mortar section follows the lead scout platoon and remains positioned to range from 3,000 to 3,500 meters forward of the lead scout elements. The mortar section avoids moving on the route and stays on the flanks near the route. The tank/AT platoons follow and support behind the second scout platoon. They key their movement on the advance of the scout platoons. The tank/AT platoons should be kept back in depth to retain flexibility, because they must be ready to react anywhere in the troop zone. The troop trains follow about 2 kilometers behind the tank/AT platoons, and bound from one covered and concealed position to another. The troop CP displaces generally along the route using terrain that affords effective and continuous communications with troop elements and the squadron. The troop commander positions himself to observe the actions of the scout platoons.

EXAMPLE OF A TROOP ROUTE RECONNAISSANCE

The 1st platoon (scout) crosses the LD (PL BOB) using the two-section organization, and reconnoiters the terrain on either side of ROUTE SABER up to PL TOM. The platoon looks for the threat around the outskirts of the village, and explores routes into the village that intersect ROUTE SABER. The scouts also search the woods near checkpoint 4 for the enemy. The platoon scans the terrain north of the stream, and then moves up. One section locates a fording site west of the bridge at checkpoint 16, and the other section swims the stream on the east. The platoon makes a quick visual inspection of the bridge, then searches for the threat in the forested terrain near checkpoints 12 and 55. The mortar section crosses the LD next, about 2 kilometers behind the 1st platoon (scout), and establishes a firing position on the outskirts of the village. The 3d platoon (scout) crosses the SP when the 1st platoon (scout) crosses PL TOM.

The 3d platoon (scout) reconnoiters and classifies the route through the village, then moves up and classifies the bridge just north of town. The troop commander bounds closely behind the 3d platoon (scout) and observes their actions. The 2d and 4th platoons (tank) follow along the flank of ROUTE SABER about 1 kilometer behind the 3d platoon (scout), then move into the outskirts of the village. They observe the dominating terrain on the far side of the stream while the 3d platoon (scout) reconnoiters the bridge. The troop CP and trains remain in concealed positions south of the LD (see Figure 3-9).

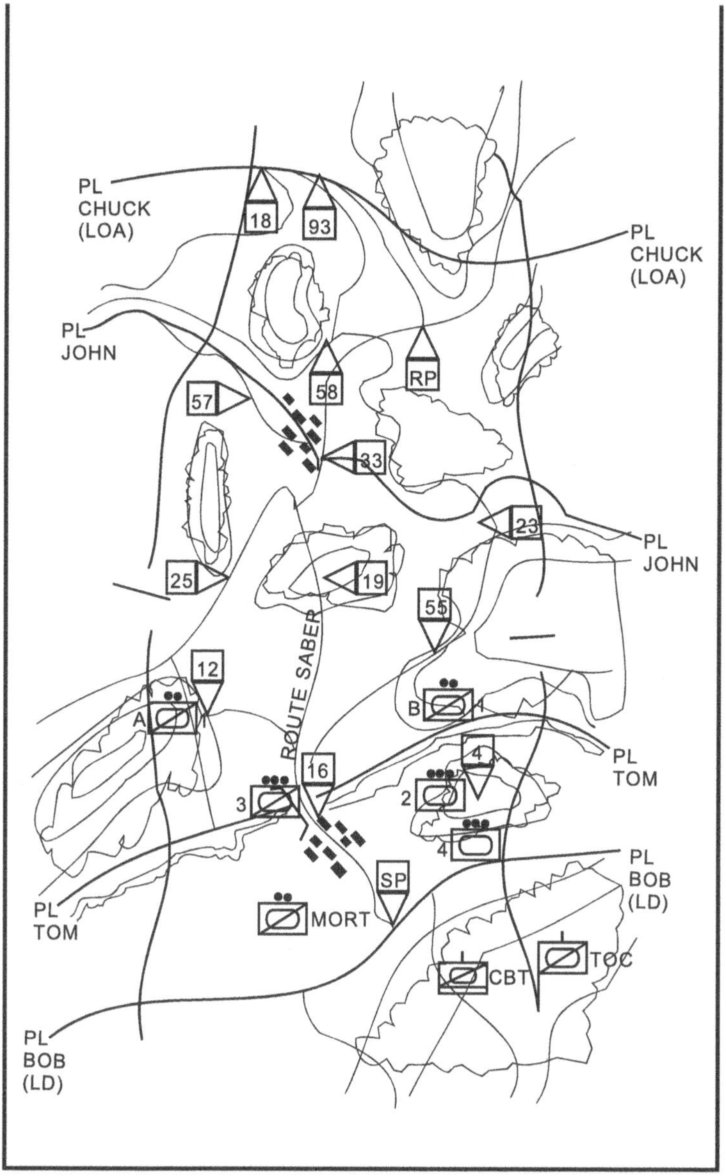

Figure 3-9. Route reconnaissance (part one).

The 1st platoon continues to reconnoiter in zone up to PL JOHN. The platoon searches the wooded areas for the threat, and explores the high-speed lateral routes to the east and west of ROUTE SABER. The platoon clears the shoulders and makes a visual inspection of the defile on ROUTE SABER at checkpoint 19, then moves forward to search the outskirts of the village at checkpoint 33. As with the first village, the platoon looks for threat forces along the roads into the village. The mortar section follows the 1st platoon and establishes a firing position on the edge of the woods southwest of checkpoint 19. The 3d platoon continues classifying the route along ROUTE SABER. At the defile near checkpoint 19, scouts using mine detectors check the route to ensure it is free of mines. The 2d and 4th platoons cross the bridge at checkpoint 16, and then move into concealed positions in the woods on either side of ROUTE SABER near checkpoint 19. The troop CP moves to a concealed position on high ground near checkpoint 12. The troop trains move into covered and concealed positions in the village south of checkpoint 16 (see Figure 3-10).

As the 1st platoon crosses PL JOHN and approaches the RP, the lead scout squad of Bravo section is hit by cannon and heavy machine-gun fire from concealed positions near checkpoint 93. The vehicle has its left track blown off and sustains external damage. The gunner and one scout are wounded by shell fragments. The PSG immediately returns fire and deploys to cover then reports the contact to the platoon leader who forwards a contact report to the troop commander. Once the PSG determines the location of the enemy force, he sends a spot report to the platoon leader and a request for immediate suppression to the troop FSO. After the call for fire is sent forward, the PSG directs his remaining scout squad to search for additional threat forces in the woods north of the RP and along the high-speed route heading northeast (see Figure 3-11). As mortar fire falls on the enemy position, the 1st platoon leader quickly moves up near checkpoint 58 to observe the threat. He directs his section to search the woods north of checkpoint 57, and to move around to the west flank of the threat and find out

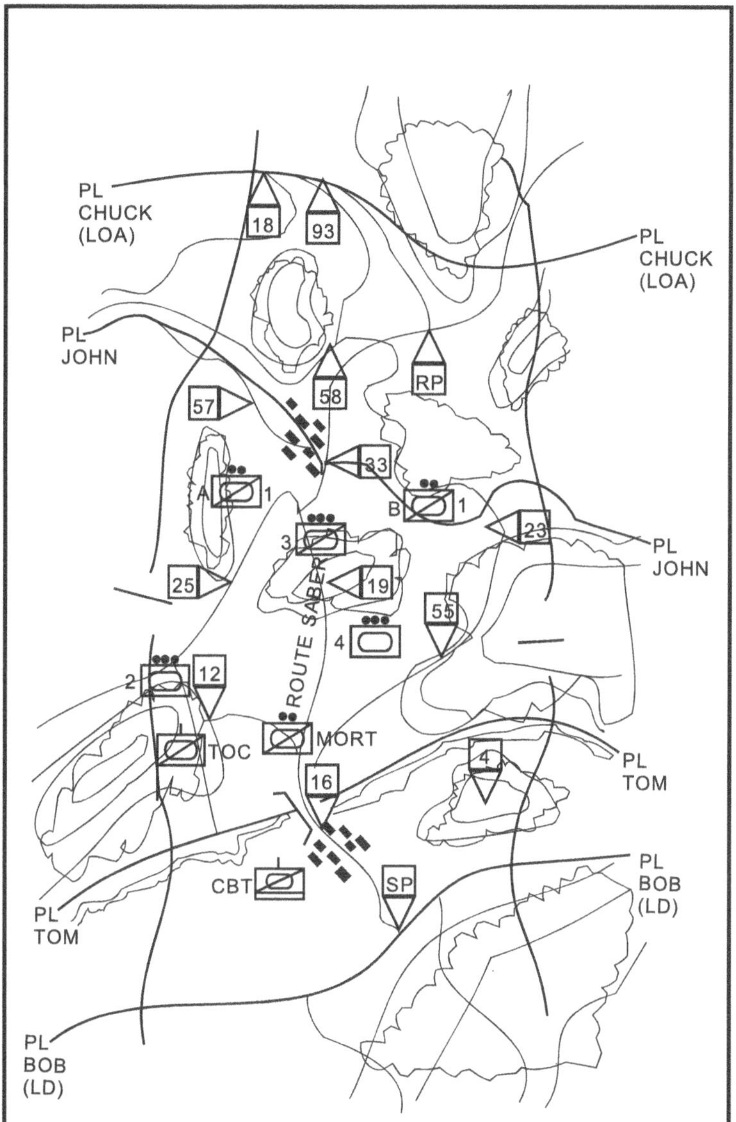

Figure 3-10. Route reconnaissance (part two).

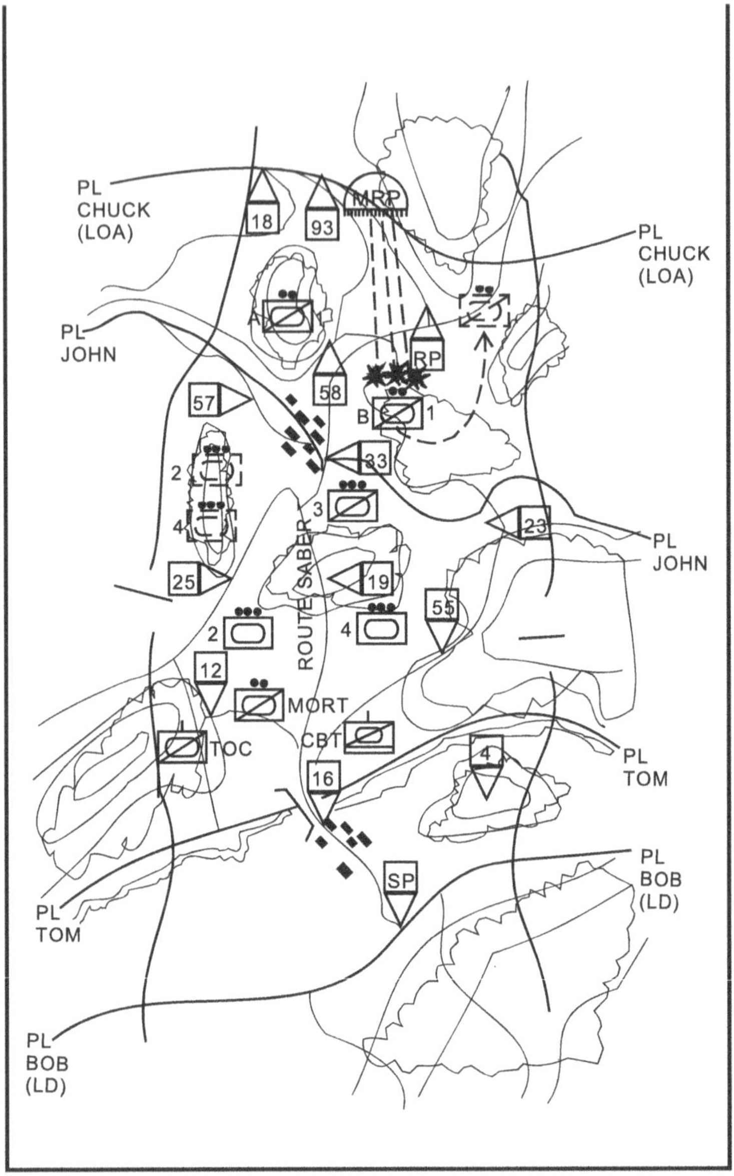

Figure 3-11. Route reconnaissance (part three).

what the platoon is up against and whether or not the threat force is supported by other elements in the village at checkpoint 18. He then reports to the troop commander. The troop commander immediately moves forward to join the 1st platoon leader and to see the situation for himself. En route, the commander tells the 2d and 4th platoons to move into concealed positions in the village at checkpoint 33, and to prepare to counterattack. The 3d platoon continues to reconnoiter the route through the village at checkpoint 33, and then moves into covered positions along ROUTE SABER just south of checkpoint 33. The troop CP holds in place and reports the situation to squadron. The 1SG quickly leads the M88 and the medic vehicle forward to treat and evacuate the injured scouts and their damaged vehicle.

Dismounted scouts from the 1st platoon determine that the threat force consists of three BMPs and one T-64 in prepared positions. Threat crews remain mounted and continue to slew turrets, scanning for more friendly elements. The threat reinforced platoon is oriented southeast toward the route. Their western flank is free of obstacles and unobserved. The troop commander decides to conduct a hasty attack (see Figure 3-12).

When the tank platoons complete their hasty attack, the troop commander gets permission to advance beyond the LOA and reconnoiter the enemy position. The 3d platoon returns to complete the reconnaissance of ROUTE SABER. As the platoon checks for mines along the route to the RP, they discover the route is mined just short of the RP. They find a bypass around the obstacle and forward a bypass report (Blue 10) to the troop XO. The platoon then organizes into two sections and moves out to occupy OPs overlooking the lateral routes at the east and west flanks of the troop near PL CHUCK. The troop CP moves into position on high ground near checkpoint 58. The 1SG directs activities of the troop trains, to include evacuation of injured soldiers and inoperable vehicles, security and evacuation of injured prisoners, requests for casualty replacements and resupply.

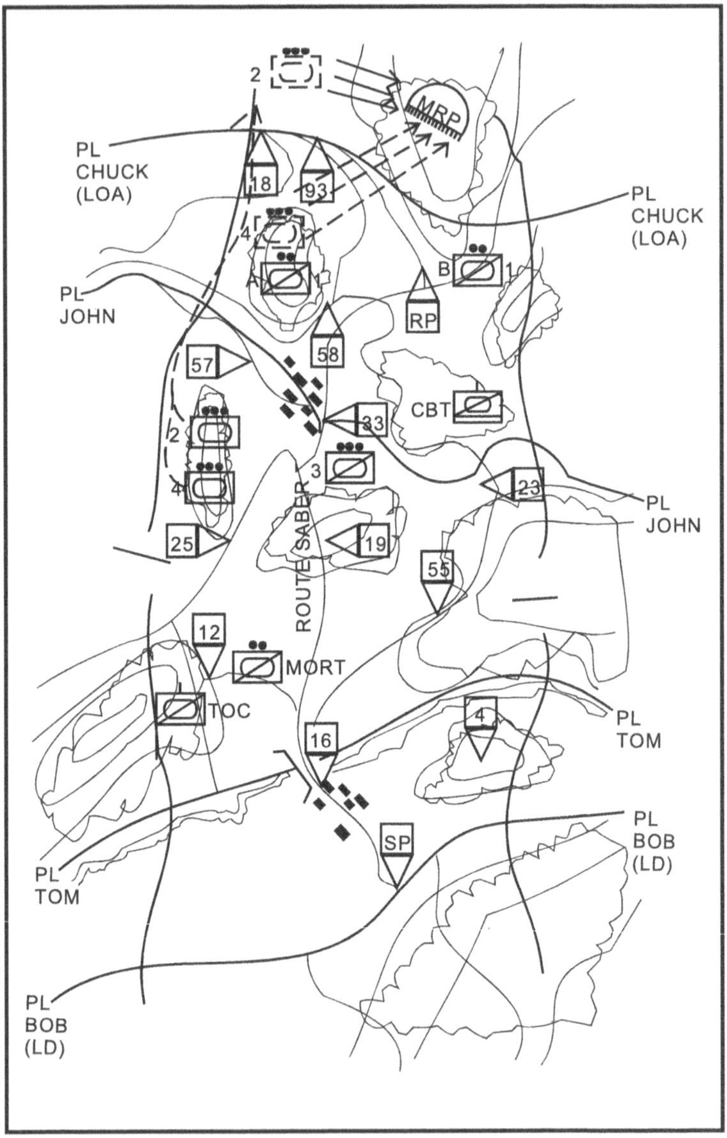

Figure 3-12. Route reconnaissance (part four).

Section IV. Zone Reconnaissance

Zone reconnaissance is the directed effort to obtain detailed information concerning all routes, obstacles, terrain, and enemy forces within a zone defined by boundaries. Obstacles include chemical and radiological contamination. A zone reconnaissance is assigned when the enemy situation is vague or when information concerning cross-country trafficability is desired. It is appropriate when previous knowledge of the terrain is limited or when combat operations have altered the terrain.

CRITICAL TASKS

Zone reconnaissance is a deliberate, time-consuming process. During a zone reconnaissance, the troop accomplishes the following critical tasks unless specifically directed otherwise by the higher commander:

- Reconnoiter all terrain within the zone.
- Inspect and classify all bridges within the zone.
- Locate fords or crossing sites near all bridges in the zone.
- Inspect and classify all overpasses, underpasses, and culverts.
- Locate and clear all mines, obstacles, and barriers in the zone within its capability.
- Locate a bypass around BUAs, obstacles, and contaminated areas.
- Find and report all enemy forces within the zone.
- Report reconnaissance information.

TECHNIQUES

The cavalry troop is usually assigned a zone reconnaissance as part of a squadron zone reconnaissance mission.

A zone reconnaissance mission is very time-consuming. Any commander who orders this mission must remember the number and complexity of the tasks to be accomplished. If he wants a faster tempo, he must prioritize reconnaissance tasks for the troop.

The cavalry troop can effectively reconnoiter a zone from 6 to 10 kilometers wide. If stretched any farther, the troop quickly exceeds its ability to accomplish the critical tasks.

Because the enemy situation is vague and knowledge of the terrain is limited to what is shown on maps or aerial photos, platoons will face unexpected situations around every corner; therefore, the course of action selected must also provide a good measure of protection for the troop as it executes the mission. When considering techniques of conducting a zone reconnaissance, the scheme of maneuver has to be flexible. The troop commander must convey his intent to subordinates so they can act quickly and without orders. He should employ forces, tank or AT platoons, in depth to give the troop flexibility and responsiveness.

When the troop receives a zone reconnaissance mission, the zone is usually identified by lateral boundaries. The line of departure and a reconnaissance objective or LOA are specified. The commander should divide the troop zone into two platoon zones for the scout platoons. Use caution when drawing the boundary. Make sure it is on easily identifiable terrain that is not a high speed avenue of approach. The number of critical tasks within the troop zone, and terrain restrictions should guide the commander in determining the location of the scout platoon boundaries. Doctrinal distances are not always the best solution if one platoon will be overtasked.

Add phase lines every 5 to 8 kilometers on easily identifiable terrain to control progress of the troop through the zone. Place contact points near the intersection of the boundary and all phase lines. Ensure the contact points provide at a minimum the potential for concealment from enemy observation to allow the exchange of information between flank units. Use checkpoints to focus the efforts of scout platoons and to assist in maneuvering the troop (see Figure 3-13). Use TIRS on maps as described in Chapter 2.

Use a troop vee or split-vee to conduct zone reconnaissance. As scout platoons reconnoiter the zone, the tank/AT platoons follow and support behind the scouts, keying their movement off the scouts' forward progress. The tank/AT platoons are kept far enough in depth to retain flexibility, and yet remain responsive to situations developed by the scout platoons. The distance from the scouts is determined by the terrain and enemy situation. Therefore, the distance the tanks/ATs move from the scouts may vary during the course of the operation. Control tank/AT platoon movement by one of the following methods:

- Give movement guidance to the platoon leaders. Have them key movement off the scouts.
- Move the platoons by CPs, battle positions, and hide positions.

Move the mortar section center of the troop zone to range forward of the scouts. The range forward of the scouts is determined by the scouts' ability to acquire the enemy. The mortars may move under the control of the FIST or eavesdrop on the troop net and key their movement off the progress of the scout platoons. Regardless of who controls their movement, the TOC and the FIST track the mortar position constantly to ensure they can support the scouts.

Note. If the mortar section is busy firing missions and computing fire data, they may not be able to stay abreast of the friendly situation. Therefore, the FIST or XO should be ready to control the mortars displacement.

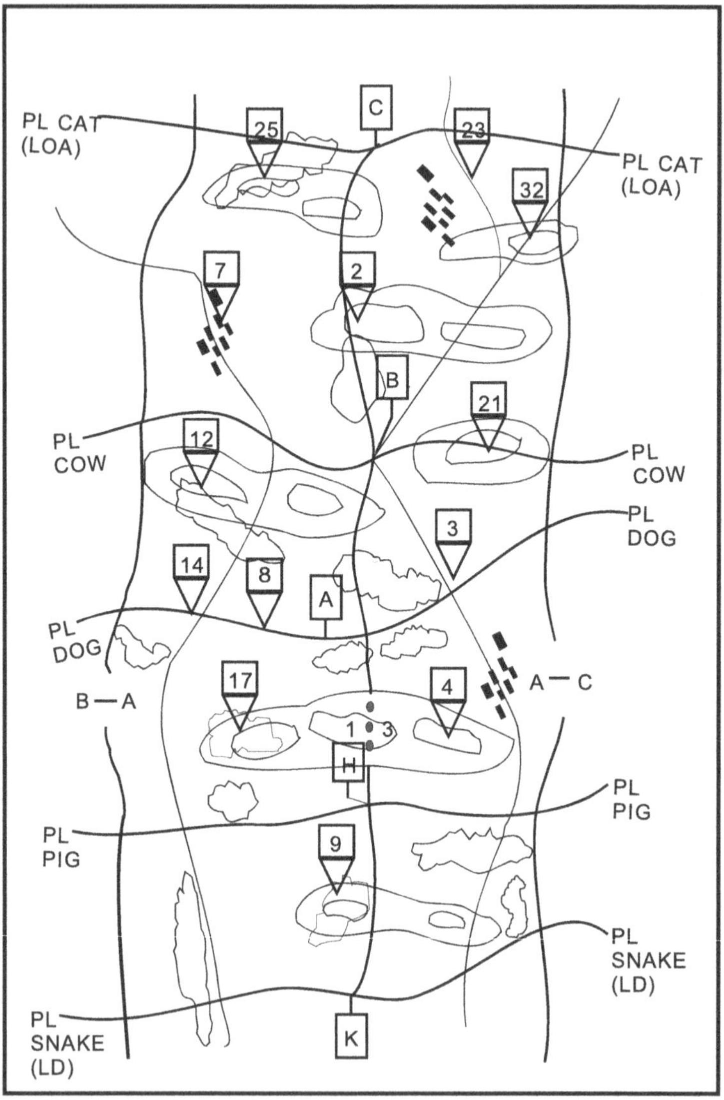

Figure 3-13. Zone reconnaissance graphic control measures.

The XO displaces the troop CP through the zone using terrain that affords effective and continuous communications with troop elements and the squadron. The 1SG moves the troop trains about 2 kilometers behind the tank platoons and bounds the trains from one covered and concealed position to another. The troop commander positions himself well forward to observe the actions of one scout platoon or the other. His position usually depends on where he expects initial enemy contact or problems to occur.

LIGHT TROOP EXAMPLE

The troop commander deploys the troop in a split-vee formation. The 1st and 3d platoons (scout) cross the LD, PL UTAH, abreast and reconnoiter the terrain in troop zone up to PL OHIO. The platoons determine both the open terrain and the trafficability of the route within their zone. Some hill slopes are greater than thirty percent and dense woods inhibit armor movement. Except for a marshy area north of checkpoint 9, the open terrain is dry and firm. Both routes will support heavy armor movement. The troop commander follows the 1st platoon. The mortar section eavesdrops on the troop command net to follow the progress of the scout platoons, and coordinates its moves with the troop FSO accordingly. As the scout platoons cross PL OHIO, the 2d and 4th platoons (antitank) cross the LD. They eavesdrop on the troop command net and key their movement on the progress of the scout platoons. The troop CP and the trains hold in place.

The scout platoons continue reconnaissance up to the river (PL BAMA). They take a close look at the dominant terrain near checkpoint 17 and at the village near checkpoint 4. As the platoons approach the river, scouts move into covered and concealed positions, dismount, and visually search the dominant terrain on the north side of the river. The 1st platoon inspects the bridge at checkpoint 14 and determines that it will support up to 40 tons. Scouts also verify that a good fording site exists near checkpoint 8. The 3d platoon confirms there is a fording site with a rock bottom just east of the blown bridge at checkpoint 3. It will support heavy armored traffic (see Figure 3-14).

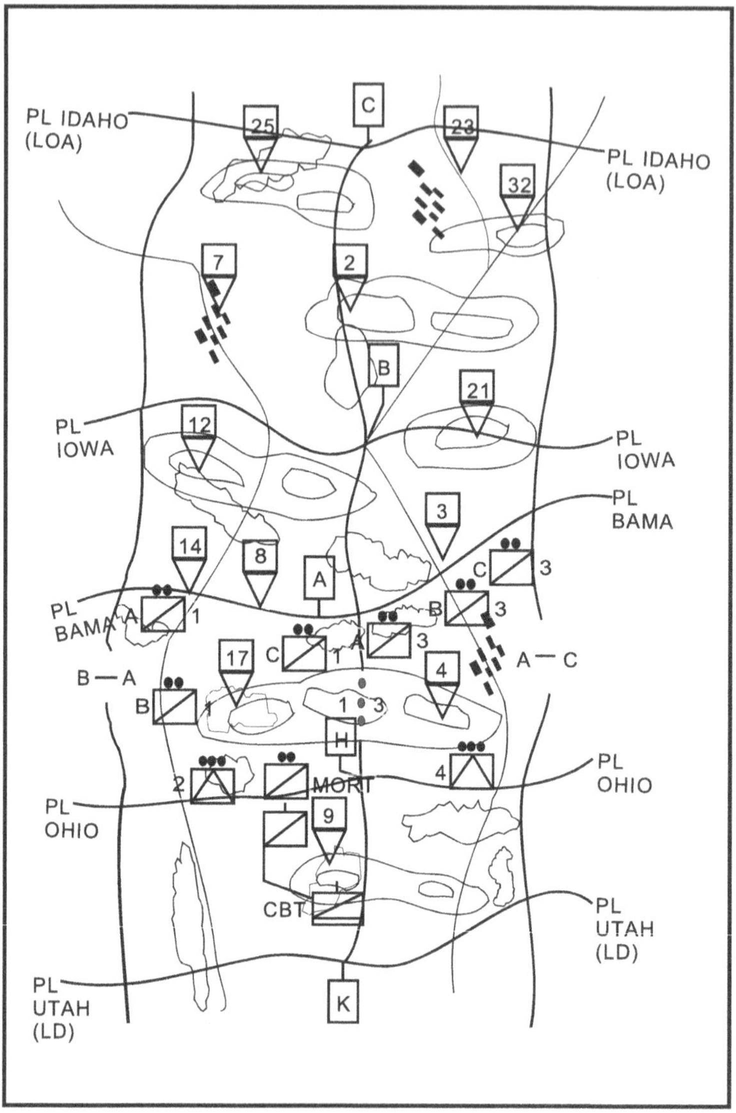

Figure 3-14. Zone reconnaissance (part one).

As the scout platoons continue reconnaissance towards PL IOWA, the 1st platoon observes two stationary BRDM-2s on the high ground near checkpoint 12. They look like a security outpost. The platoon engages with immediate suppression from the troop's 120-mm mortars. As the rounds impact, the outpost withdraws behind the ridgeline to the north. The 3d platoon determines the dominant terrain near checkpoint 21 is clear of threat forces. The mortar section crosses the river at checkpoint 8. The 2d platoon follows, crossing at checkpoint 8, while the 4th platoon uses the ford near checkpoint 3. The troop CP bounds forward to high ground near checkpoint 17. The troop trains cross the river and move into the woods on the northern bank.

The scout platoons continue reconnaissance while moving toward the troop LOA. As the 3d platoon crosses the ridge (PL IOWA), Bravo section receives heavy machine-gun fire from concealed positions south of checkpoint 2. The platoon immediately deploys, reports contact to the troop commander, and calls for indirect fire to suppress the threat (see Figure 3-15).

Under this protection, the platoon, using primarily a dismounted technique, reconnoiters to the flanks and rear to develop the situation. The troop commander acknowledges the report, and moves immediately to link up with the 3d platoon leader. The troop commander orders the 1st platoon to continue its reconnaissance to locate the enemy's flank and coordinate with the FIST to switch indirect fires to the squadron howitzer battery. The 3d platoon is organized in a three-section, three-vehicle configuration; the platoon leader is with the Charlie section. The platoon leader sends his Bravo section forward on the enemy's left and his Alpha section on the right to determine if the enemy is mutually supported by other threat forces from the flanks or rear. The Charlie section sergeant and a two-man dismount team move in closer and determine that the enemy consists of an MRP with three BTR 70s and one BMP-2. The Alpha scout section and scouts from 1st platoon identify the right flank of the threat and discover the flank is unprotected by obstacles and is exposed to direct fires.

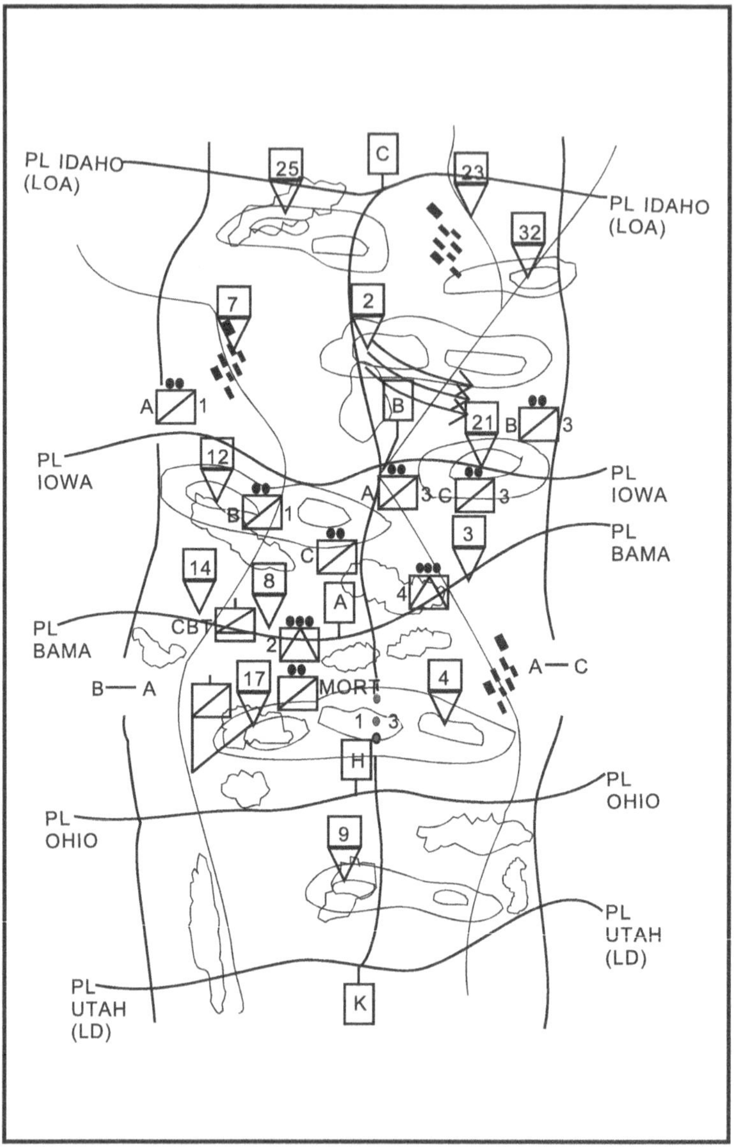

Figure 3-15. Zone reconnaissance (part two).

While en route to the 3d platoon's zone, the troop commander tells the 1st platoon to continue reconnaissance forward to PL IDAHO and determine if other enemy forces are providing mutual support to the enemy in the 3d platoon's zone. At PL IDAHO, he tells the platoon to establish a screen. The troop commander tells the 4th platoon to move over and join the 3d platoon Charlie section in overwatch of the enemy MRP, and prepare to overwatch with fires. The troop commander tells 2d platoon to prepare to conduct a hasty attack of the enemy MRP. The 1st platoon scout squad moves to checkpoint 12, links up with the 2d platoon, and guides them into position to attack the threat platoon.

The 2d platoon reports they are set in the assault position. The 4th platoon reports they are set in the overwatch position with the 3d platoon Charlie section. The 3d platoon leader reports that his dismount teams from his Alpha and Bravo sections are out of the target area and set. The 1st platoon leader reports he is set in a screen along PL IDAHO. The troop commander orders the FSO to fire for effect on the enemy position at his command. The commander issues orders and guidance for limits of fire and signals for the initiation, shifting, and lifting of fires. The troop commander issues the fire command for indirect fire over the command net and follows with a fire command for direct fire for the 4th platoon and the Charlie section of 3d platoon. The indirect fire from the squadron's howitzer battery and the troop mortars impact on the enemy position and is immediately followed by the fire from four TOWs, MK-19s, and M2 machine guns. The commander then issues the orders for 2d platoon to initiate its move to the attack-by-fire position followed by the order for the support element (4th platoon, Charlie section of 3d platoon) to shift fires. The troop FSO shifts fires to the rear of the enemy position. The 2d platoon moves forward and engages the enemy target from the flank to complete the destruction of the enemy force (see Figure 3-16).

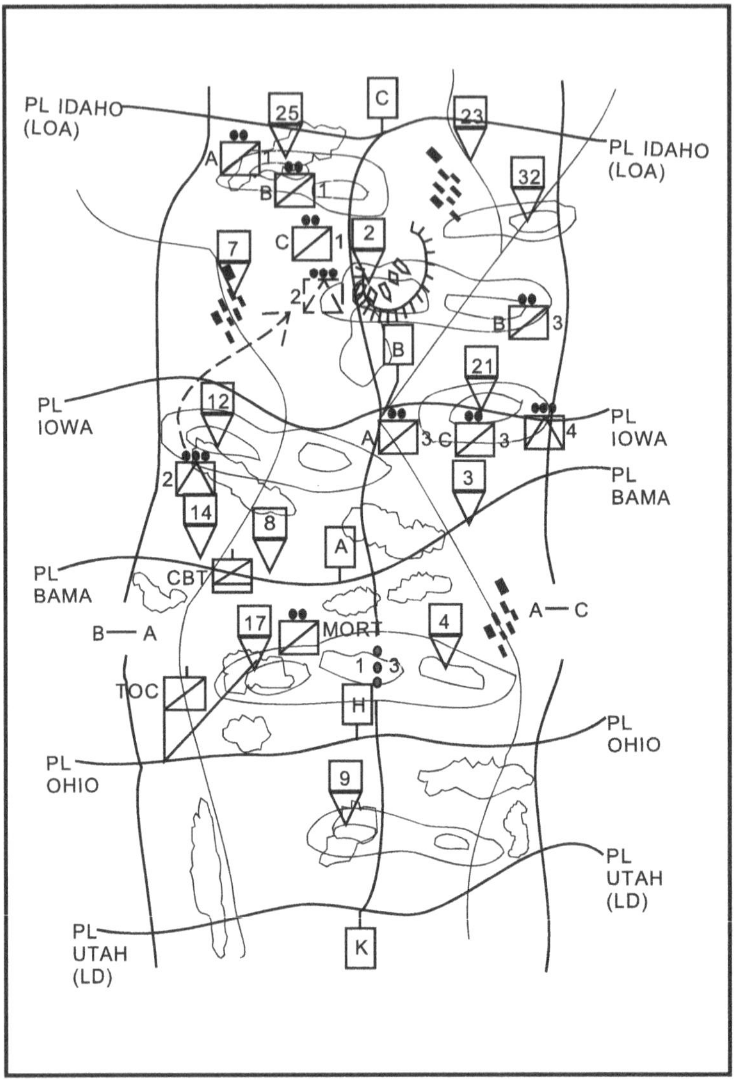

Figure 3-16. Zone reconnaissance (part three).

The antitank platoons consolidate in covered and concealed positions west of checkpoint 2. Scout dismount teams from the 3d platoon move in quickly, capture two wounded prisoners, and search the vehicles and personnel. The 3d platoon leader orders one scout squad to stay and secure the prisoners, and then completes his reconnaissance up to the LOA. The 1SG leads the medics to the location of the 3d platoon scout squad and the EPWs. The XO collects and transmits final reconnaissance reports to the squadron.

AIR-GROUND COORDINATION

Cavalry troops will often conduct a coordinated reconnaissance with an air cavalry troop. Because of the air cavalry troop's ability to maneuver faster and look deeper than ground cavalry, they are often placed to the flanks or forward of ground cavalry. The air cavalry troop should focus its reconnaissance efforts on areas that impede ground cavalry movement.

The air cavalry troop quickens the pace or tempo of the ground reconnaissance because it provides the ground unit with added security during movement and increased situational awareness once contact is made. The air cavalry troop can gain initial contact with the enemy or a critical piece of terrain. Once the air troop develops the situation with his assets, he can pass off the information/contact to the ground troop (see Figure 3-17).

- Ground troop and air troop coordinate zone reconnaissance operations.
- Air troop divides the zone into platoon zones.
- Air troop will use checkpoints to identify critical areas. Checkpoints should enhance ground troop movement throughout the zone.
- Platoon on the right reconnoiters forward of ground troop, identifies platoon ford site, and conducts route reconnaissance.
- Platoon on the left reconnoiters forward of ground troop, supports the other platoon in conducting route sweep, checks lateral routes in the zone, and looks for sites along the obstacle (river/road).

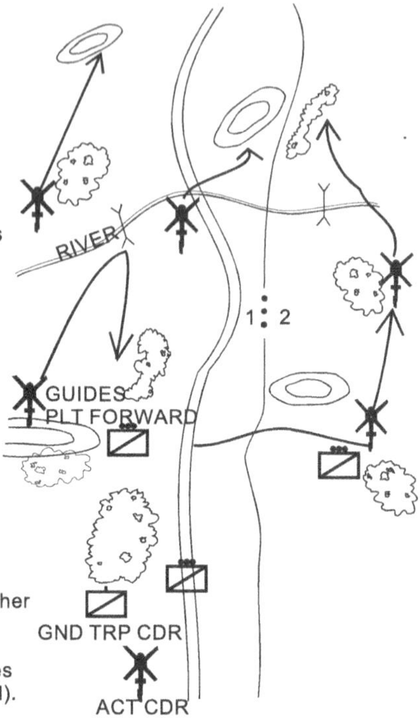

- ACT commander and ground troop commander identify critical areas where air-ground coordination is required or needed.
- ACT commander positions himself close to the ground troop commander (if needed).
- If the ford site or bridge site is blocked, ACT elements reconnoiter the site forward of the location. They also find a bypass to the obstacle and lead ground troops to bypass the location. The ACT will start the screen forward of ground elements.

Figure 3-17. Air-Ground troop coordinated reconnaissance.

Section V. Area Reconnaissance

An area reconnaissance is a specialized form of zone reconnaissance. It is a mission conducted to gain detailed information about terrain features and enemy forces within a specified area or point that other forces intend to occupy, pass through, or avoid. A commander usually calls for area reconnaissance before he sends his forces into or near an area to avoid being surprised by actual terrain conditions or unexpected enemy forces.

CRITICAL TASKS

During an area reconnaissance, the following critical tasks must be accomplished unless the SCO directs the troop to do otherwise:

- Reconnoiter all terrain within the area.
- Inspect and classify all bridges within the area.
- Locate fords or crossing sites near all bridges within the area.
- Inspect and classify all overpasses, underpasses, and culverts.
- Locate and clear all mines, obstacles, and barriers in the area within its capability.
- Locate a bypass around BUAs, obstacles, and contaminated areas.
- Find and report all enemy within the area.
- Report reconnaissance information.

TECHNIQUES

An area reconnaissance is conducted like a zone reconnaissance. When the troop receives an area reconnaissance mission, the assigned area is identified as the terrain inside a solid, continuous boundary. Planning the movement to the area is the first step. Select the route(s), establish a march order on each route, and specify a start point, checkpoints, and a release point. Use a movement technique that keeps the troop moving quickly and securely. If possible, avoid contact with the enemy while en route. Report and bypass. Be careful when approaching the area to begin the reconnaissance. Use common sense. Avoid known enemy forces outside of the area where reconnaissance will be conducted; start somewhere else.

As in a zone reconnaissance, enclose the area within a troop zone. Draw a line of departure, an LOA, and lateral boundaries. Divide the troop zone into two platoon zones. Add phase lines along identifiable terrain to control movement through the area. Place contact points at the intersection of the platoon boundary and all phase lines. Place TIRS on the map. To identify specific areas or features, use checkpoints for reference (see Figure 3-18).

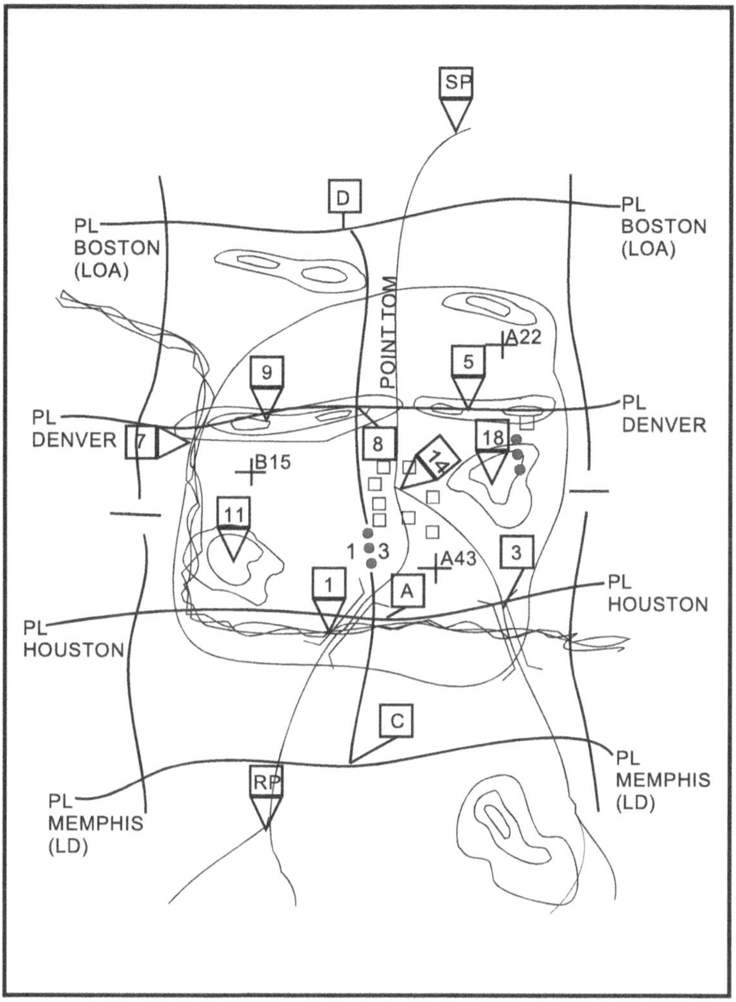

Figure 3-18. Area reconnaissance.

Use a troop vee or split-vee to conduct an area reconnaissance. Scout platoons deploy abreast from the release points to accomplish all the reconnaissance tasks. They move out across the line of departure first. The mortar section moves through the center of the troop zone, remaining in position to range from 3 to 3.5 kilometers forward of the lead elements in the scout platoons. The mortar section could also follow one of the scout platoons, depending on where initial enemy contact is expected. The tank platoons are kept back far enough in depth to retain flexibility, but remain responsive to situations developed by the scout platoons. They key their advance on the progress of the scout platoons. The tank platoons are told how far behind the scout platoons they should stay as the troop deploys across the line of departure. The distance between the tank platoons and the scout platoons will usually be adjusted as the terrain or enemy situation changes. The troop CP displaces through the zone using terrain that affords effective and continuous communication with troop elements and squadron. The troop trains follow about 2 kilometers behind the tank platoons and bound from one covered position to another. The troop commander positions himself well forward to observe the action of one scout platoon or the other. The location usually depends on where initial enemy contact or problem situations are expected.

Chapter 4

Security

Security is an essential part of all offensive and defensive operations. Cavalry provides security for the commander along an exposed front, flank, or rear of the main body where a threat may exist. Surveillance is continuous during security operations. Even during security missions that involve fighting the enemy, the scouts' primary task remains gathering information. Scouts do this by establishing OPs, conducting patrols, and performing reconnaissance.

Counterreconnaissance is an inherent task in all security operations. Counterreconnaissance is not a mission. It is the sum of actions taken at all echelons to counter enemy reconnaissance and surveillance efforts through the depth of the area of operations. Counterreconnaissance denies the enemy information about friendly units. It is both active and passive and includes combat action to destroy or repel enemy reconnaissance elements.

CONTENTS

Section I. Purpose, Fundamentals, and Capabilities

Security includes screening operations, guard operations, covering force operations, area security operations, convoy security operations, and route security operations. The squadron performs screen, guard, and route security missions. Covering force operations are normally the mission of a cavalry regiment. Separate brigades or task organized divisional brigades may perform cover operations as well.

Both the heavy and light cavalry troops perform two security missions (screen and convoy security) independently or as part of their parent squadron. Both also participate in guard, cover, and route security missions as part of their parent squadron or regiment. Either troop will normally perform reconnaissance, screen, defend, delay, attack, or a combination of these missions in support of their parent squadron or regiment (see the applicable section of Chapters 3, 5, and 6 for specifics).

PURPOSE

Security operations are designed to obtain information about the enemy and to provide reaction time, maneuver space, and protection to the main body. Security operations are characterized by conducting continuous reconnaissance to reduce terrain and enemy unknowns, gaining and maintaining contact with the enemy to ensure continuous information, and providing early and accurate reporting of information to the protected force.

FUNDAMENTALS

Five fundamentals are common to all cavalry security operations.

Orient on the Main Body. As a security force, the troop will be operating at a specified distance from a main body, between it and a known or suspected enemy force. If the main body moves, the troop also moves. The troop commander must know how the main body commander intends to maneuver his forces and where he wants the troop in relation to his movement. The troop commander maneuvers his troop to positions where he can provide the needed security.

Perform Continuous Reconnaissance. The troop's security and the security of the main body come in large measure from knowing everything about the terrain and the enemy within the troop's area of operations (AO). Reconnaissance and continuous patrolling go hand in hand with security operations. Cover all the ground in the AO. Determine what the terrain will allow the troop *and* the enemy to do.

Provide Early and Accurate Warning. Early and accurate warning of enemy approach is the cornerstone of security operations. The main body commander needs as much time as possible to shift and concentrate his forces to meet and defeat an unexpected enemy attack. Put observers in positions that afford long-range observation of expected enemy avenues of approach. Use aeroscouts and ground surveillance radar (GSR), if available, to enhance their ability to see. Place remote sensors in the ground to monitor avenues of approach that cannot be easily observed. If possible, send dismounted or mounted patrols forward of OPs to extend their ability to see, providing additional reaction time for the main body commander.

Provide Reaction Time and Maneuver Space. All security operations are designed to provide reaction time and maneuver space for the main body so it can deal effectively with an unexpected enemy attack. The troop provides early warning to the main body commander and sometimes has to fight hard to buy time and space so the main body's combat power can be concentrated on defeating the enemy.

Maintain Enemy Contact. Once visual or physical contact with the enemy is gained, do not allow him to break the contact. Maintain enemy contact and continue to report his activities until told to stop.

CAPABILITIES

Capabilities of the heavy troop include—

- Screen up to a ten-kilometer-wide sector.
- Maintain continuous surveillance of up to six battalion-size avenues of approach.*
- Can establish up to 12 short-duration OPs.

Capabilities of the light troop include—

- Screen up to a ten-kilometer-wide sector.
- Maintain continuous surveillance of up to six battalion-size avenues of approach.*
- Can establish up to 16 short-duration OPs.

*The maximum six long-duration OPs either cavalry troop can occupy is a function of personnel required to perform the following tasks at each OP:

- Man the actual OP.
- Maintain radio communications with the OP and with the platoon leader.
- Provide local security for the vehicles.
- Conduct dismounted patrols as required.
- Conduct resupply.
- Perform maintenance.
- Sleep/rest.

The ability to perform the above tasks simultaneously for periods in excess of 12 hours requires at least 9 to 10 personnel (collocation of two scout squads for M3-mounted scouts, or three scout squads for HMMWV-mounted scouts).

Section II. Screen

Screen is the most common security mission heavy and light cavalry troops conduct. Both troops conduct screen missions for their parent squadrons or other forces to—

- Provide early warning of enemy approach.
- Provide real-time information, reaction time, and maneuver space to the protected force.
- Destroy enemy reconnaissance elements within their capability (perform counterreconnaissance).
- Impede and harass the enemy.

The screen mission provides the least amount of protection of any security mission, and is appropriate when operations have created extended flanks, when gaps between forces exist and cannot be secured in force, or when required to provide early warning over gaps that are not considered critical enough to require security in greater strength. A commander normally assigns cavalry this mission when he needs time to respond to an unexpected enemy attack, and cannot afford to commit other forces to the task.

The screen mission is defensive in nature. As such, both heavy and light cavalry troops screen the front, flanks, and rear of a stationary force but only to the flanks or rear of a moving force. Screening operations are not performed forward of a moving force. Zone reconnaissance or movement to contact is the appropriate cavalry troop mission suited to the requirements of the offensive force.

CRITICAL TASKS

A screen mission has certain critical tasks that guide planning. To achieve the intent of a screen mission, the troop must accomplish the following critical tasks:

- Maintain continuous surveillance of all battalion-size avenues of approach into the troop sector under all visibility conditions.
- Destroy or repel all enemy reconnaissance patrols (if within the unit's capability and directed by the higher commander).
- Locate lead elements of enemy order of battle and determine the direction of movement of each.
- Maintain contact with, report the activity of, and impede and harass the enemy while displacing.

EXTENDED SCREENS

Both the heavy cavalry troop and the light cavalry troop normally screen a sector up to ten kilometers in width. However, METT-T may dictate an extended screen across frontages in excess of the norm. Either troop's ability to accomplish its critical tasks, or its ability to screen in depth, can diminish rapidly as frontages increase.

Examples of extended screens are described below.

- Heavy cavalry troop screens 20 kilometers of southern bank of unfordable river crossed by four bridges in sector.
- Light cavalry troop screens 25 kilometers of desert terrain, from dominant ridge.

DEPTH

Depth is also important in a screen. The term *"screen line"* is descriptive only of the forward trace along which security is provided. Depth allows an enemy contact to be passed from one element to another without requiring displacement. Depth is advantageous to—

- Destroy an enemy reconnaissance patrol without compromising critical OPs.
- Prevent an enemy from penetrating the screen line too easily.
- Prevent gaps from occurring when OPs displace or are lost.
- Maintain contact with moving enemy without compromising OPs.
- Prevent enemy templating of the screen line.

Depth is achieved primarily by positioning OPs, particularly where there are limited avenues of approach. Tank platoons, AT platoons, the mortar section, and attached elements positioned behind the screen line establish local security and provide surveillance. The degree to which depth can be attained is a function of many factors, which include—

- Higher commander's intent and concept as expressed in—
  - Graphical trace of the screen line (LOA).
  - Engagement criteria.
  - Destruction criteria.
  - Displacement/disengagement criteria.
- Width of the sector.
- Depth of the troop sector.

- Terrain and avenues of approach it will support.
- Attachments and detachments.

Screening is largely accomplished by establishing a series of OPs and conducting patrols to ensure adequate surveillance of the assigned sector. Screens are active operations. Stationary OPs are only one part of the mission. Employing patrols (mounted and dismounted), aerial reconnaissance, ground-based sensors, intelligence from space-based sensor systems, and OPs relocated on an extended screen ensure that continuous overlapping surveillance occurs. Inactivity in an immobile screen promotes complacency.

PLANNING CONSIDERATIONS/GUIDANCE TO SUBORDINATES

The enemy situation is often vague when planning a screen. The troop should develop plans that are flexible enough to react to any enemy course of action, particularly the worst case. Planning considerations for a screen should include a detailed description of how contact with the enemy reconnaissance will be gained then how and where it will be destroyed. Planning should also cover the method of displacement once the main body of the enemy force has been identified and how that force will be handed off to the main body in the main defensive belt. Because of the need for flexibility, screen operations will often begin to inherit the characteristics of defense or delay missions. To cover the displacement of scouts, some elements of the troop may be required to execute missions such as delay or defend.

Screen operations at troop level usually occur in four phases.

- Movement to and occupation of the screen line.
- Surveillance and counterreconnaissance.

- Gaining and maintaining contact with the enemy main body displacement of screen.
- Rearward passage of lines.

Higher command guidance should address each phase of the operation and cover at least the following:

- Location/orientation/width of the screen.
- Duration of the screen.
- Method of movement to and occupation of the screen line.
- Location and disposition of the friendly force being screened.
- Engagement/destruction criteria.
- Displacement/disengagement criteria.
- Follow-on missions.
- Positioning and orientation guidance for GSRs (if attached).

The following must be considered when developing and completing the plan and executing the screen mission:

Time Screen Must Be Established. The time the screen must be set and active will influence the troop's method of deploying to and occupying the screen line.

Movement to the Screen Line. If the screen mission is the result of a previous tactical maneuver such as zone reconnaissance, the troop will essentially be postured to begin screening from present positions. This situation occurs frequently, and may be the result of a FRAGO to halt at a specified phase line.

If the troop is not currently sited on the screen line, obviously, deployment to the screen line must occur before actually beginning the screen mission. Time determines the method of occupying the screen line. Thorough analysis of METT-T will determine which deployment technique or combination of techniques best meets mission requirements.

Trace and Orientation of Screen Line. The initial screen line is depicted as a phase line and often represents the forward line of own troops (FLOT). As such, the screen line may be a restrictive control measure for movement (limit of advance); coordination/permission would be necessary to move beyond the line to establish OPs or to perform reconnaissance. When occupied, OPs are sited on or behind the phase line. OPs should be given specific orientation and observation guidance.

Initial OP Locations. The squadron or troop commander may determine tentative initial OP locations to ensure effective surveillance of the sector and designated named area(s) of interest (NAI). At a minimum, the troop commander designates a primary orientation for the scouts during the conduct of the screen. Scouts, once set on the screen line, will report their location to the troop TOC and verify they are in compliance with the commander's orientation guidance. The scouts who occupy each OP always retain the responsibility to modify the location to achieve the commander's intent and guidance for orientation. The OPs are positioned along or behind the screen line.

OPs are generally categorized based on their expected duration of employment as either—

- Short-duration (less than 12 hours)

  or
- Long-duration (more than 12 hours).

OPs may be either mounted or dismounted. Mounted OPs maximize use of vehicular optics, weapon systems, and speed of displacement, but are more readily detected by the enemy. Dismounted OPs provide maximum stealth at the expense of speed of displacement, and vehicle-mounted optics and weapons.

A heavy cavalry troop can occupy up to 12 short-duration OPs (one per scout squad, six per scout platoon). For extended periods of time, the heavy troop can occupy six OPs (one per scout section, three per scout platoon).

A light cavalry troop can occupy up to 16 short-duration OPs (one per scout squad, eight per scout platoon). For extended periods of time, the light troop can occupy six OPs (one per scout section, three per scout platoon).

Width and Depth of the Screened Sector. The troop sector is defined by lateral boundaries extending out to a limit of advance (the initial screen line), forward of a rear boundary. The troop sector is established by the squadron or unit being screened. The troop rear boundary may be a squadron phase line and may serve as a battle handover line (BHL) to control passing of responsibility for the enemy to the protected force. The troop's ability to gain depth decreases as screened frontage increases.

Locations of Subsequent Screen Lines. The squadron or troop commander uses additional phase lines to control the operation. These phase lines may serve as subsequent screen lines. Displacement to the subsequent screen lines is event driven.

Scout Platoon Sectors. Assign clear responsibility of identified avenues of approach and designated NAIs. The nature of a screen normally requires both scout platoons to deploy abreast.

Tank/AT Platoon Sectors. Position the tank (heavy troop) or AT (light troop) platoons in the scout platoons' sectors. They may occupy hide or battle positions along avenues of approach. The tank/AT platoons remain responsive to the troop commander. They are the primary direct-fire killing asset.

Force to be Screened. The troop must orient on the force it is securing. If the main body is moving, the troop must move to maintain the screen's position relative to the main body.

Reinforcements. Any unique requirement posed by the mission may require assets not organic to the troop. GSR and engineers are common attachments at troop level.

- **GSR.** During screen operations, GSR is used to augment scout OPs and to add depth to the screen. GSRs should be attached to scout platoons, and the commander should provide the scout platoon leader with positioning and orientation guidance.
- **Engineers.** If engineers are attached to the troop, the troop commander should assign them with priority of mission and priority of effort in support of commander's guidance. During screen operations, engineers will normally dig survivability positions for scouts and tanks. Also engineers may emplace obstacles in support of the counterreconnaissance battle or assist the troop with displacement of the screen once contact has been established with the enemy body.

Special Requirements or Constraints. Specify all requirements for observing any NAI identified during the intelligence preparation of the battlefield (IPB). Task subordinate platoons as required. Specify the following in the troop OPORD:

- Engagement criteria.
  - What size force will scouts engage and destroy (if any)?
  - Where will this action occur?
  - What size force will the tank or AT platoons engage and destroy?
  - Where will this action occur?

- Disengagement criteria.
  - What event will cause scouts to displace from the initial screen line?
  - How will scouts maintain contact with the enemy while displacing?
  - What event will cause the tank or AT platoons to displace to subsequent or alternate positions?

Indirect Fire Planning. Fire planning integrates artillery and mortar fires. Position the troop mortars to fire up to two-thirds of their maximum range forward of the initial screen line. A wide sector may require the troop commander to position them to provide effective coverage of the most likely avenue of approach determined by IPB. The troop FSO plans artillery fires to adequately cover any gaps in mortar coverage.

Direct Fire Planning. Based on his analysis of the terrain, the troop commander determines where to engage the enemy (engagement areas). He also determines the location of battle positions that provide observation, fields of fire, and cover and concealment that support each engagement area.

Positioning of C2, CS, and CSS Assets. The troop commander positions himself to observe the most dangerous enemy avenue of approach. The troop TOC positions itself in depth to provide continuous control and reporting during initial movements. After the screen line has been reestablished in depth following displacement from the initial screen line, the TOC can reposition. Combat trains position behind masking terrain close enough for rapid response. They are best sited along routes providing good mobility laterally and in depth.

Patrol Requirements. Patrols may be required to cover gaps between OPs. The troop commander tasks the scout platoon leaders to perform specific patrols.

Coordination. The troop commander coordinates his concept with air cavalry troop (ACT) commanders who may be operating the same ground, flank troop commanders, tank company (heavy cavalry) and AT company (light cavalry) commanders, and other unit commanders as appropriate.

TECHNIQUES

Graphics

The following is a list of common graphic control measures used to control screen missions:

- Boundaries.
- Phase lines.
- Checkpoints.
- Contact points.
- NAIs.
- OPs.
- Mortar firing positions.
- Battle positions.
- Hide positions.
- TIRS.
- Unit symbols.

Movement to the Screen Line

In deploying to the screen line, the troop commander must deal with the competing requirements to establish the screen quickly to meet mission requirements and to provide the necessary level of security for the troop in doing so. The troop moves to the screen line using one of three basic methods—a tactical road march, zone reconnaissance, or movement to contact.

Tactical Road March. The troop conducts a tactical road march to a release point behind the screen line. From the release point, platoons deploy to occupy initial positions. This method of deploying to the screen line is the fastest, but least secure. It is appropriate when enemy contact is not expected and time is critical, or when an air cavalry troop is conducting zone reconnaissance forward of the ground troop.

Movement to Contact. The troop conducts a movement to contact from a line of departure to the initial screen line. This method is slower than a tactical road march, but more secure. It is appropriate when enemy contact is likely, time is limited, or when an air cavalry troop is conducting zone reconnaissance forward of the ground troop.

Zone Reconnaissance. The troop conducts a zone reconnaissance from a line of departure to the initial screen line. Given adequate time, this method is preferred as the troop can clear the zone of any enemy and enables platoons to become thoroughly familiar with the terrain. The troop can reconnoiter potential subsequent OP locations, battle and hide positions, and mortar firing positions, for example, as they move to the screen line. A zone reconnaissance is appropriate when time is available and information about the enemy or terrain is unknown.

Security Drill

A security drill is a series of rehearsed actions (battle drills) a scout platoon or cavalry troop takes to maintain contact with the main body of an advancing enemy force. It is used when collapsing the screen line to subsequent OP positions or when transitioning from a screen mission to a delay or defend mission.

At platoon level, OPs gain contact with the enemy main body, then report and prepare to displace to a subsequent position. When the enemy force reaches the OP's break point (point where the OP must displace or his position/movement will compromise him to the enemy), the OP passes off the responsibility to track the enemy to another OP in depth. The platoon displaces its OPs to subsequent positions in depth while maintaining contact with the enemy.

At troop level, the security drill combines the collapse of the initial screen line with the actions of organic tank or AT platoons. Scout platoons may perform platoon security drills initially, consolidating some or all of their combat power at a battle position to aid execution of a troop-level engagement.

At platoon and troop levels, conduct of security drills is tempered by the commander's overall concept, intent, and scheme of maneuver. Enemy actions (events) drive security drill execution (response) (see Figure 4-1).

STATIONARY SCREEN

Take a close look at the high-speed avenues of approach into the sector. Divide the sector into two platoon sectors. Make sure the platoon boundary is on easily identifiable terrain. Do not split avenues of approach with a platoon boundary or place the boundary on a road. Place NAIs, TIRS, or checkpoints beyond the screen line to focus surveillance from OPs. If needed, add additional phase lines to control withdrawal of the troop at 5- to 8-kilometer intervals. Place contact points at the intersection of the platoon boundary and all phase lines. Place TIRS on the map or overlay as described in Chapter 2.

Deploy the scout platoons abreast and establish a series of OPs along or behind the initial screen line, as terrain allows, but never forward of it without permission. Make it clear to the scout platoon leaders which avenues of approach (depicted as NAIs or checkpoints) they are to observe. If unobservable areas between OPs need to be routinely checked, have the scout platoon leaders prepare patrol plans for approval and subsequent execution.

Position the mortar section to fire 3 to 4 kilometers forward of the initial screen line, oriented on the expected enemy avenue of approach. Establish subsequent firing positions for the mortar section back through the sector. Position each tank (heavy) or AT (light) platoon in one of the scout platoon sectors, or have the tank/AT platoons consolidate. Position the tank/AT platoons in hide positions or battle positions in depth behind the scout platoons, oriented on the expected enemy avenue of approach. Establish subsequent positions for the tank/AT platoons back through the sector to support the scheme of maneuver.

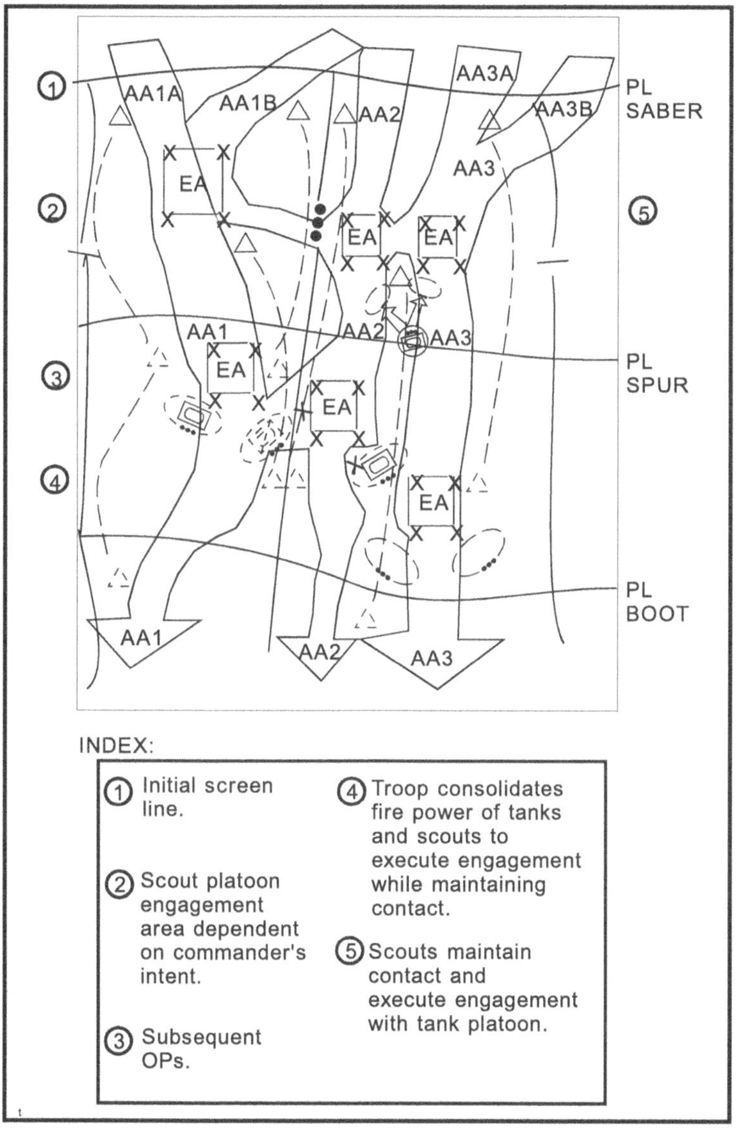

Figure 4-1. Scout platoon/cavalry troop security drill.

Have the XO position the command post on terrain that affords good FM radio communications with troop elements and squadron headquarters. If possible, the command post should be positioned behind the subsequent screen positions. This allows the TOC to remain in position during the initial collapse of the screen line. Establish tentative subsequent command post sites back through the troop sector. The XO retains the authority to adjust the actual TOC location to maintain effective communications.

The 1SG first positions the troop trains within 5 to 8 kilometers of the initial screen line, then he establishes subsequent locations for the trains to bound back through the troop sector. The troop commander positions himself well forward where he can best observe and control the actions of the troop (see Figure 4-2).

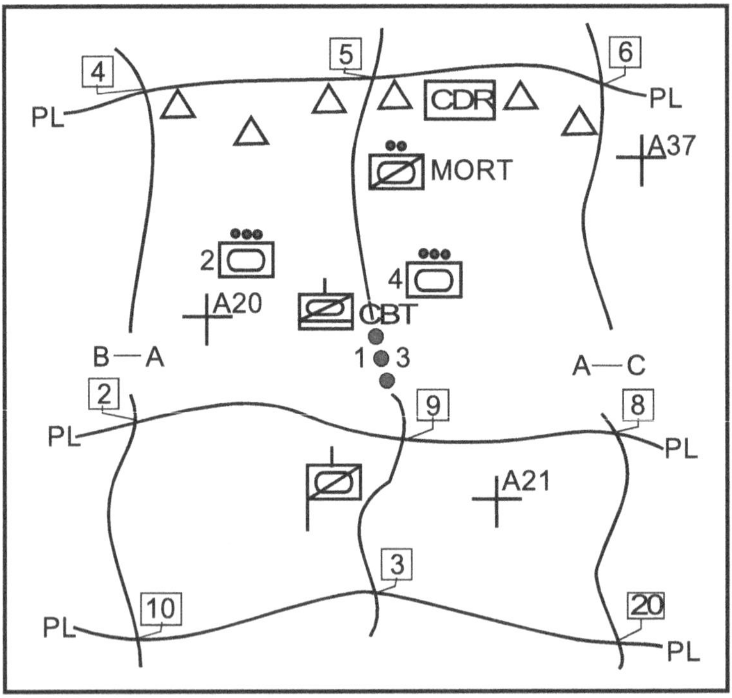

Figure 4-2. Troop screen positioning.

MOVING FLANK SCREEN

The troop may conduct a moving flank screen by itself or as part of the squadron. It may be tasked to screen the exposed flank of the squadron while the squadron conducts a movement to contact, a hasty attack, or a zone reconnaissance (see Figure 4-3). The troop may also participate in a squadron mission to screen or guard another combined arms force (see Figure 4-4).

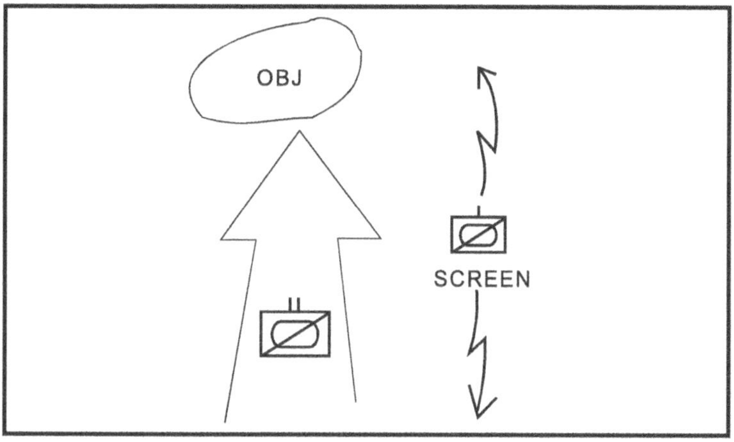

Figure 4-3. Squadron movement to contact; troop flank screen.

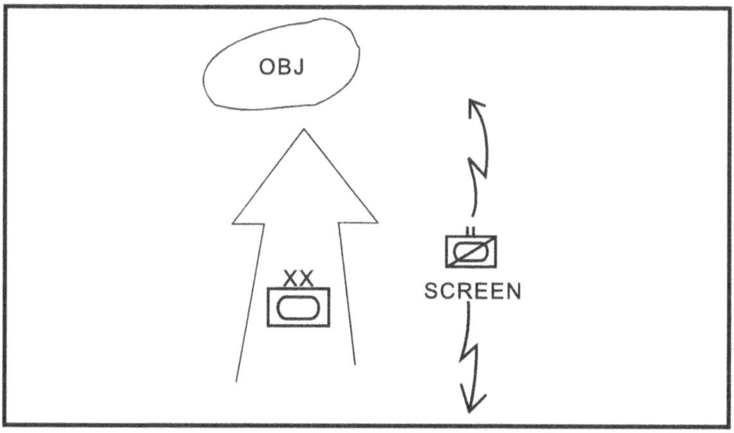

Figure 4-4. Squadron flank screen.

A moving flank screen uses the same techniques as when screening a stationary unit. Position the scout platoons to maintain continuous surveillance on the avenues of approach, the mortar section to cover likely avenues of approach with indirect fire, and the tank/AT platoons in depth to destroy or repel enemy reconnaissance units.

Applying Graphic Control Measures

Because of the inherent dual orientation of a moving flank screen (direction of movement versus orientation of the screen line), control of the operation poses numerous challenges. Control measures must facilitate both orientations (see Figure 4-5).

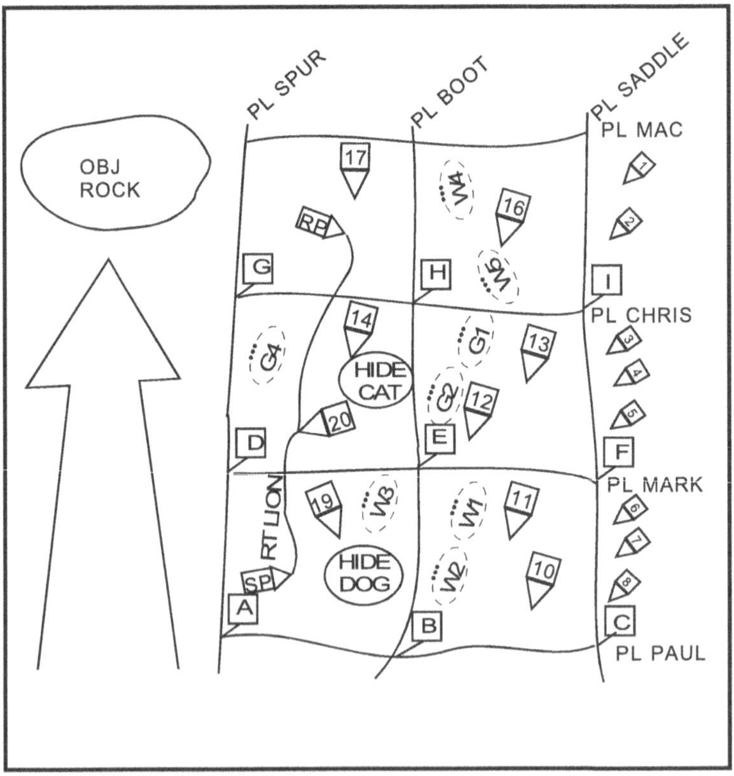

Figure 4-5. Troop moving flank screen graphics.

Follow the procedures below when applying graphic control measures to a moving flank screen.

- Use phase lines to control the scout platoon's movement (placed perpendicular to the screen line). Plan to use these phase lines as on-order boundaries for subordinate platoons if enemy contact is gained. Place phase lines no more than 5 kilometers apart (corresponding to the width of a scout platoon's screen frontage). Do not divide avenues of approach with them.

- Use additional phase lines rearward of (parallel to) the initial screen line to control retrograde movement (toward the protected force). Plan to use these phase lines as subsequent (on-order) screen lines.

- Use objectives, checkpoints, or axes of advance to control the movement of the tank/AT platoons for movement. Position and plan to use these objectives or checkpoints as battle positions for the tank/AT platoons if the scouts make enemy contact at the screen line. Plan subsequent (on-order) battle positions between the screen line and the protected force.

- Use mortar firing positions or checkpoints to control movement of the mortar section. Position these firing positions rearward of the screen line where they allow the mortars to fire two-thirds maximum range forward or cover likely avenues of approach. Plan subsequent (on-order) mortar firing positions between the screen line and the protected force.

While this number of graphic control measures and required planning may seem excessive, they provide maximum flexibility in terms of mission execution. The troop commander can issue simple FRAGOs to adjust the plan to the enemy situation.

Repositioning the Screen

The troop must reposition to stay oriented on the force it is securing. Movement along the screen line is determined by the speed of the protected force. Movement is conducted by one of three techniques—continuous marching, bounding by platoons (alternately or successively), and bounding by OPs (alternately or successively).

Continuous Marching. This technique is appropriate when the protected force is moving quickly and contact is not likely. It is the least secure movement technique.

Deploy both scout platoons abreast with the two tank or AT platoons, the mortar section, and the remainder of the troop in depth (between the screen line and the protected force). The trace of the screen line is essentially the route of advance for the two scout platoons in column. Have the tank/AT platoons and the remainder of the troop move along a designated route or axis of advance (see Figure 4-6).

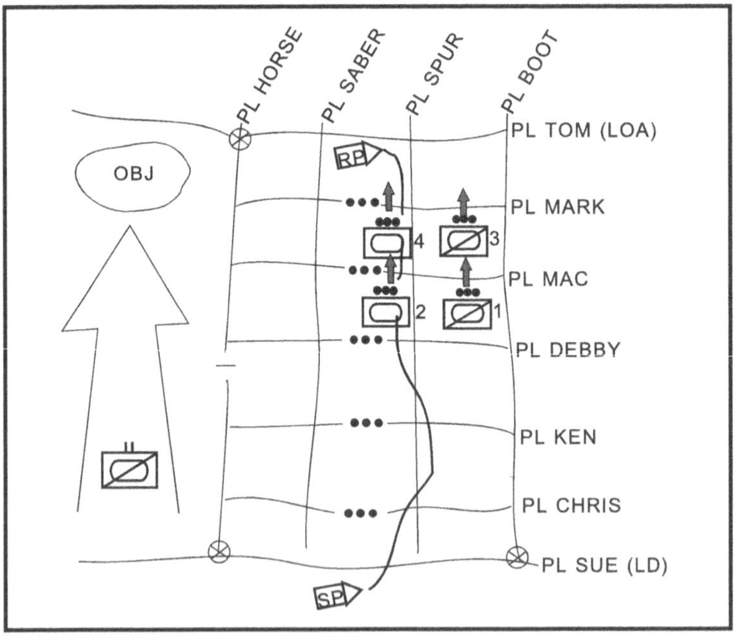

Figure 4-6. Troop moves by continuous marching.

Bounding by Platoons. This technique is appropriate when the protected force requires greater protection than afforded by continuous marching, is not moving quickly, or knows enemy contact is possible. Bounding platoons alternately may leave temporary gaps in the screen line as they move. Bounding platoons successively is more secure but slower than bounding platoons alternately.

Deploy both scout platoons abreast with the two tank/AT platoons, the mortar section, and the remainder of the troop in depth (between the screen line and the protected force). Alternately bound one scout platoon around (to the rear of) the other to assume new positions along the screen line (see Figure 4-7), or successively bound the scout platoons along the screen line (see Figure 4-8). Have the tank/AT platoons occupy designated positions sequentially or alternately.

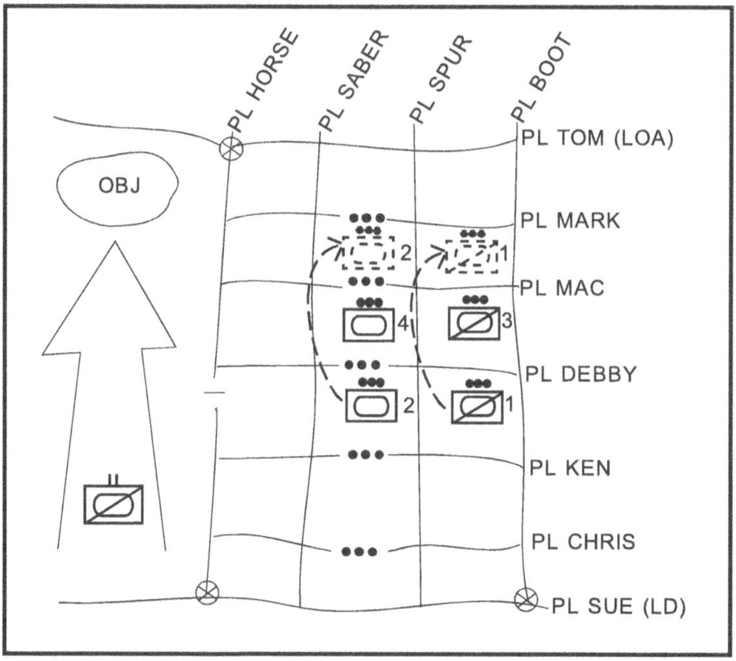

Figure 4-7. Troop moves by alternately bounding platoons.

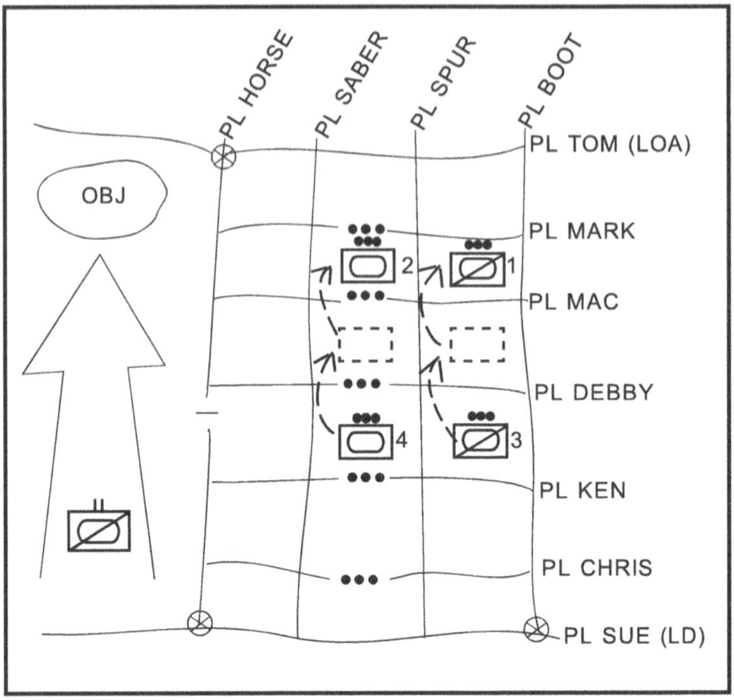

Figure 4-8. Troop moves by successively bounding platoons.

Bounding by OPs. This technique is appropriate when the main body is moving slowly, contact is possible, and maximum security is required. Bounding OPs alternately will disrupt the integrity of the scout platoons as OPs bound to their next position. Bounding OPs successively is easier for the scout platoons to control.

Deploy both scout platoons abreast with the two tank/AT platoons, the mortar section, and the remainder of the troop in depth (between the screen line and the protected force). Alternately bound the rearmost OP around (to the rear of) the other OPs (both platoons) to assume a new position along the screen line (see Figure 4-9), or successively bound the OPs along the screen line (see Figure 4-10). The number of OPs on the screen line may be reduced, as two or more may be bounding at any given time. The rate of advance of the protected force will determine this. Have the tank/AT platoons occupy designated positions sequentially or alternately.

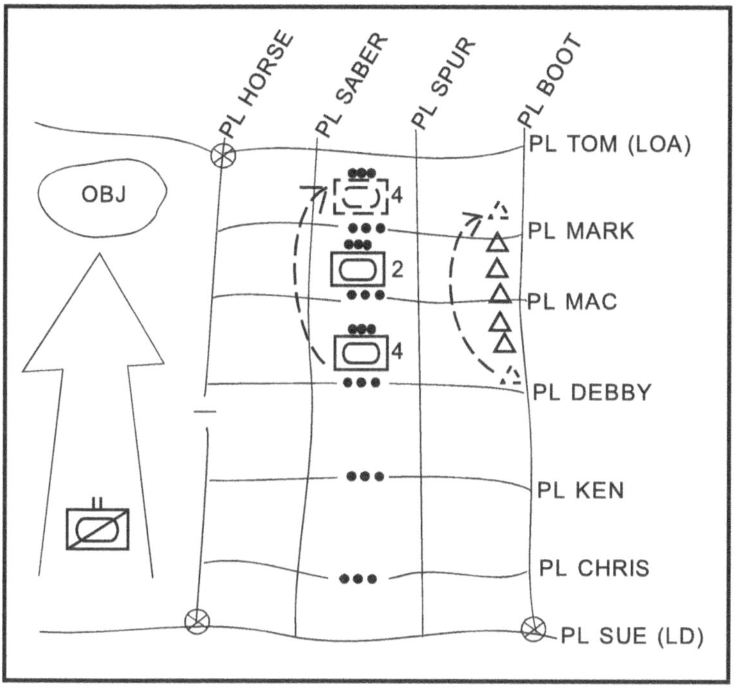

Figure 4-9. Troop moves by alternately bounding OPs.

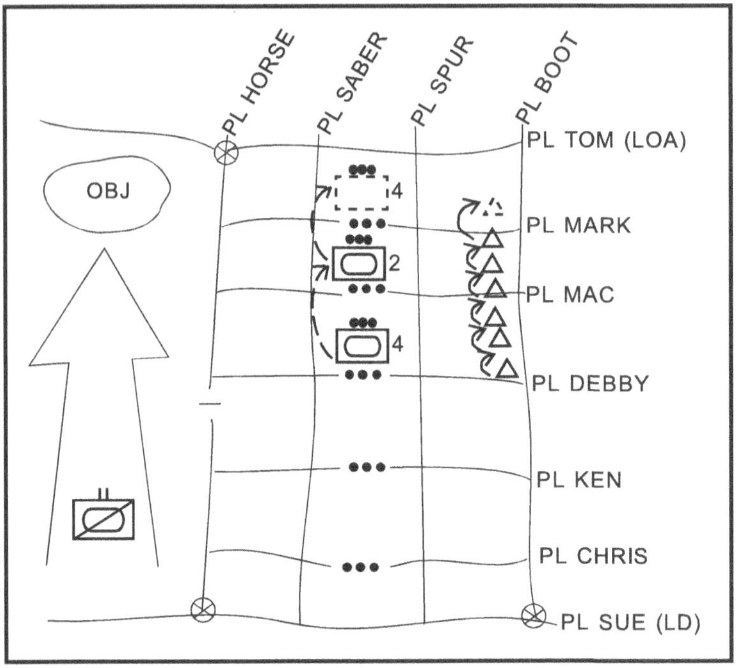

Figure 4-10. Troop moves by successively bounding OPs.

Example of a Screen Forward of Stationary Main Body (Heavy Troop)

The 1st and 3d platoons deploy abreast along PL BOOT, the initial screen line, and establish OPs to maintain continuous surveillance of the high-speed avenues of approach into the troop sector. Route 220 in the 1st platoon sector is the most likely route of enemy approach. Scout platoons execute their approved patrol plans between OPs, and periodically check the areas that cannot be observed. The mortar section lays its guns in a centrally located position about 3 kilometers behind the initial screen line and orients on Route 220. The 2d platoon occupies a hide position behind the 1st platoon. The 4th platoon occupies a hide position north of checkpoint F. Both positions provide good lateral movement and several maneuver options. The troop CP positions south of TIRS A21. The 1SG positions the troop trains in the woods just north of contact point 8, which is near a good lateral road network. The troop commander and FIST collocate well forward with the 1st platoon leader, overlooking Route 220 (see Figure 4-11).

The 1st platoon leader reports two BMPs and one T-64 bounding across the open terrain, moving southwest astride Route 220 about 2,000 meters east of PL BOOT. He maintains contact with the patrol as it crosses PL BOOT and enters the troop sector. The troop commander orders 2d platoon to attack the patrol by fire. The 2d platoon moves into hull-down positions and ambushes the patrol. All three vehicles are hit and damaged. The 1st platoon leader sends his PSG to capture the surviving BMP crew members and to search the vehicles for vehicle unit symbols, maps, sketches, and any other items of intelligence value. The PSG links up with the 1SG, who has moved forward. He hands over the prisoners and material, and returns to his position on the screen line. The 1SG evacuates the prisoners and captured material to the squadron combat trains. The 3d platoon reports no contact in its sector. The XO reports the situation to the squadron.

The troop commander tells the troop to be alert for other divisional reconnaissance patrols in the sector, and to expect regimental reconnaissance within the next hour. The XO eavesdrops on the squadron OI net and monitors reports from Troop B that indicate it has encountered several divisional reconnaissance patrols. The XO passes this information to the troop commander.

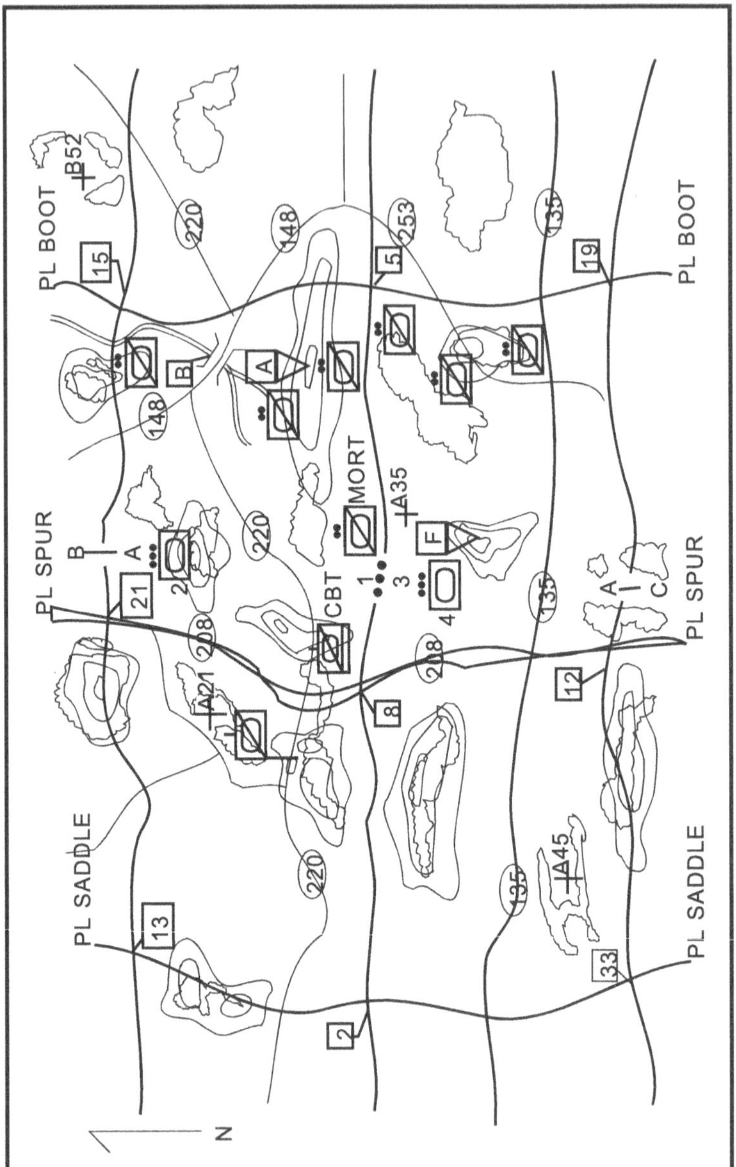

Figure 4-11. Troop screen.

Within the next 45 minutes, both scout platoons report reconnaissance patrols of two BMPs each in their sectors. The troop commander maneuvers a tank platoon into position to ambush each patrol. Three of the four BMPs are quickly destroyed. One BMP in the 1st platoon's sector is damaged, but is able to move into a covered and concealed position in the wooded area west of checkpoint A. Artillery begins to fall on the dominant terrain east of PL SPUR. The XO reports the situation to squadron.

The troop commander orders the 1st platoon to regain contact with the damaged BMP. The 1st platoon leader sends Bravo section to locate the damaged BMP. Bravo section moves into the wooded area and quickly finds and destroys the BMP.

The troop commander orders the scout platoons to move one of their scout sections back in sector in anticipation of the arrival of enemy reconnaissance elements. Both scout platoons reposition a scout section back towards PL SPUR while the troop commander shifts his tank platoons, mortar section, and trains.

The 1st platoon later reports a reinforced MRP moving southwest along Route 220. The 1st platoon continues to observe the enemy reconnaissance elements and reports their progress to the troop. The FSO engages the enemy with mortar high explosive (HE) and smoke rounds to disrupt their movement and keep them buttoned up. The troop commander tells the 4th platoon to move north and be prepared to counterattack the enemy by fire. He quickly outlines its route using TIRS. The enemy reconnaissance elements rapidly move in an attack formation toward the terrain previously occupied by the 2d platoon. It appears that the enemy elements are attempting to secure the dominant terrain in the 1st platoon sector and fix the tank platoon so the lead company of the enemy advance guard can maneuver around it.

The FSO continues to suppress the enemy with smoke and HE while it attempts to maneuver toward its objective. The troop commander directs 2d platoon into position to establish a base of fire and fix enemy elements while 4th platoon counterattacks by fire from the south. The FSO coordinates mortar and artillery fires to suppress and isolate enemy elements as the tank platoons execute their counterattack.

While the tank platoons complete the destruction of the enemy, heavy artillery begins to impact along and east of PL SPUR. The 1st platoon scouts report a tank platoon followed by six BTR-60s moving west along Route 220. The troop commander reports this information directly to the squadron commander. He orders the troops to fall back to PL SPUR, and to begin coordination for battle handover and passage of lines. The troop commander orders the 3d platoon to begin to bound back to PL SPUR. He tells the 1st platoon to maintain contact with the enemy lead company as it falls back to PL SPUR. The troop commander maneuvers the tank platoons west of PL SPUR.

The 1st platoon leader bounds his scout sections back to PL SPUR as he maintains contact with the lead company. The FSO engages the MRC with mortar and artillery fires to slow and disrupt its advance. The XO keeps squadron informed of the situation in the troop sector.

The troop XO moves back to the contact point to begin coordination for battle handover and passage of lines. The troop continues to maintain contact with the lead MRC as it maneuvers back in sector, while the FSO engages the MRC with mortar and artillery fires.

Section III. Guard

Cavalry squadrons conduct guard missions to provide another unit early warning of enemy approach, and to prevent enemy ground maneuver forces from coming within direct-fire range of the protected unit as it performs other missions or tasks. A guard is conducted to the front, flanks, or rear of a stationary or moving force. Many of the considerations for conducting a moving flank screen apply to conducting a moving flank guard.

A cavalry troop mission unique to moving flank guard operations is assigned to the lead troop of the squadron. This three-fold mission consists of the following tasks:

- Maintain contact with the protected force.
- Reconnoiter the zone between the protected force and the squadron route of advance.
- Reconnoiter the squadron route of advance.

The lead troop accomplishes these tasks by performing zone reconnaissance. The speed of the protected force determines how thoroughly the reconnaissance is performed. Assistance is required if the zone is too wide for the lead troop. An air cavalry troop may maintain contact with the protected force, or a following troop may perform route reconnaissance along the squadron's route of advance.

Note. For a detailed description of techniques for performing a squadron guard mission, see Chapter 4, FM 17-95.

The cavalry troop does not conduct guard operations independently. It normally performs reconnaissance, screen, defend, delay, and attack missions as part of a squadron guard operation. Review the appropriate chapter in this manual for specific mission details. See Chapter 3, Sections III and IV, Route and Zone Reconnaissance; Chapter 5, Section III, Hasty Attack; and Chapter 6, Sections II, III, and IV, Defend From a Troop Battle Position and Defend/Delay in Troop Sector.

Section IV. Cover

A covering force operates apart from the main body to develop the situation early and deceives, disorganizes, and destroys enemy forces. It accomplishes all the tasks of screening and guard forces. Unlike screening or guard forces, a covering force is a tactically self-contained force (that is, it is organized with sufficient combat support and combat service support assets to operate independently of the main body). Covering force operations can be either offensive or defensive in nature and can be conducted to the front, flanks, or rear of a stationary or moving force.

Cover operations are performed by the regiment (ACR or LACR) for the corps commander to provide the time and space he needs to make decisions and to achieve his operational objective.

Defensive cover operations, whether forward, flank, or rear, prevent the enemy from attacking the corps main body at the time and place and with the combat strength he chooses. Defensive cover operations are generally intended to destroy the enemy's initiative and set him up for defeat. Both heavy and light cavalry troops participate in defensive cover operations as part of their parent squadrons. Depending on the squadron commander's scheme of maneuver, they either screen, defend, delay, or attack.

Offensive cover operations are conducted by the regiment to seize and retain the initiative for the corps commander, and to allow him to attack decisively with the main body of the corps at the time and place he chooses. In offensive cover operations, the troop usually conducts movement to contact or zone reconnaissance missions. See the appropriate chapters in this manual for mission details.

Section V. Route Security

Cavalry squadrons and regiments conduct route security missions to prevent enemy ground maneuver forces from coming within direct-fire range of the protected route. A route security force operates on and to the flanks of a designated route. Route security operations are defensive in nature and, unlike guard operations, are terrain oriented. A route security force conducts reconnaissance, screens, attacks, defends, and occupies key locations along the route to prevent an enemy force from impeding, harassing, containing, seizing, or destroying traffic along the route.

Route security operations are not conducted independently by a cavalry troop. Both the heavy and light cavalry troops participate in route security operations as part of their parent squadrons and may conduct reconnaissance, screen, defend, delay, attack, or convoy security missions throughout the area of operations.

Section VI. Area Security

An area security force neutralizes or defeats enemy operations in a specified area. It operates in an area delineated by the headquarters assigning the area security mission. It screens, reconnoiters, attacks, defends, and delays as necessary to accomplish its mission. Area security operations may be offensive or defensive in nature and focus on the enemy, the force being protected, or a combination of the two.

Area security operations are conducted to deny the enemy the ability to influence friendly actions in a specific area or to deny the enemy use of an area for his own purposes. This may entail occupying and securing an area before the enemy can, or taking actions to destroy enemy forces already present.

The area to be secured may range from specific points (bridges, defiles) to areas such as terrain features (ridgelines, hills) to large population centers and adjacent areas. The factors of METT-T and unit capability will determine specific unit missions.

Area security missions are conducted by cavalry troops, squadrons, and regiments who employ the techniques of screen, guard, offense, and defense, depending on the nature and purpose of the mission.

Section VII. Convoy Security

Convoy security operations are performed as a minimum by a cavalry troop or a company team. Both heavy and light cavalry troops are suited to the requirements of protecting a convoy due to their organic reconnaissance capability and combat power. Both the cavalry troop and company team should be reinforced with engineers. METT-T considerations, such as restrictive terrain and limited time, may dictate a coordinated effort with air cavalry assets.

Convoy security operations are conducted when insufficient friendly forces are available to continuously secure lines of communication in an area of operations. They may also be conducted in conjunction with route security operations. A convoy security force operates to the front, flanks, and rear of a convoy element moving along a designated route. Convoy security operations are offensive in nature and orient on the force being protected.

A convoy security mission has certain critical tasks that guide planning and execution. To protect a convoy, the security force must accomplish the following critical tasks:

- Reconnoiter the route the convoy will travel.
- Clear the route of obstacles or positions from which the enemy could influence movement along the route.

- Provide early warning and prevent the enemy from impeding, harassing, containing, seizing, or destroying the convoy.

The convoy security force is organized into three or four elements to accomplish the following (see Figure 4-12):

- **Reconnaissance element.** The reconnaissance element performs tasks associated with zone and route reconnaissance forward of the convoy.
- **Screen element.** The screen element provides early warning and security to the convoy's flanks and rear.
- **Escort element.** The escort element provides close-in protection to the convoy. May also provide a reaction force to assist in repelling or destroying enemy contact.
- **Reaction force.** Provides firepower and support to the elements above in order to assist in developing the situation or conducting a hasty attack. May also perform duties of the escort element.

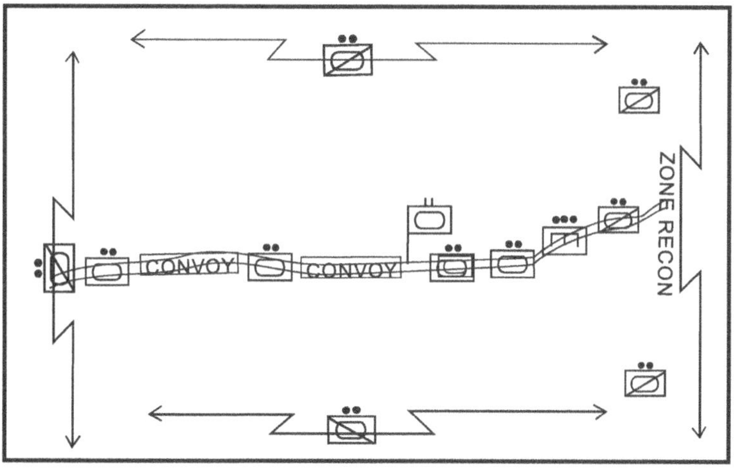

Figure 4-12. Cavalry troop conducts convoy security.

TECHNIQUES

The troop commander organizes and coordinates the efforts of his unit to fulfill the critical tasks associated with the convoy security mission.

Tasks of the reconnaissance element can usually be fulfilled by a single scout platoon. The troop commander ensures the reconnaissance element focuses on trafficability of the route and enemy forces that may influence movement along the route. METT-T may dictate the use of engineers to assist in reconnoitering and clearing the route. Convoy speed is determined by the pace of reconnaissance (METT-T). As a guide, the reconnaissance element should operate a minimum of 3 to 5 kilometers ahead of the main body of the convoy.

Tasks of the screen element can usually be fulfilled by a single scout platoon also. The troop commander ensures the screen element establishes OPs to provide early warning on critical portions of the route or key avenues of approach to the route. OPs have a limited ability to destroy enemy forces; therefore, their primary purpose is to acquire the enemy and direct reaction forces or indirect fire to destroy it.

Tasks of the escort element are best performed by tank platoons in a heavy troop. A light troop should consider a scramble of platoons (MK-19/M2/TOW—see Chapter 2, Figure 2-13) to gain the suppressive firepower necessary to protect the convoy. The troop commander ensures the escort element is positioned to provide security throughout the length of the convoy. This requires elements of the two tank (heavy troop)/mixed (light troop) platoons be dispersed throughout the convoy order of march. If there is no reaction force available or designated, a task of the escort element may be to provide reaction forces that respond to enemy forces identified by the reconnaissance or screening elements. Depending on the length of the convoy and METT-T considerations, the troop commander may keep one tank/mixed platoon consolidated or specifically designated as the reaction force.

Chapter 5

Offense

Offense is the decisive form of war. Brigades, battalion/task forces, and company teams are the principal offensive force for the corps or division. Cavalry units normally perform reconnaissance and security missions in support of corps and division offensive operations. Cavalry troops may perform certain offensive missions as part of a squadron, regiment, or other combined arms force. These offensive missions are normally performed during reconnaissance or security operations.

If required, cavalry troops may perform offensive operations within an economy-of-force role for a higher headquarters.

CONTENTS

Section I. Purpose and Fundamentals

PURPOSE

The main purpose of the offense is to defeat, destroy, or neutralize the enemy force. Offensive operations are also undertaken to secure decisive terrain, to deprive the enemy of resources, to gain information, to deceive and divert the enemy, to hold the enemy in position, to disrupt his attack, and to set up conditions for future successful operations.

FUNDAMENTALS

Successful offensive operations have four fundamentals in common.

- **Surprise.** Strike the enemy at the time and place or in a manner that he least expects.
- **Concentration.** Mass available forces, strive for overwhelming superiority in men, weapons, and firepower. With concentration, however, vulnerability becomes a factor. A force that is dispersed is much more survivable. The commander must maintain a high sense of situational awareness to anticipate the conditions of battle that will allow him to mass at the critical point, kill the enemy, and quickly disperse to survive.
- **Tempo.** Tempo is the rate of speed of military action. Controlling or altering the rate is essential for maintaining the initiative. Tempo can be fast or slow dependent on the capabilities of the troop relative to that of the enemy. Commanders must adjust tempo to ensure synchronization.
- **Audacity.** Boldness in the plan's execution is key to success in offensive operations. Commanders should understand when and where they are taking risks, but must not become tentative when executing their plan.

Section II. Movement to Contact

A movement to contact is a mission undertaken to gain or regain contact with the enemy. Unlike a zone reconnaissance, which is focused on reporting detailed information on the terrain and enemy within a given zone, movement to contact is focused on finding the enemy. The critical tasks are geared for achieving fast movement and rapid location of enemy forces.

The cavalry troop normally conducts a movement-to-contact mission as the lead element of a squadron or combined arms force hasty attack. The troop may also conduct the mission as part of a squadron movement to contact when the squadron is leading the advance of another combined arms force, such as a brigade or division.

A movement to contact is characterized by rapid aggressive action. The troop commander must rapidly develop the situation and may be permitted to bypass enemy forces to maintain momentum.

CRITICAL TASKS

During a movement to contact, certain critical tasks will be accomplished. Unless the squadron commander gives guidance otherwise, the troop will—

- Reconnoiter and determine the trafficability of all high-speed routes within the zone.
- Inspect and classify all bridges, culverts, overpasses, and underpasses along high-speed routes. Identify all bypasses and fords that cannot support rapid heavy-armor movement.
- Clear all high-speed routes of mines, obstacles, and barriers within its capability, or locate a bypass.
- Find and report all enemy forces within the zone and determine their size, composition, and activity.

TECHNIQUES

The cavalry troop can perform movement to contact in a zone up to 10 kilometers wide. The size of the zone depends on the terrain and enemy situation. Scout squads, and in some cases sections, must be mutually supporting during movement based on the enemy situation. Mutual support between squads and sections is gained by reducing the size of the zone based on the terrain. The more restrictive the terrain the less mutual support becomes available.

When considering techniques for conducting a movement to contact, remember the basics.

- Always retain the ability to maneuver.
- Make contact with the smallest force possible.
- Employ forces in depth; stay flexible.
- Develop the situation rapidly once contact is made with the enemy force. Operate at a tempo that forces him to react to you, not you to him.

When the troop receives a movement-to-contact mission, the zone of action is normally identified by lateral boundaries. A line of departure and an objective or limit of advance are specified to orient the troop's movement. The objective is usually a terrain feature placed deep enough to ensure contact with the enemy. Divide the troop zone into two platoon zones. Think about where to draw the boundary. Make sure it does not split enemy avenues of approach and that it is on easily identifiable terrain. Do not use a road for a boundary. To control progress through the zone, add phase lines every 5 to 8 kilometers on easily identifiable terrain. Place contact points near the intersection of the platoon boundary and all phase lines. If specific areas need to be checked, identify them using checkpoints for reference. Place TIRS on the map.

Use a troop vee or split vee formation when conducting a movement to contact. Deploy scout platoons abreast and have them accomplish all the related reconnaissance tasks. Move them out first across the line of departure. The mission of the scouts is to find the enemy, develop the situation, and recommend to the troop commander the best course of action for the employment of the tank/AT platoons.

The mortar section, under control of the troop fire support officer or mortar section sergeant, follows about 2 kilometers behind the scouts, centered in the zone, to provide continuous coverage forward of the scouts. The troop command post, under control of the XO, displaces in the zone, positioning on terrain that affords effective and continuous communication with troop elements and the squadron. The troop trains, under control of the 1SG, follow about 2 kilometers behind the tank/AT platoons, and bound from one covered and concealed position to another. The troop commander positions himself well forward to observe one scout platoon or the other. His position is usually located where he expects initial enemy contact or difficult situations.

Heavy Troop

Due to the survivability and firepower of scout platforms (M3) in the heavy troop, the tank platoons may be kept in considerable depth (1,000 to 3,000 meters) behind the scouts until enemy contact is made. Once contact is made and the scouts have developed the situation/recommended a course of action, the troop commander may then employ his tanks, using firepower and maneuverability to fix the enemy or destroy it by hasty attack.

Light Troop

In the light cavalry troop, the AT platoon(s) should be kept close (500 to 1,500 meters) to the scout platoons. Upon contact, the HMMWV scouts will require more time to develop the situation than Bradley-mounted scouts. The AT platoon(s) can assist the scouts in developing the situation by overwatching the scouts. Also the TOW takes time to employ, so reaction time is slower than with the tank. During movement to contact, the light troop commander must be able to execute his actions on contact very quickly to maintain the momentum of the operation. A technique the light troop might use is outlined below.

- Scouts make contact with an enemy force.
- The troop commander immediately moves one AT platoon forward to overwatch the scouts as they develop the situation.

- If the situation dictates, the AT platoon suppresses with direct fires to allow the scouts to maneuver.

- Once the situation is developed, the troop positions to either fix the enemy for the follow-on force or destroy it by a hasty attack.

EXAMPLE OF MOVEMENT TO CONTACT (HEAVY TROOP)

The troop is ordered to reestablish contact with elements of a withdrawing enemy mechanized infantry platoon. As the troop deploys in a split-vee formation, the 1st and 3d platoons cross the line of departure abreast, north and south respectively, and begin to reconnoiter in troop zone up to PL BOOT. The scout platoons determine the trafficability of high-speed routes within their zones, and search for the enemy in areas that dominate the routes. The bridge across the creek at checkpoint B will support only 20 tons, so the 1st platoon finds a solid-based fording site 100 meters south of the bridge. Thus far, the high-speed routes are not damaged or blocked. The 3d platoon reports finding tracks of what appears to be three BMPs heading northwest from Hill 244. The 1st platoon reports tracks of an enemy company-size force heading west along Route 220. Track patterns indicate BMPs and T-64s/T-72s. The mortar section eavesdrops on the troop command net, and moves with the scout platoons, staying back about 1.5 kilometers, centered in the troop zone. The 2d and 4th platoons are paired with their sister scout platoons and key their movement on the scout platoon's progress, staying back about 1 kilometer and using terrain that affords good cover and concealment. The troop CP and trains hold in place (see Figure 5-1).

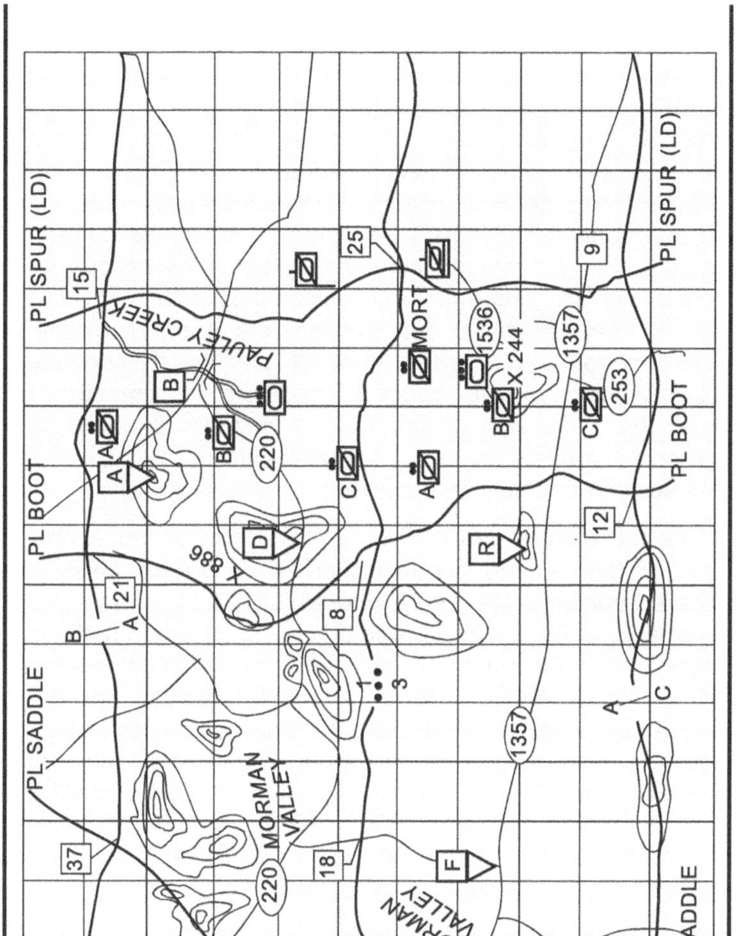

Figure 5-1. Movement to contact.

As the 1st platoon approaches PL BOOT, it reports two BRDM-2s withdrawing west at high speed on Route 220. The 3d platoon reports no contact in zone and moves forward to PL BOOT. Routes 1536, 1357, and 253 are reported trafficable. The 1st platoon reconnoiters the ridgeline in the vicinity of checkpoint D. As a scout squad crosses the southwestern side of the ridge, it receives heavy machine gun and cannon fire from an enemy position at the base of Hill 886. The 1st platoon immediately executes actions on contact. The troop commander quickly moves to a covered position from which he can observe the enemy position. While en route, he orders the 3d platoon to continue reconnaissance forward about 2 kilometers and establish a screen line. He also orders the 2d and 4th platoons to prepare for a hasty attack. The mortar section moves into a firing position and lays on the enemy (see Figure 5-2).

The 1st platoon maneuvers to the flanks of the enemy position. Dismounted scouts report a reinforced MRP in prepared positions oriented on Route 220. Enemy infantry is dug in around the buildings, and dug-in BMPs are in defilade behind them. The enemy's left flank appears to be ignored and weakly defended. Based on this information, the 1st platoon identifies a good covered and concealed armor approach that swings north around Hill 886 into the vulnerable flank. The 3d platoon reports no enemy contact in the south. The troop commander decides to conduct a hasty attack with both tank platoons to eliminate this enemy force. The hasty attack succeeds (see Figure 5-3).

The troop commander quickly orders the 1st platoon to continue its reconnaissance, and sends the 4th platoon back to the 3d platoon zone. The 1st and 3d platoons continue their reconnaissance. As scout elements of both platoons approach PL SADDLE, they observe what appears to be minefield belts and AT ditches astride Routes 220 and 1357 in the Morman Valley. Dismounted reconnaissance determines the extent of the obstacles, and reveals heavy concentrations of enemy forces in several company strongpoints along the dominating ridgeline by Highway 920.

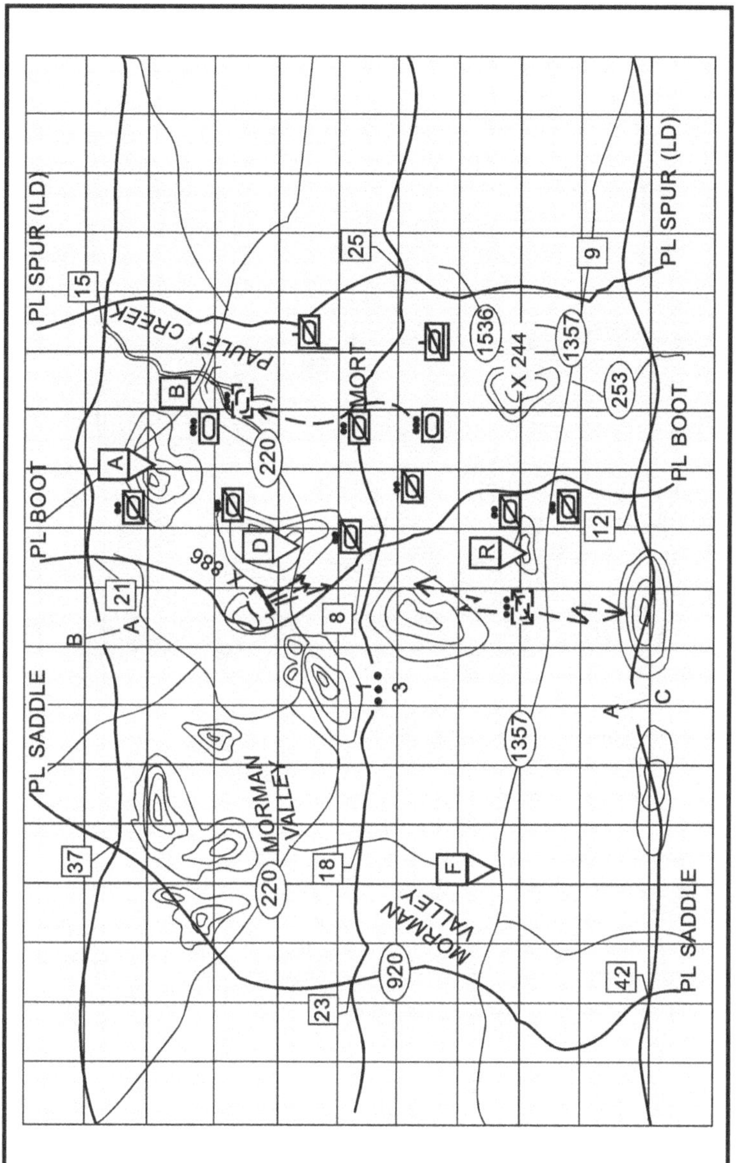

Figure 5-2. Actions on contact.

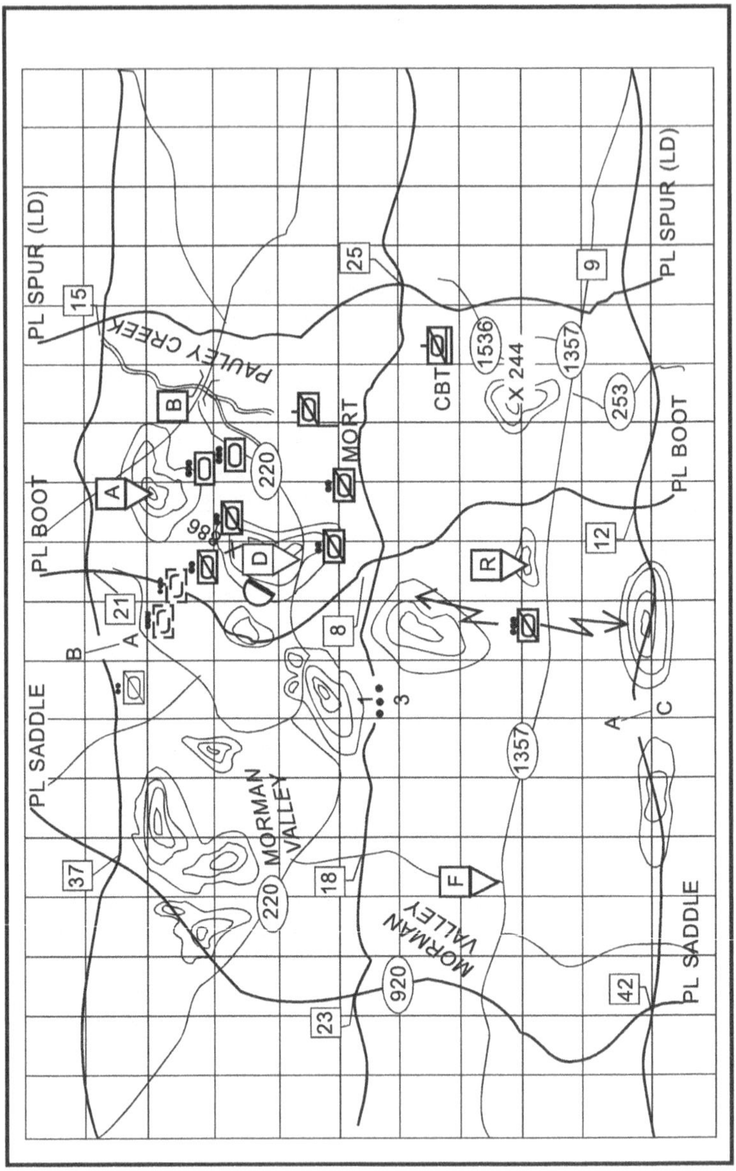

Figure 5-3. Hasty attack.

The troop commander takes a look. En route, he orders both scout platoons to determine if any gaps or boundaries exist between the enemy positions. After seeing the situation, the troop commander calls his squadron commander and recommends the troop establish a hasty defense and prepare to assist in battle handover and forward passage of following attack forces (see Figure 5-4).

Section III. Hasty Attack

A hasty attack is conducted with a minimum of preparation to defeat an enemy force that is not prepared or deployed to fight. It is a course of action routinely employed in cavalry operations to seize or retain the initiative, or to sustain the tempo of operations. A hasty attack can be executed while the troop is engaged with a zone reconnaissance mission or movement to contact. (See Chapter 3, Section II, and Section II of this chapter for other examples of hasty attack.)

CRITICAL TASKS

To execute a hasty attack, the following critical tasks must be accomplished:

- Reconnoiter and determine the size, composition, and orientation of the enemy force.
- Determine if the objective enemy force is supported by other units nearby.
- Find a high-speed covered and concealed approach into the enemy's flank(s).
- Establish a maneuver element (usually one or both tank platoons) to move to a position of advantage over the enemy and attack him by fire.

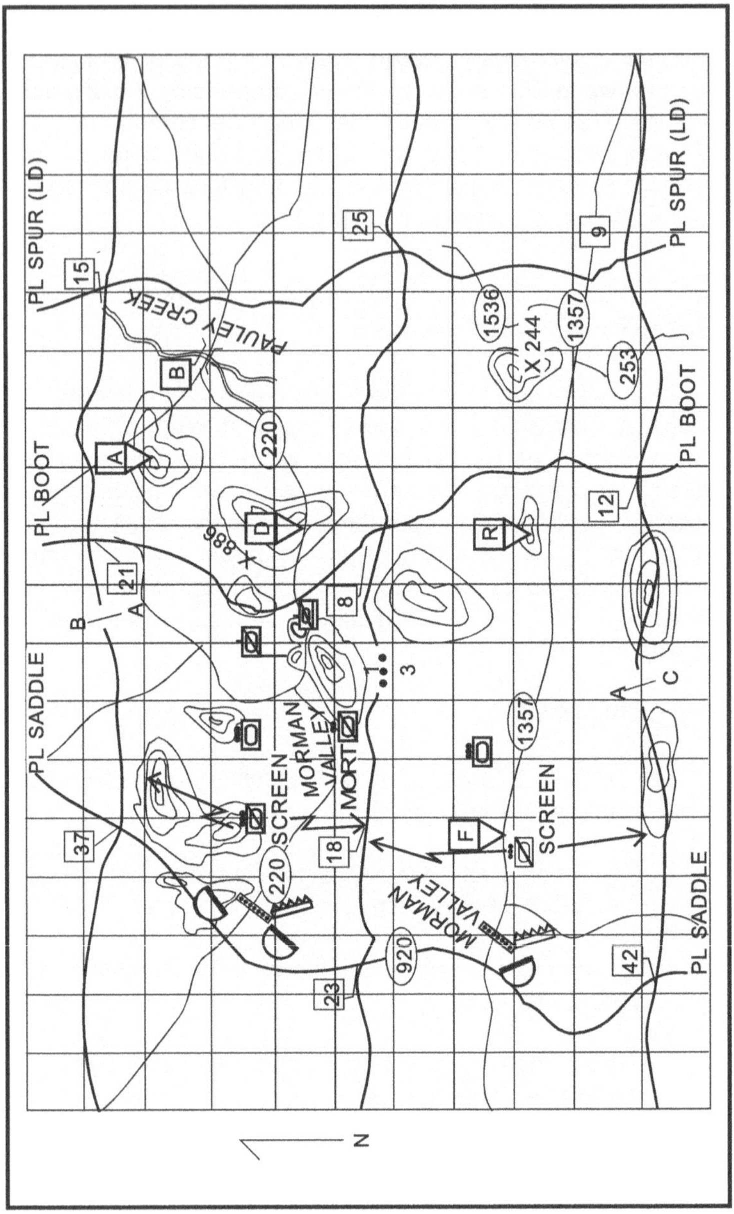

Figure 5-4. Hasty defense.

- Establish a base-of-fire element (usually one or both scout platoons) to defeat or suppress all observed enemy AT weapons with long-range direct and indirect fires before the maneuver force deploys into its attack.

- Isolate the objective enemy force from other mutually supporting units with indirect fires (usually with smoke and HE ammunition).

- Attack the enemy by fire or by fire and movement, and defeat him.

- Once the attack is completed, immediately establish hasty defensive positions and OPs on high-speed avenues of approach into the troop position.

TECHNIQUES

Each critical task has a time at which it will be accomplished in relation to all other critical tasks. A good hasty attack depends on the commander's sense of timing and on his ability to employ his forces to accomplish the tasks in the proper sequence. The commander has to synchronize—to concentrate and apply different forms of combat power against the enemy at the right times and places. The decision to conduct a hasty attack is usually made after an enemy force's reconnaissance and dispositions show that winning requires a quick strike, with little preparation. Tactics for conducting a hasty attack have three common features.

- Known or suspected enemy AT weapons are suppressed and destroyed with direct and/or indirect fires before the maneuver force is committed.

- The enemy is forced to fight in two directions.

- The enemy is suppressed and unable to react.

SETTING UP THE TROOP FOR A HASTY ATTACK

While conducting other missions, scouts will often make contact with an enemy force. In developing the situation, a scout platoon may recommend hasty attack as a course of action to the troop commander, who decides to execute the recommended course of action. The troop commander issues FRAGOs that will position forces to execute an attack simply and effectively.

The scout platoon in contact continues to reconnoiter the enemy's position and to accomplish its reconnaissance tasks. One section of the scout platoon remains in contact with the enemy. The other scout platoon continues its reconnaissance up to a limit of advance established by the troop commander. Both scout platoons continue to develop the situation further by looking for the presence of other enemy units, to the flanks or rear, supporting the enemy contact.

The FIST moves to a good position to see the battlefield and to control the indirect fires. The FSO places the mortar section on terrain where it establishes a firing position and prepares to suppress the enemy position.

The size and strength of the enemy may require the use of one or both tank/AT platoons. The commander may choose to do one of the following options:

- Move one tank/AT platoon into an overwatch position and attack by fire with one.

- Keep one in reserve behind the scout platoon not in contact and attack by fire with one.

- Attack by fire with both platoons.

The scouts determine a good attack position and attack-by-fire position for the tank/AT platoon(s). A scout squad from the scout platoon in contact moves to a link-up point with one or both tank/AT platoon(s) to guide them into the attack position.

The troop commander moves to collocate with the scout platoon in overwatch. The 1SG moves medics and recovery vehicles close to the battlefield. The XO assists the commander in control of the troop and keeps the higher commander informed.

INDIRECT FIRES IN SUPPORT OF A HASTY ATTACK

Indirect fires should complement the troop's scheme of maneuver. In the light troop, dependent on availability of supporting indirect-fire systems, indirect fires may prove to be the best weapon of destruction. The troop can acquire and engage the enemy with indirect fire from positions offering good protection from enemy direct/indirect fire. Because the light troop is in covered and concealed positions, the survivability of the troop is enhanced. Indirect fires must also be controlled to prevent fratricide. The troop commander must determine the following:

- Who will control the indirect fires during the hasty attack?
- Who will initiate indirect fires onto the objective?
- Who will lift and shift the indirect fires to subsequent targets?
- What will the signal be for lifting and shifting indirect fires?
- Are there any restrictive fire measures or restrictive fire areas?

There are many different answers to these questions. METT-T will determine which answer works best. Under most conditions the commander or the FSO will be in the best position to control the engagement of indirect-fire systems.

The troop commander should use available indirect fires from mortars and supporting artillery to—

- Suppress the enemy while scouts are maneuvering to develop the situation.
- Screen enemy observation of scouts or assault element during the conduct of the hasty attack.
- Isolate the enemy contact by firing HE and smoke between the enemy force and any possible supporting positions.
- Shift indirect fires off the objective to block enemy withdrawal routes.

DIRECT FIRES DURING HASTY ATTACK

When executing the hasty attack the troop must combine indirect fires with direct fires. The direct-fire systems available to the heavy troop commander are highly destructive. The high rate of fire from the M3 and M1 allows the commander to quickly place effective fires into the objective area while stationary or moving.

The systems available to the light troop commander can be just as destructive when used in combinations and against a comparably equipped enemy. The light troop commander should strive to engage the enemy with combinations of weapons. Using the bounding technique and move-set drills by platoons, the troop commander can place effective suppressive and destructive fires into the objective area. (The high rates of fire from the M2 caliber .50 and MK-19 should be used to suppress the enemy and cover the TOW systems as they engage.) See Figure 5-5.

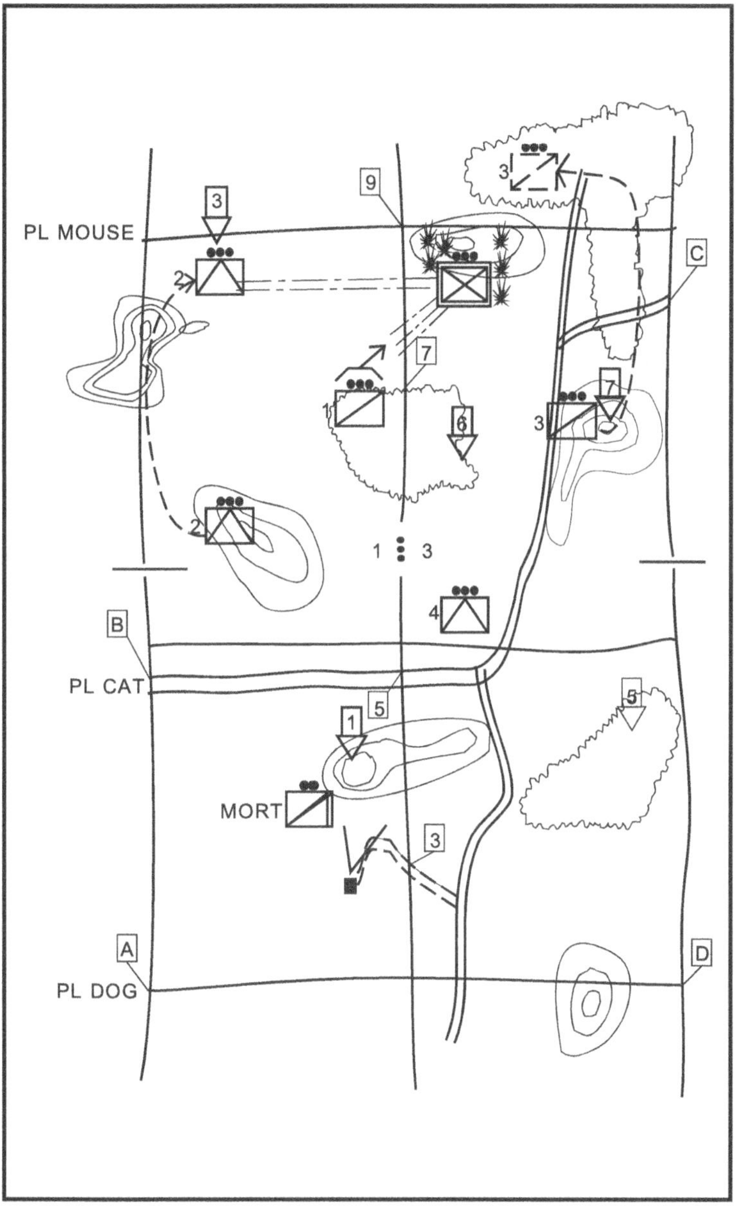

Figure 5-5. Light cavalry troop using fire and maneuver.

Like indirect fires, direct fires must be controlled. The commander must determine the following before executing a hasty attack:

- Who will initiate direct fires into the objective area and from where will they be initiated?
- What is the aim point for the support element? When do they shift fires and where do they shift fires to?
- What is the limit of advance for the assault force?

To determine the answers to these questions the commander must look at the enemy he is facing, the terrain he is operating in, and the forces he has to bring to bear against the enemy.

THE OBJECTIVE

Once the troop has initiated its attack and enemy resistance in the objective area has ceased, the troop may begin consolidating in preparation of continuing its mission. Have one or both scout platoons dismount scouts to sweep the area quickly for prisoners and other items of tactical value. Immediately establish a hasty defense with tank/AT platoons oriented on high-speed approaches into the troop position, and with scout platoons oriented on expected high-speed enemy approaches at OPs out from the defensive position. Position the mortar section and have the guns lay on the most likely enemy avenue of approach. Redistribute ammunition and balance crews as time allows. Replace key leaders quickly. If time is available and enemy counterattack unlikely, begin resupply operations.

EXAMPLE OF A HASTY ATTACK

A troop is conducting a zone reconnaissance moving toward its limit of advance (LOA). As the 3d platoon crosses the ridge (PL SABER) in its zone, Charlie scout section is hit by enemy tank cannon and heavy machine-gun fire from concealed positions near checkpoint 2. The platoon immediately takes cover and reports contact to the troop commander. The troop commander acknowledges and immediately moves to link up with the platoon leader. The platoon sergeant calls for indirect fire to suppress the enemy force. Under this protection, the platoon leader orders dismounted reconnaissance through the woods on the enemy's right to find his flank. He then sends his Bravo scout section forward on the enemy's left to find his other flank and to determine if the enemy is mutually supported by other enemy forces near checkpoint 2. The platoon leader and two scouts dismount, move in closer, and determine the enemy force consists of a reinforced MRP, two tanks, and three BMPs, in a hasty defensive position. Scouts report what appear to be minefields forward and to the flanks of the enemy position. The Alpha section identifies the right flank of the enemy position and discovers the flank is not protected by obstacles and not covered by direct-fire weapons (see Figure 5-6).

En route to the 3d platoon's zone, the troop commander tells the 1st platoon to continue reconnaissance forward to the LOA and find out if other enemy forces are providing mutual support to the enemy platoon in the 3d platoon's zone. He also tells the platoon leader to establish a screen along the LOA in his platoon zone. He then tells the 4th platoon to join the 2d platoon near checkpoint 12, using the ridgeline to cover its move, and prepare to conduct a hasty attack. Using TIRS, he outlines a route that runs from checkpoint 12 around the western edge of the village to checkpoint 7, then to a position from which it can engage the enemy on his most vulnerable flank (see Figure 5-7).

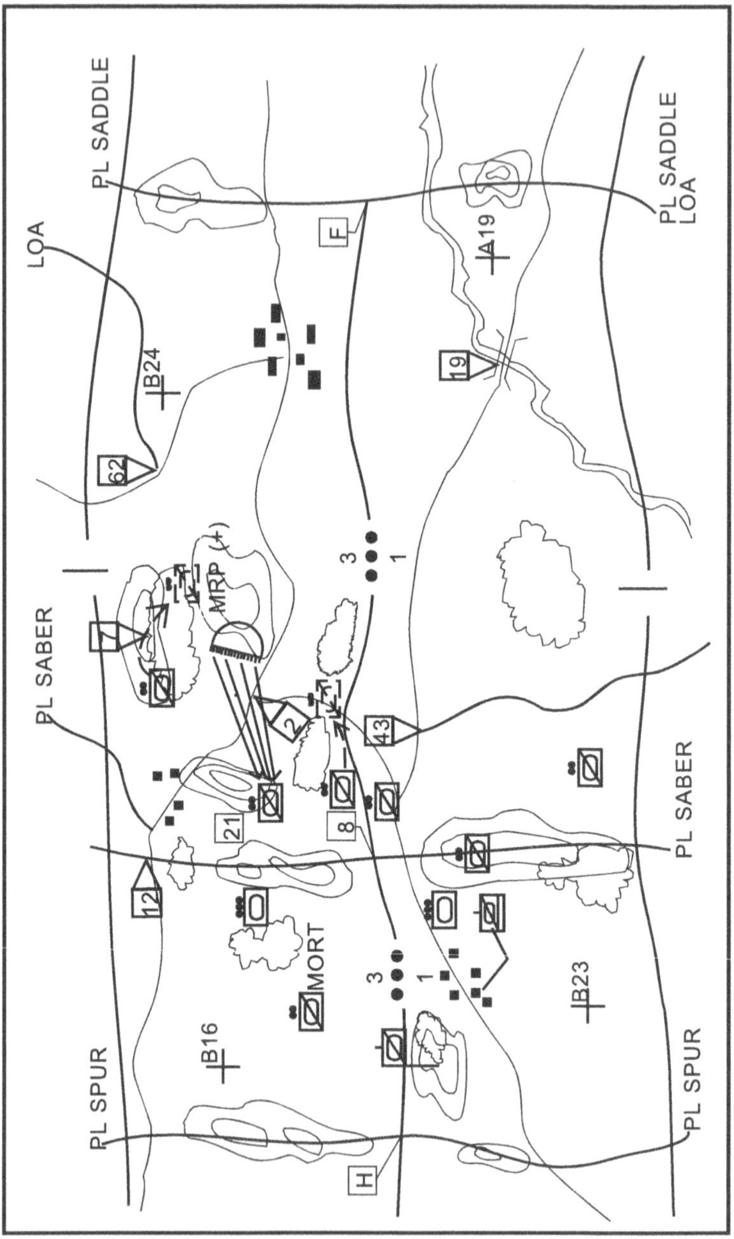

Figure 5-6. Develop the situation.

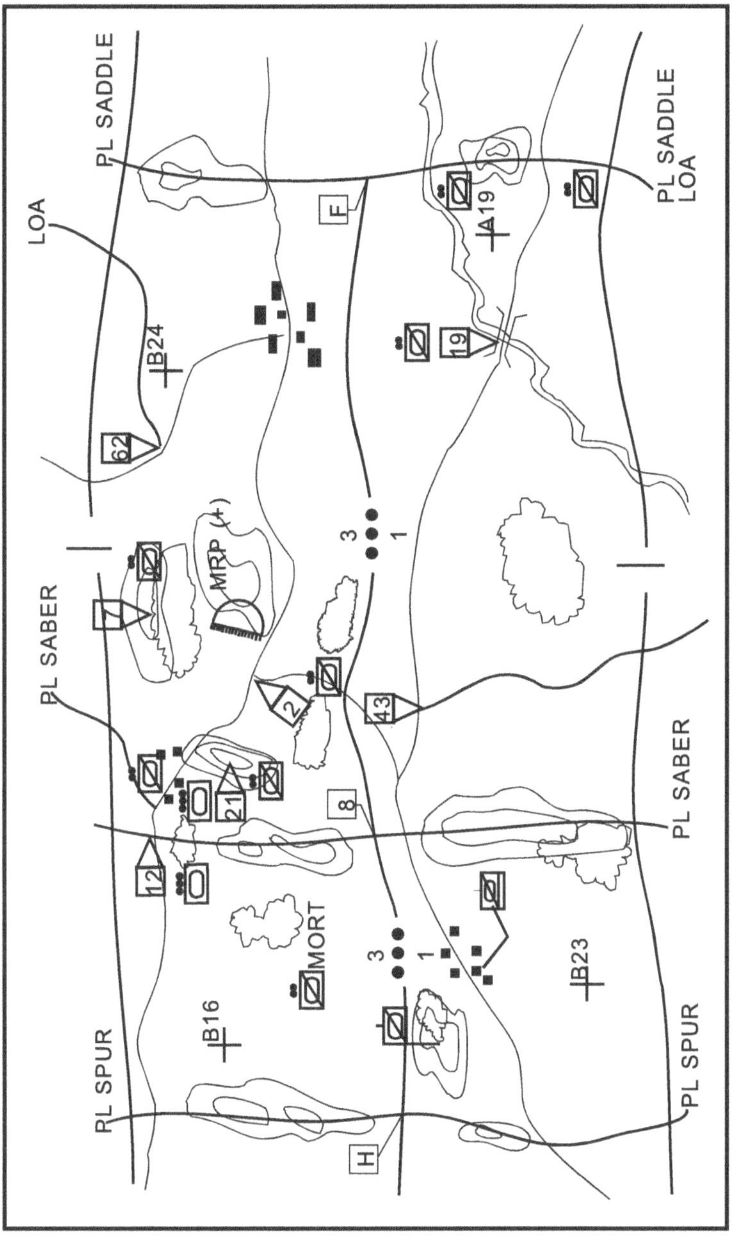

Figure 5-7. Set for hasty attack.

After a first-hand look at the situation in the 3d platoon's zone, the troop commander decides to attack. He tells the FSO to continue suppressing the enemy. Next, he tells the 3d platoon leader to establish a base of fire from covered positions near checkpoint 21. The 3d platoon is tasked to destroy or suppress all AT weapons observed. While heavy suppressive fires are raining on the enemy force, the commander joins the tank platoons at checkpoint 12 and leads them into the attack. After passing checkpoint 7, the tank platoons begin maneuvering. The troop commander tells the FSO to shift his fires on routes of withdrawal behind the enemy position. The tank platoons go around the woodline, deploy abreast, and attack by fire (see Figure 5-8).

Two tanks in the 4th platoon are hit and damaged; one soldier is killed and three are injured. The tank platoons consolidate in covered and concealed positions near checkpoint 7. Scouts from the 3d platoon move in quickly, capture several wounded prisoners, and search the vehicles and dead for anything of intelligence value. The 3d platoon leader orders one scout squad to stay and secure the prisoners, and then takes the rest of the platoon forward to finish zone reconnaissance up to the LOA.

The 1SG leads the medics and M88 to the 4th platoon's casualties and picks up the prisoners the 3d platoon secured. The XO coordinates with the S4 for immediate resupply, then collects and transmits final reconnaissance reports to the squadron.

Section IV. Deliberate Attack

A deliberate attack is an offensive mission conducted to defeat enemy forces in dug-in and prepared defensive positions or strongpoints. It is preceded by intensive intelligence gathering using sources such as patrols, aerial photographs, reconnaissance by fire, radio intercepts, radar surveillance, satellite reconnaissance, and prisoner interrogation. Large volumes of indirect preparatory fires are normally used. Detailed tactical planning and rehearsal of the attack are crucial to success.

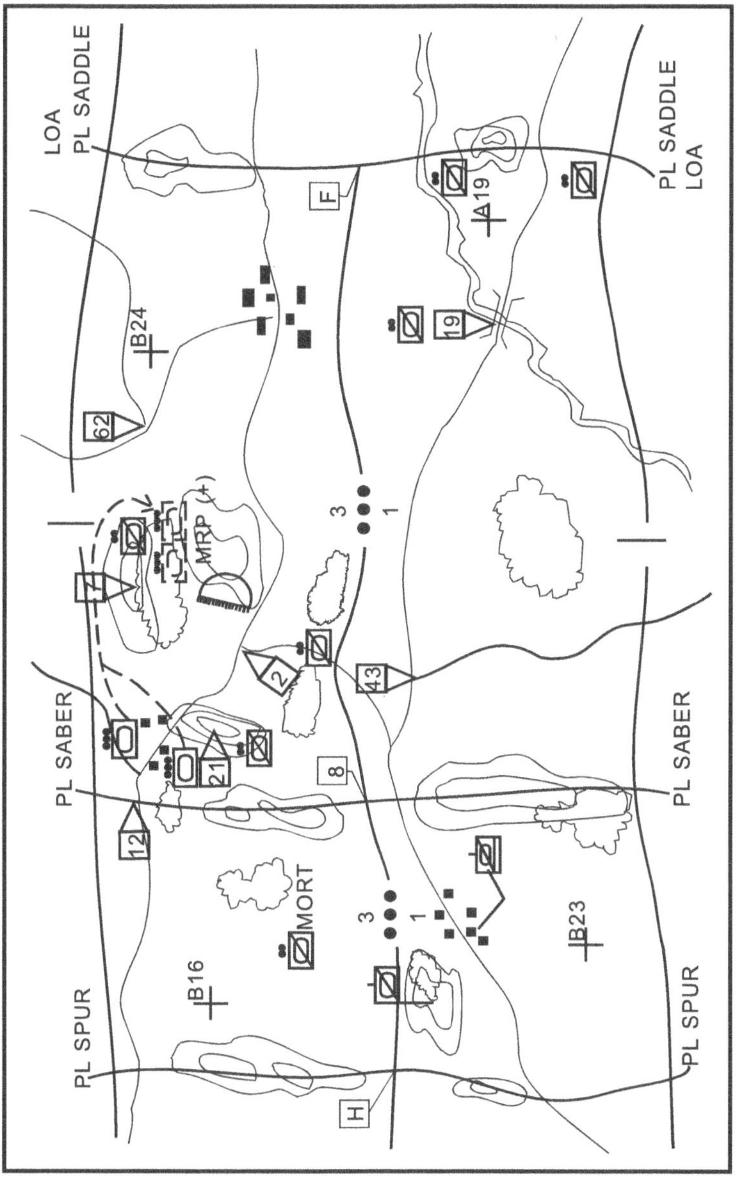

Figure 5-8. Tank platoons move into position for the hasty attack.

The cavalry troop alone cannot conduct a deliberate attack against prepared defensive positions. It usually participates with its parent squadron or other combined arms force as one of the three major elements in a breaching and assault operation. The troop can serve as the *overwatch/support element,* the *breaching element* (if reinforced with engineers), or the *assault element.* Doctrine for attacking a complex obstacle or a strongpoint is found in FM 71-1 and FM 17-95-10.

Section V. Raid

A raid is an attack into enemy territory for a specific purpose, with no intent to gain or hold terrain. The troop returns to friendly lines after the attack.

The cavalry troop normally conducts a raid against clear, well-defined targets whose destruction will have an adverse effect on an enemy force. Good targets for a troop raid are enemy artillery, rear service units, or command and control headquarters.

CRITICAL TASKS

During a raid, the following critical tasks must be accomplished:

- Coordinate and establish a restricted fire area around the raid area.
- Reconnoiter the troop's direction of attack from the point of departure to the objective.
- Establish visual contact with the enemy. Reconnoiter and determine the best place to initiate the attack.
- Establish a base of fire element to suppress the enemy force.

- Establish a maneuver element (usually one or both tank platoons) to move to a position of advantage over the enemy and attack them by fire.
- Suppress the enemy force with long-range direct fires while the maneuver element moves up to begin its attack.
- Establish surveillance on enemy avenues of approach to the objective, where follow-on enemy maneuver units would most likely appear.
- Attack the enemy by fire or by fire and movement to defeat them. Finish quickly.
- Establish security to protect the withdrawal of the remainder of the troop.
- Return rapidly to friendly lines and enter at the designated passage point.

TECHNIQUES

A raid achieves its purpose through speed, violent execution, and surprise. The troop's maneuver should produce that effect. There will probably not be a chance to reconnoiter before the attack; speed and surprise generate the security needed. Consequently, the troop might not always take the mortar section, combat trains, or command post. Vehicles disabled on a raid may be left behind in the interests of speed and security for the entire troop. When the troop receives a FRAGO to conduct a raid, the commander should receive the following graphic control measures from the S3 or squadron commander: a point of departure along the line of departure or line of contact, a direction of attack, an objective, a route of withdrawal, and a passage point to enter friendly lines. Analyze the terrain and the enemy situation, then issue the troop FRAGO.

The FSO should coordinate with the squadron FSO to establish a restricted fire area around the raid area. The 1SG should coordinate for immediate resupply of Class III and Class V when the troop returns. Have one scout platoon take the lead and reconnoiter the troop direction of attack from the point of departure on the line of departure to the objective. The lead platoon should establish visual contact with the raid objective to confirm the troop's scheme of maneuver and establish surveillance of the objective. The other scout platoon may act as the base-of-fire element, and follow about 1 kilometer behind the lead platoon. Have the lead platoon guide it into an overwatch position to deliver long-range suppressive fires on the enemy. Consolidate the tank platoons, move forward to join them, then follow about 1 kilometer behind the second scout platoon.

Once the base-of-fire element is set, have the platoon suppress the enemy with direct and indirect fires. Guide the maneuver element into a position to attack by fire. Finish quickly and move out.

Tell the scout platoon performing reconnaissance and surveillance tasks to screen and protect the withdrawal of the remainder of the troop back to friendly lines. Have the other scout platoon lead the way back on the designated route of withdrawal, and have the tank platoons follow. Stay back and supervise the conduct of the rear screen until all troop elements have entered friendly lines. Consolidate and resupply the troop with fuel and ammunition as soon as possible. Get ready to fight again.

EXAMPLE OF A RAID (HEAVY TROOP)

The squadron is defending in sector astride a likely Threat regimental zone of attack. The three armored cavalry troops are deployed abreast with the tank company in reserve. The squadron frontage is about 10 kilometers wide. Threat forces attack from the march. Squadron elements in the northern half of the sector are softened with long-range preparatory fires delivered by rocket and artillery forces. As artillery fires increase in intensity, the fire support element and the advance guard battalion of the regiment deploy from the march into prebattle formation and then attack. The advance guard battalion is quickly defeated by the concentrated fires of the northern cavalry troop and a surprise counterattack by the tank company. The squadron rapidly repositions to meet the assault of the regimental main body. Within one hour, the regimental main body approaches the line of contact with two MRBs abreast. Their axis of advance indicates most of their combat power will be concentrated in the northern half of the squadron's sector. Artillery suppressive fires increase as the Threat battalions deploy into attack formation. Aeroscouts report two 122-mm artillery battalions of the regimental artillery group moving into firing positions about 4 kilometers behind the attacking combat maneuver units. In this situation, with good defensible terrain, the squadron commander decides to concentrate the combat power of the troops, the tank company, and his available artillery in the north to defeat the advancing maneuver forces. He orders the southern cavalry troop to conduct a raid behind enemy lines to destroy the regimental artillery group simultaneously.

The troop commander quickly analyzes the terrain, and then issues a FRAGO to the troop. He tells the 1st platoon, which is currently on the screen line, to take the lead and reconnoiter the troop's direction of attack from the point of departure on the LD to the objective area. The platoon is tasked to establish visual contact with the Threat artillery units and to locate the best covered and concealed approach into their flanks or rear. The troop commander tells the 3d platoon to follow about 1 kilometer behind the 1st platoon and establish a base of fire to suppress and fix the artillery units. He orders the 2d and 4th platoons to consolidate rapidly near the point of departure. He tells the mortar section to remain in the troop sector. Priority of fires goes to the 1st platoon. As the 1st platoon approaches the point of departure, the troop FSO coordinates with the squadron FSO to establish a restricted fire area around the raid area. Once it is established and confirmed by the FSO, the troop commander orders the 1st platoon forward (see Figure 5-9).

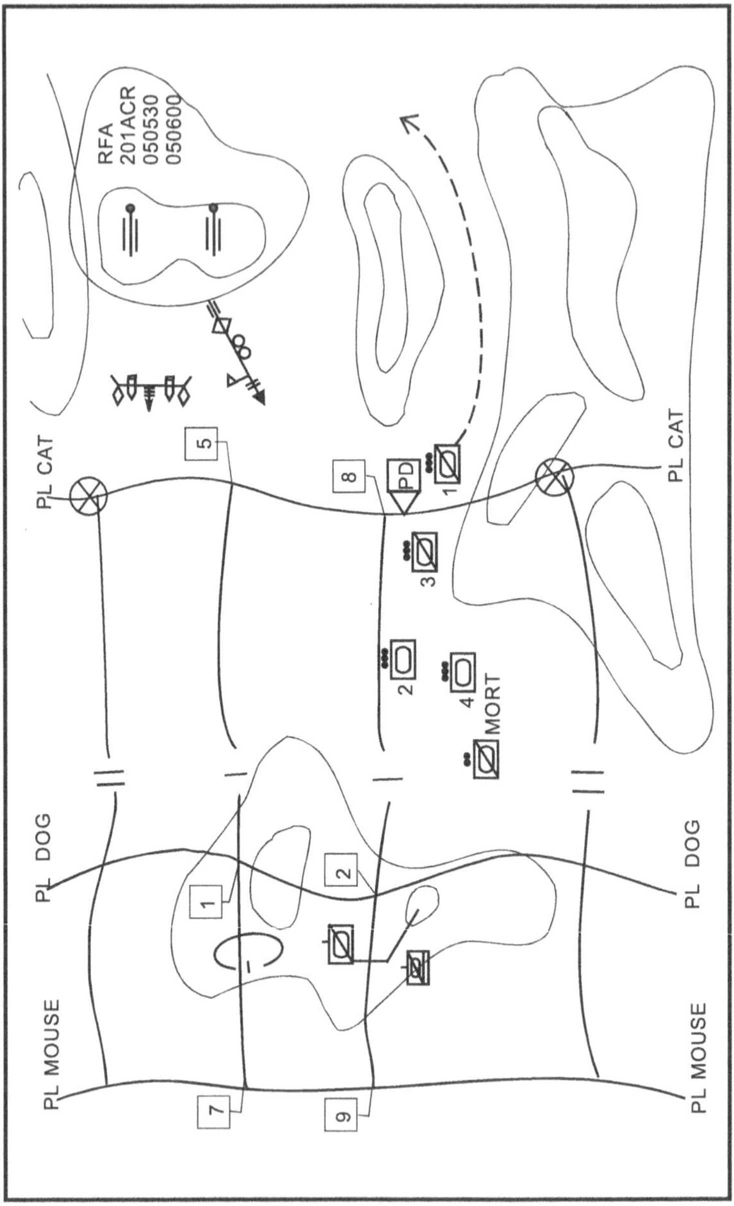

Figure 5-9. Heavy troop raid (part one).

The 1st platoon makes visual contact in about 10 minutes, and continues to reconnoiter for a good axis of attack. The 3d platoon moves into an overwatch position about 1,800 meters away and lays down a curtain of suppressive fire with chain guns, TOW missiles, and mortar fire. While suppressive fires are delivered, the tank platoons maneuver into the left rear of the artillery units, then deploy abreast and attack the Threat artillery battalions by fire (see Figure 5-10).

After silencing the battalions, the troop commander immediately orders the 3d platoon to lead the troop back to friendly lines along the designated route of withdrawal. The 2d and 4th platoons follow while the 1st platoon screens the withdrawal of the troop. The troop commander joins the 1st platoon and supervises the conduct of the rear screen until all troop elements have reentered friendly lines (see Figure 5-11). The troop returns to defensive positions, and the 1SG begins resupply actions.

EXAMPLE OF A RAID (LIGHT TROOP)

The light troop should consider the raid target when planning its operation. A combination of weapon systems may best suit the destruction of the raid objective rather than operating in pure platoons. A mix of weapon systems may also assist in security of the objective during the raid. The troop scramble should be considered as a possible task organization when planning a raid.

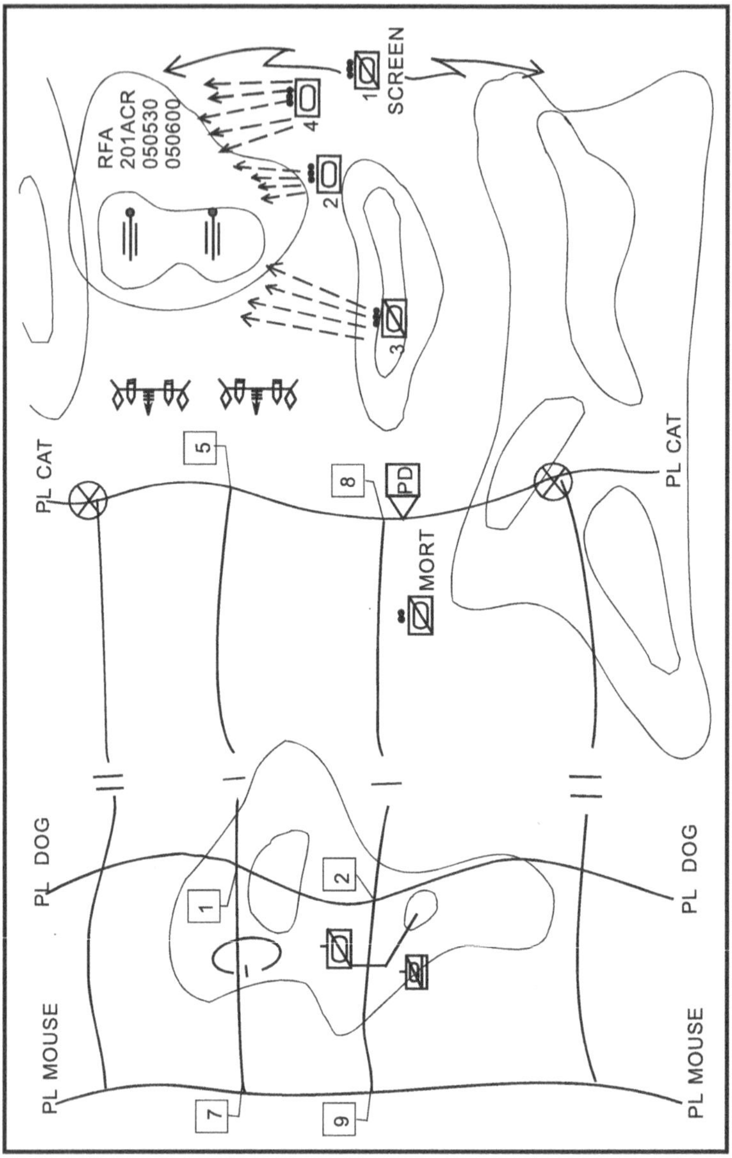

Figure 5-10. Heavy troop raid (part two).

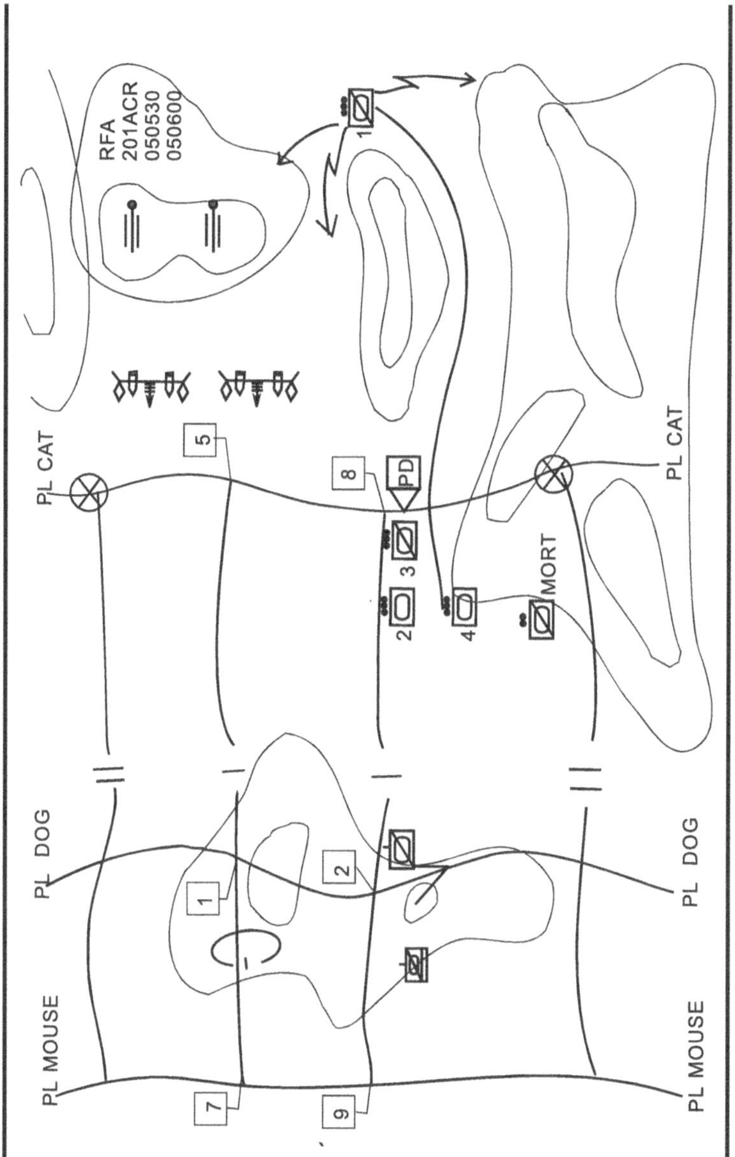

Figure 5-11. Heavy troop raid (part three).

Example steps of a light troop conducting a raid are as follows:

- Troop moves in column, platoons in a vee. Troop is a scramble organization with platoons integrating the AT vehicles into the scout sections. Each scout platoon is running two sections of three vehicles plus the platoon leader. See Figure 5-12.
- Lead platoon (1st platoon) conducts reconnaissance forward of the troop.
- 1st platoon establishes contact with the objective, then secures the flanks and conducts surveillance of the objective.
- 2d platoon, second in the order of movement, moves through 1st platoon, establishes security on the far side of the objective, orienting on the mounted avenues of approach into the objective area.
- 3d platoon establishes an attack position vicinity of BP1. 4th platoon establishes an attack position vicinity of BP2.
- Troop commander initiates the attack with indirect fires. 3d and 4th platoons move and attack by fire from BP1 and BP2.
- Troop engages enemy position using TOWs to destroy hard targets and fires from caliber .50 and MK-19s for soft targets and suppression of enemy defenses.
- Troop uses indirect fires to cover their withdrawal.
- 1st platoon covers the troop withdrawal.
- 4th platoon displaces first, followed by 3d platoon, then 2d, then 1st.
- Troop moves in column, platoons in vee with 1st platoon providing rear security. 4th platoon conducts a rapid zone reconnaissance forward of the troop as it moves toward friendly lines. See Figure 5-13.

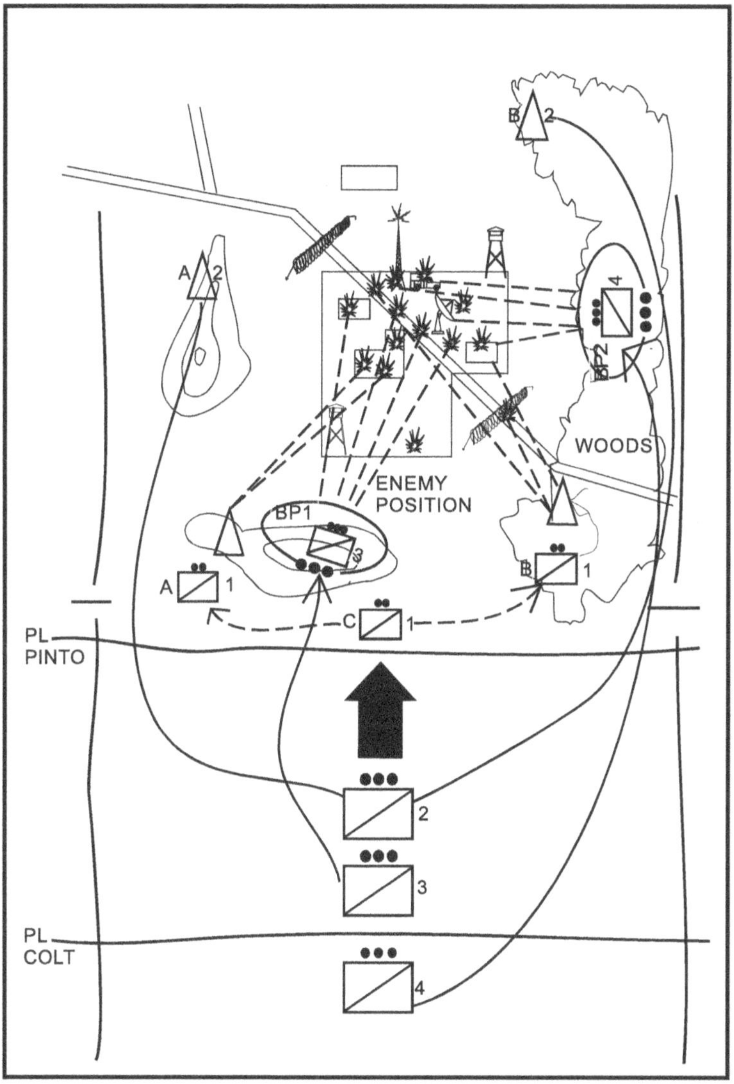

Figure 5-12. Light cavalry troop conducts a raid using a troop scramble.

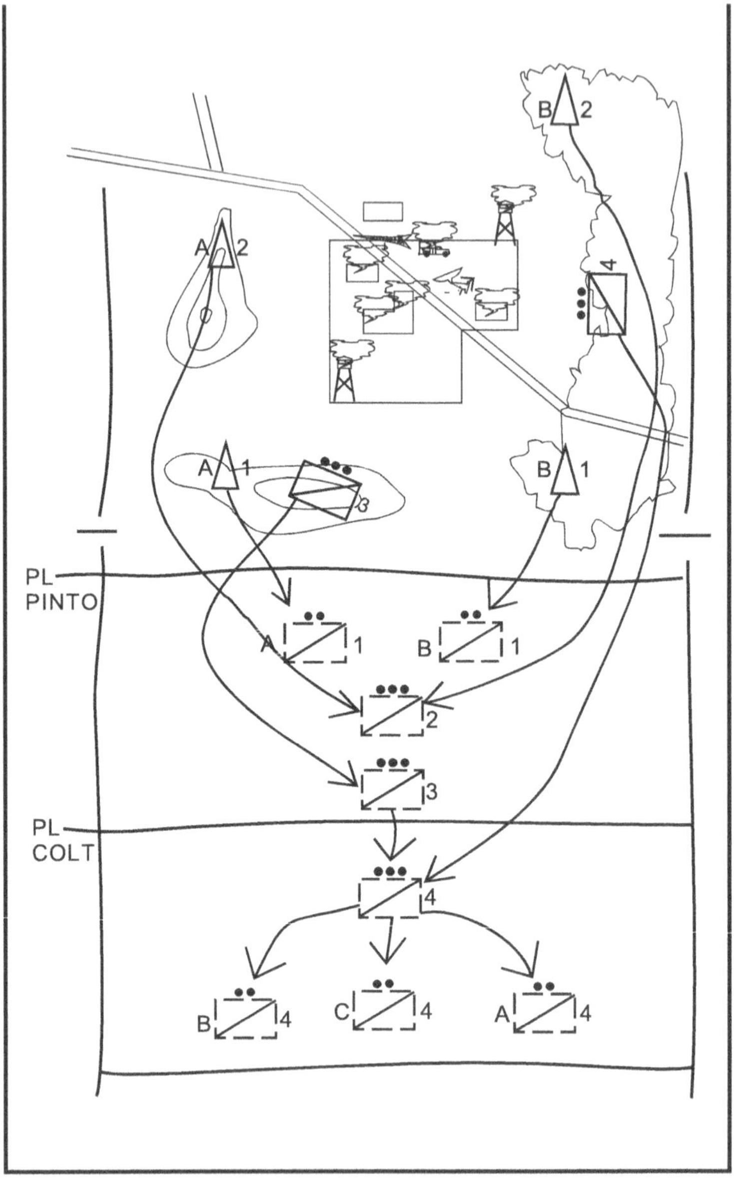

Figure 5-13. Light cavalry troop completes the raid and returns to friendly lines.

Chapter 6

Defense

Defense is not the decisive form of war. While defense can deny success to the enemy, it seldom assures victory. The defense is, however, the stronger form of war due to the inherent advantages of the defender. Army operations recognize the strength of the defense, but emphasize the necessity to transition to the offense quickly.

Brigades, battalion task forces, and company teams are the principal defensive forces for the corps or division. Cavalry units normally perform security missions for the defense or reconnaissance missions to support attacks. Cavalry units frequently perform defensive operations as a part of security missions, or when required to defend in an economy-of-force role for a higher headquarters.

Contents

Section I. Purpose, Fundamentals, and Schemes of Maneuver

PURPOSE

The immediate purpose of defensive operations is to defeat an enemy attack and create opportunities to go on the offense. For both the heavy and light cavalry troops, defensive operations may also achieve one or more of the following:

- Gain time.
- Concentrate forces elsewhere.
- Attrit enemy forces as a prelude to offensive operations.
- Control key or decisive terrain.
- Retain tactical, strategic, or political objectives.

In some cases, the troop may be required to defend because it cannot muster enough combat power to attack.

FUNDAMENTALS

Eight fundamentals are common to cavalry troop defensive operations.

Depth. Position platoons in depth, and place obstacles in depth. Depth allows the troop to—

- Gain enemy contact early.
- Perform counterreconnaissance tasks.
- Ascertain enemy direction of attack/intentions.
- Develop the situation, providing reaction time and maneuver space to concentrate combat power when and where it is needed.

Dispersion. Deploy subordinate elements as far apart as possible without losing their ability to concentrate (mass) firepower against the enemy. The more dispersed the troop, the harder it is for the enemy to mass fires against it as a whole. However, do not allow the enemy to concentrate its forces or fires against isolated elements.

Security. The troop may employ passive or active measures, or a combination of techniques. All must be considered in the defensive plan. Examples of passive and active security measures follow:

Passive	Active
Disperse vehicles and platoons.	Screen/Establish OPs.
Use camouflage/ cover and concealment.	Perform mounted/ dismounted patrols.
Impose radio listening silence.	Establish GSR posts.
Use hide positions.	Establish M8 chemical alarm net.
Enforce noise and light discipline.	
Minimize movement.	
Do not position in likely target areas.	

Maximize Terrain Advantages. Study the terrain. Reconnoiter it from both the troop commander's and the enemy's view, if possible, to determine—

- Avenues of approach.
- Reconnaissance avenues of approach.
- Restricted/severely restricted areas.
- Defiles (canalizing terrain).

- Engagement areas.
- Battle positions.
- Subsequent and alternate battle positions.
- Hide positions to support battle positions.
- OP positions forward of defensive positions.
- Subsequent and alternate OP positions.
- Positions where obstacles can be tied in with natural obstacles to turn, disrupt, or block the enemy.
- Positions that facilitate counterattacking by fire or by fire and maneuver into the flanks and throughout the depth of the enemy.
- Routes to and from each position.

Stop Enemy Rate of Advance. Offense is based on two principles—speed and mass. Develop a defensive plan that blunts the momentum of the enemy attack through the use of obstacles and fires (direct and indirect). If most of his combat power is killed, the enemy will be forced to dig in and establish a hasty defense or withdraw from combat.

Mass Combat Power at the Right Place and Time. In order to defeat a massed attack, the troop must mass fires against the enemy where and when he is least able to escape its effects. The ability of the troop to mass combat power when and where it is needed is a function of—

- Early warning/reaction time.
- Responsive/rapid maneuver by subordinate elements.

Force the Enemy to Fight in Two Directions. When engaging the attacking force, maneuver platoons into positions that force the enemy to turn and fight in two or more directions. This will force him to split his fires, preventing him from concentrating fires, and to expose his vulnerable flanks.

Counterattack. Draw the enemy into structured engagement areas and attack him *en masse* with overwhelming firepower to destroy him quickly and decisively. Wrest the initiative from him. Maneuver forces to exploit the situation. Attack by fire and by fire and maneuver into his vulnerable flanks and throughout the depth of his formation.

DEFENSIVE SCHEMES OF MANEUVER

There are three basic schemes of maneuver the commander can use in designating a course of action for a defensive mission. These schemes of maneuver center on the use of battle positions and sectors for subordinate platoons, or a combination of the two.

Defend from a Troop Battle Position. Both heavy and light cavalry troops may defend from a battle position. This method is be used when the enemy situation is clear, when there is only one avenue of approach, or when the troop commander must coordinate subordinate unit fires. In this scheme of maneuver the troop commander retains most of the authority for fighting the battle. The troop commander must understand his squadron commander's intent and concept to prevent holding the troop in place and risking its destruction.

Defend in Troop Sector. Both the heavy and light cavalry troops may defend in troop sector. This method may be used when the enemy situation is vague, when there is more than one avenue of approach, or when subordinate platoons require more freedom or action. In this scheme of maneuver, the troop commander delegates much of the responsibility for fighting the battle to his subordinate platoon leaders. He maintains control through the use of effective control measures and a clearly understood intent and concept.

Defend by Combination of Battle Positions and Sectors. Both heavy and light cavalry troops may defend using a combination of battle positions and sectors. This is the most common method of defense for a cavalry troop as it provides the troop commander maximum flexibility to maneuver and mass fires throughout the depth of the sector when changes in METT-T dictate.

Section II. Defend From a Troop Battle Position

Either the heavy or light cavalry troop may defend from a troop battle position. This mission is normally assigned when the squadron commander elects to concentrate the direct fires of the troop or squadron within an engagement area. The troop cannot maneuver outside the position without the squadron commander's permission. Within the battle position, the troop commander positions his platoons to concentrate all direct fires where the squadron has specified. The troop fights to retain the position unless ordered by the squadron commander to counterattack or withdraw. The troop may still retain the task to screen/maintain contact with the enemy forward of the battle position, depending on the squadron commander's intent. If so, the tank/AT platoons will initially be positioned within the troop battle position, and the scout platoons may assume adjacent positions within the battle position following execution of a security drill (see Security Drill paragraph in Chapter 4, page 4-15).

CRITICAL TASKS

To defend a troop battle position, the following critical tasks must be accomplished:

- Decide where the enemy will be killed and designate the engagement area.
- Establish OPs oriented forward and to the flanks of the BP to gain contact with the enemy force and provide early warning.
- Establish primary and alternate platoon BPs to concentrate direct fires within the engagement area as directed by the squadron commander.

- Designate supplementary platoon BPs to cover other routes of enemy approach.
- Establish sectors of fire for each platoon.
- Reconnoiter and establish platoon routes from hide positions to platoon BP, and for withdrawal to subsequent platoon/troop BPs.

TECHNIQUES

The squadron commander assigns troop battle positions when he can mass the fires of two or more troops in a squadron engagement area. The size of a troop battle position can vary, but it should provide enough depth and maneuver space for platoons to maneuver into alternate/supplementary positions and execute local counterattacks.

The troop commander conducts a thorough terrain study before the troop occupies the position, keeping in mind where the squadron commander wants the troop to concentrate its fires.

- Designate primary and alternate positions for each platoon.
- Position platoons to achieve flanking fires along the avenue of approach. Consider the effective range of each platoon's weapon systems.
- Position platoons to mass direct fires within the engagement area and to provide mutual support.
- Position platoons to cover any dead space in the engagement area. If not possible, plan indirect fires to cover the dead space.
- Use TRPs to control fires and orient weapon systems for each platoon.
- Occupy the BPs from the rear. Establish OPs to support the squadron plan. Allow platoon leaders time to reconnoiter and position vehicles to place effective direct fires within their sectors of fire. If better

defensive terrain lies outside the BP, call the S3 or commander for approval to adjust the boundaries.

- Once platoons are set, inspect them to make sure each platoon is properly oriented and has good fields of fire. When satisfied, have the platoons reconnoiter routes to subsequent positions and select firing positions for each vehicle. If time and engineer support are available, dig prepared positions for vehicles. Have the platoon leaders prepare platoon fire plans for approval.

- After preparing the BP, have all elements except OPs move to hide positions to reduce the risk of enemy observation and to decrease their vulnerability to enemy fires.

- Maintain security. Position the command post where it has FM communications with subordinate elements and the squadron. Position the troop trains behind good cover that is out of direct-fire range and allows quick access to each platoon position. The troop commander is positioned where he can observe the engagement area and control the troop. The FSO should be nearby to ensure coordinated fire support.

- Identify trigger points/lines.
  - Upon the enemy reaching what terrain feature does the troop initiate indirect fires? Direct fires?
  - Upon the enemy reaching what terrain feature does the troop displace to prevent decisive engagement?

- If the troop has to disengage and displace under fire to a subsequent position, bound the troop back by platoon(s), consistent with the squadron scheme of maneuver. If the troop disengagement and displacement are covered by another element, the troop may move as a whole to a subsequent position.

See Figure 6-1 for an illustration of the techniques described above.

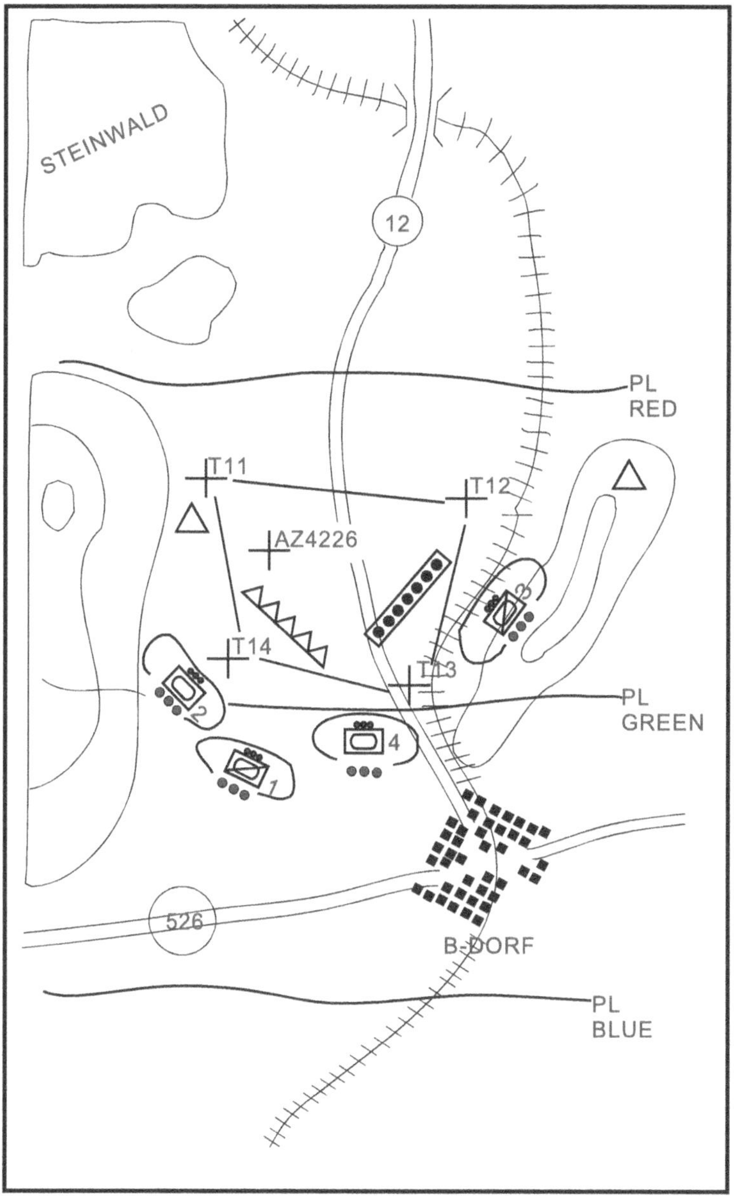

Figure 6-1. Troop engagement area.

Section III. Defend in Troop Sector

Both heavy and light cavalry troops may defend in sector. METT-T considerations determine optimal troop sector width; however, the troop is normally allocated a sector oriented on a single battalion-size avenue of approach.

Either troop may defend in sector when—

- The squadron cannot concentrate its fires due to—
  - extended frontages.
  - defending along a cross compartment.
  - multiple avenues of approach.
- Retention of specific terrain features is not necessary.
- The troop may use the depth of the sector to dissipate the enemy's attack.
- Maximum flexibility to maneuver is desired.

CRITICAL TASKS

- Maintain continuous surveillance of high-speed routes or avenues of approach into the troop sector (screen).
- Destroy or repel all enemy reconnaissance elements forward of the troop's initial defensive positions (counterreconnaissance).
- Structure engagement areas.
- Position platoon battle positions to support engagement areas.
- Engage the enemy from more than one direction.
- Determine criteria for initiating fires, counterattack, and disengagement.
- Prevent the enemy from penetrating the troop rear boundary or designated NPL (no penetration line).

TECHNIQUES

When given the order to defend in sector, the squadron will usually provide the following graphic control measures: *troop boundaries, an initial screen line, a rear boundary, phase lines, contact points between troops, TIRS, and TRPs that support any squadron engagement areas.*

Study the terrain in sector. Identify the terrain near the initial screen line from which OPs can maintain continuous long-range surveillance of enemy avenues of approach.

Determine where platoons can be positioned astride or on the enemy avenues of approach. Look for positions that provide good observation and fields of fire into the avenue of approach, and good cover and concealment for hide and defilade positions. Take a look at proposed platoon battle positions, and determine where troop fires can be massed on the avenue of approach. Use this portion of the avenue of approach to structure a troop engagement area(s).

The engagement area is where the troop will destroy an advancing enemy force. Establish a series of TRPs and use them to assign sectors of fire to each platoon. This allows the commander to control the fires of the troop and to achieve overlapping platoon fires.

With the assistance of supporting engineers, plan obstacles within the sector to support the defensive plan. Reinforce existing obstacles within engagement areas, and plan more obstacles to slow, canalize, or turn the enemy. Obstacles can buy the troop time to engage the enemy, and increase the effectiveness of indirect fires in the engagement area by compressing threat formations, slowing them down, and detaining them in the engagement area. Obstacles can give the commander time to maneuver platoons to counterattack or to move to subsequent positions. Plan obstacles in depth so the enemy gets bottled up in the engagement area and is confronted with a series of breaching operations. Make sure the troop can observe and place fires on all obstacles in the sector. Place the obstacles to achieve different effects. Obstacles emplaced on the reverse side of a hill or depression will cause the enemy to pile into them before he sees them. Strategically emplaced obstacles will cause the enemy to turn when he sees them, exposing his flanks to direct fires.

Give the FSO planning guidance so he can develop the troop indirect-fire support plan for the mission. Plan fires to support the scouts on the screen line forward of the troop. The scouts need indirect-fire support to engage enemy reconnaissance forces, to disrupt enemy lead echelon formations, and to attack follow-on forces. Plan indirect fires to engage enemy forces in the engagement area, when they are slowed by obstacles. These fires suppress, disrupt, and confuse the enemy and allow platoons to set up the direct-fire engagement. Plan indirect fires behind the engagement area to isolate the enemy. In addition, plan fires forward of the troop's positions to help disengage from the enemy in case the troop cannot stop him from initial positions. Identify rally points in sector behind the battle positions. Crews or troop elements that become separated or disorganized during battle move to these identified rally points to reassemble or reorganize.

Position platoons to maximize their weapons' effectiveness and crew/vehicle survivability based on the given terrain and the capabilities of the enemy. Platoons positioned at the base and along one or both flanks of the engagement area will force the enemy to fight in two or more directions.

Determine how to employ the scout platoons. Although the primary role of scout platoons is to conduct reconnaissance and screening in support of the troop, the troop commander may need to use their firepower to support troop defensive missions. Depending on sector width and number of avenues of approach, one scout platoon may be employed in a screen mission forward of the troop during a mission to defend in sector. It may fall back to a battle position after identifying the attacking enemy force (security drill), or it may stay forward of the troop, continuing to screen to identify follow-on forces. The other scout platoon may fight the attacking force from a battle position. The placement of the scout platoon's BP depends on the role the commander wants the platoon to play in the troop fight.

After making final adjustments to initial battle positions with the platoon leaders, plan alternate positions and subsequent positions in depth. Give platoon leaders time to reconnoiter covered and concealed withdrawal routes to their alternate subsequent positions.

Position the troop mortars where they can support the scouts on the screen line. Ensure they can engage targets from 3 to 3.5 kilometers beyond the screen line, or as far as the scouts can observe. Plan other mortar positions so that they can support the troop fight as the threat enters the engagement area. Also, plan positions through the sector.

The first sergeant positions the troop trains behind the initial troop battle positions, where they are responsive to troop needs but not vulnerable to direct fires. The XO positions the command post behind the initial platoon BPs on terrain that affords good FM radio communications with the troop elements and squadron headquarters. If possible, the XO positions the command post behind the subsequent troop positions; this reduces its vulnerability to fires, and allows it to remain stationary and maintain good FM communications while the troop displaces to other positions. The XO and first sergeant plan subsequent positions through the sector.

Position the FIST where it can maintain good digital FM communications with the supporting artillery unit. If possible, keep the FIST where it can use the laser designator to designate high-priority targets in the engagement area for Copperhead or other laser-guided munitions.

The troop commander must be in position to see the battlefield. To control the troop fires, he must understand which areas can and cannot be engaged by platoon battle positions.

The first critical task for a defend mission is to destroy or repel the threat reconnaissance. The scout platoon(s) on the screen line is responsible for identifying enemy reconnaissance forces, engaging them with indirect fire, and defeating them if possible. The other platoons may need to assist the scout platoon(s) in defeating the reconnaissance forces after the scouts identify them.

After destroying or repelling enemy reconnaissance forces, the troop is prepared to take on the lead echelon of the enemy force. Remember the scheme of maneuver. Let the enemy enter the engagement area and then mass the direct fires of the troop to strike a decisive blow. Depending on how the obstacles are set up, the troop commander may want to strike the enemy just before he reaches the obstacles; then, as the enemy deploys in reaction to troop fires, he hits the minefields and tank ditches. The commander may wish to wait until the enemy gets into the obstacles, and strike when he is confused and his formations are compressed.

Continue the fight by maneuvering platoons into alternate or supplementary positions while counterattacking by fire to complete the destruction of the lead echelon forces. If the troop is unable to defeat the enemy in the initial engagement area, it must be prepared to displace to subsequent positions. Do not allow the troop to become decisively engaged. Use the disengagement criteria from the scheme of maneuver to ensure adequate time to bound the troop back to subsequent positions by platoon(s). Keep one or two platoons in contact with the enemy, engaging him with direct and/or indirect fires. The other platoons move back in sector to subsequent positions. Once they are set, they engage the enemy with indirect fires or long-range TOW missile shots or tank main gun fire so the remainder of the troop can break contact and move to its subsequent positions.

Once the platoons are in their subsequent positions, make any needed adjustments to their positions or orientations based on what the enemy is attempting to do. Finish off the enemy from these positions, and then move forward and reoccupy initial or alternate positions, if possible.

EXAMPLE - DEFEND IN TROOP SECTOR (HEAVY TROOP)

The troop is given a mission to defend in sector and hold the threat forward of PL SABER. The troop commander completes his plan and issues the order. The 1st platoon moves forward and establishes a screen along PL SPUR. The 3d platoon prepares BP 20 at the base of the initial engagement area and occupies a hide position to the rear. BP 20 is about 2,500 meters away from the base of EA BEAR. The platoon's fires are concentrated in the direction of TRP 25. The 2d and 4th platoons occupy BP 30 and BP 40, respectively, along the eastern flank of EA BEAR. The 2d platoon orients on TRP 21, and the 4th platoon orients on TRP 27. The western flank is protected by a steep ridgeline. The troop command post is on high ground about 2 to 3 kilometers behind the initial defensive positions. The troop trains are just south of BP 30. The troop commander positions between BP 20 and BP 30, where he has a good view of all platoon positions and the engagement areas (see Figure 6-2).

A threat reconnaissance patrol of two BMPs approaches the screen line moving along the flanks of the high-speed route into the troop sector. As the patrol crosses PL SPUR, a 1st platoon scout element unmasks and quickly ambushes the patrol with cannon fire. One scout squad immediately searches the threat vehicles for intelligence information and captures two wounded prisoners. The troop commander tells the 1st platoon leader to expect a CRP within an hour. The rest of the troop stays in hide positions away from terrain most likely targeted by threat artillery units (see Figure 6-3).

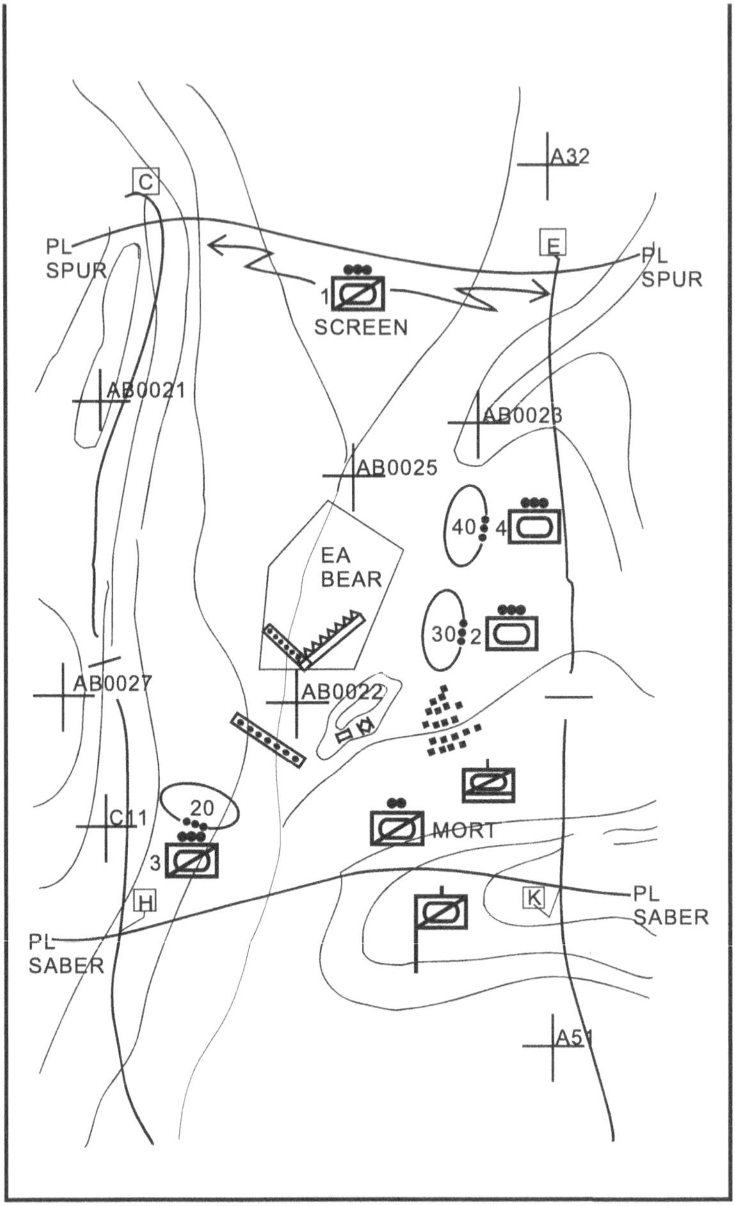

Figure 6-2. Defend in sector (part one).

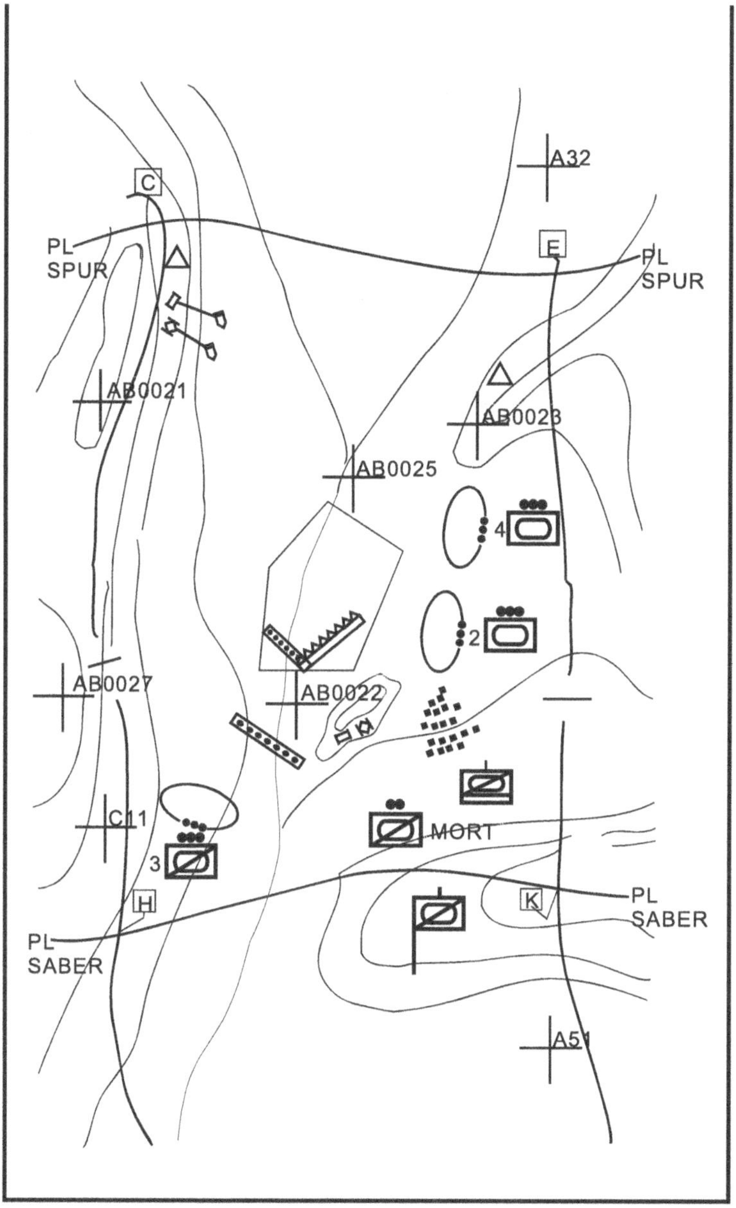

Figure 6-3. Defend in sector (part two).

About 45 minutes later, heavy concentrations of artillery and rocket fire begin falling along areas of high ground that dominate the avenue of approach near EA BEAR. Troop elements button up. When the suppression lifts, they immediately test for chemical agents. No chemicals are detected. The TCs unbutton. The 1st platoon reports two MRCs advancing abreast in prebattle formation about 2 kilometers forward of the screen line. The platoon sergeant contacts the FSO. Using the technique of fire "AT MY COMMAND," the platoon sergeant times the impact of artillery to coincide with the arrival of Threat formations at preplanned TRPs. Threat formations are disrupted, several vehicles sustain suspension damage, and the advance slows down. Threat leaders scramble to restore order, company formations are reformed, and the advance continues. The 1st platoon does not engage the advancing force, but maintains contact and reports the Threat's location and activity to the troop commander. The 1st platoon maintains its positions along PL SPUR to identify follow-on forces. It reports seeing a third MRC about 1,500 meters behind the lead companies. The troop commander now has a fairly clear picture of the Threat situation.

The platoon leaders and troop commander move into hide positions to observe the threat approach. The FSO continues to smoke and to suppress the threat lead companies with mortar fire, which keeps them buttoned up and slows their rate of advance. The troop commander orders the 3d platoon to move into firing positions and prepare to fire at his command. The 3d platoon moves into hull-down positions. The 2d and 4th platoons remain in hide positions. Their platoon leaders stay up and continue to observe. The lead threat MRCs appear about 2,800 meters away, with tanks leading platoon columns. The troop commander orders the 3d platoon to engage. The 3d platoon sends six TOW missiles down range, targeting mine-roller tanks and the threat platoon leader's BMPs (see Figure 6-4). The threat force detects the antitank guided missile (ATGM) fire and moves toward the 3d platoon, which quickly backs into defilade.

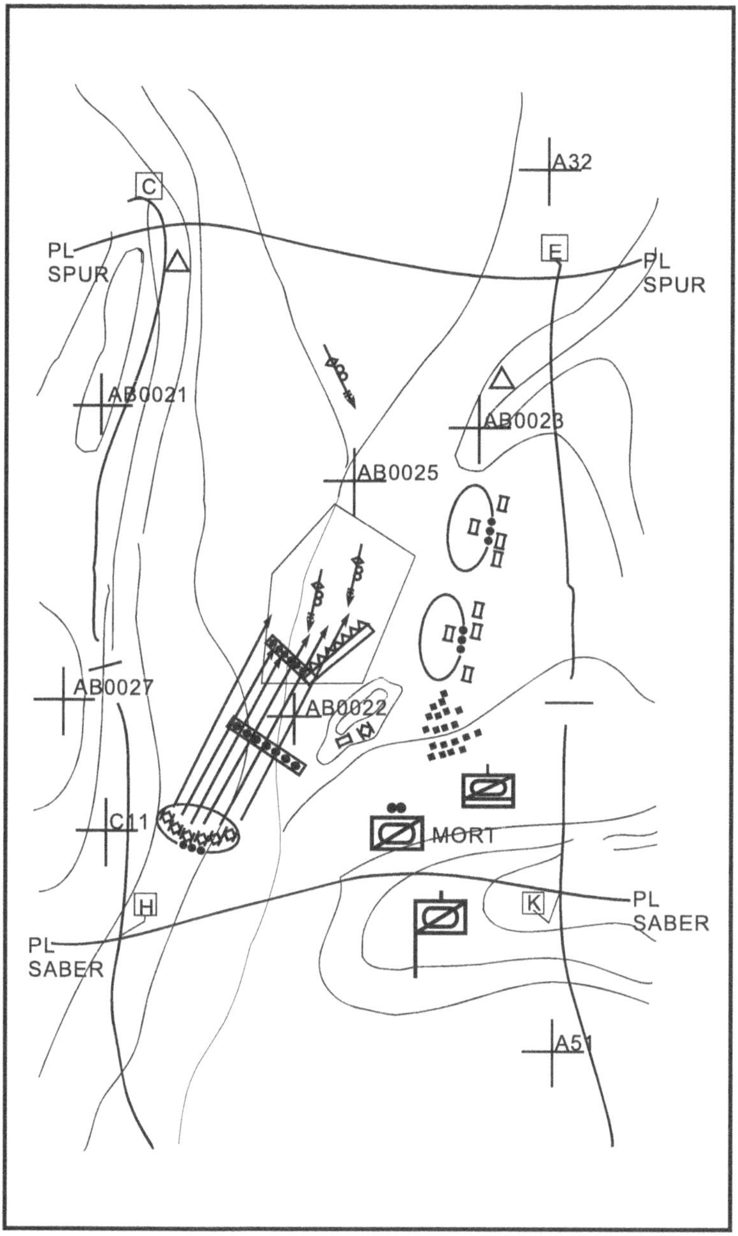

Figure 6-4. Defend in sector (part three).

The threat continues to advance through EA BEAR, piling into a minefield and tank ditch hidden on the reverse slope of a long, shallow draw. Several threat vehicles are caught in the obstacle. As the FSO observes this development, he concentrates all available fires on group targets that cover the obstacle. With the advance disrupted and stalled at the obstacle, the troop commander orders the 2d and 4th platoons to attack by fire into the flanks of the threat's lead echelon (see Figure 6-5). The concentrated cannon fire of both tank platoons quickly destroys most of the two lead companies.

The threat's third company comes into view in attack formation. The troop commander orders the 3d platoon to engage this force with long-range TOW missile fires. Survivors of this battalion begin to withdraw from the battlefield. The troop commander quickly orders the 2d and 4th platoons to counterattack and destroy the remnants of the battalion. The 3d platoon provides overwatch. Moving to alternate positions, the tank platoons counterattack by fire to finish off the remaining battalion vehicles. The troop commander then orders all platoons to reoccupy their initial hide positions and redistribute ammunition. He anticipates the second echelon will arrive within 30 minutes. The 1st platoon maintains its positions on the screen line. The first sergeant moves to each of the platoons to resupply. He collects the PWs from the 1st platoon. The XO collects all routine logistics reports and forwards them to squadron headquarters. The troop commander checks the status of leaders within the troop and designates replacements and cross-levels within the troop as necessary.

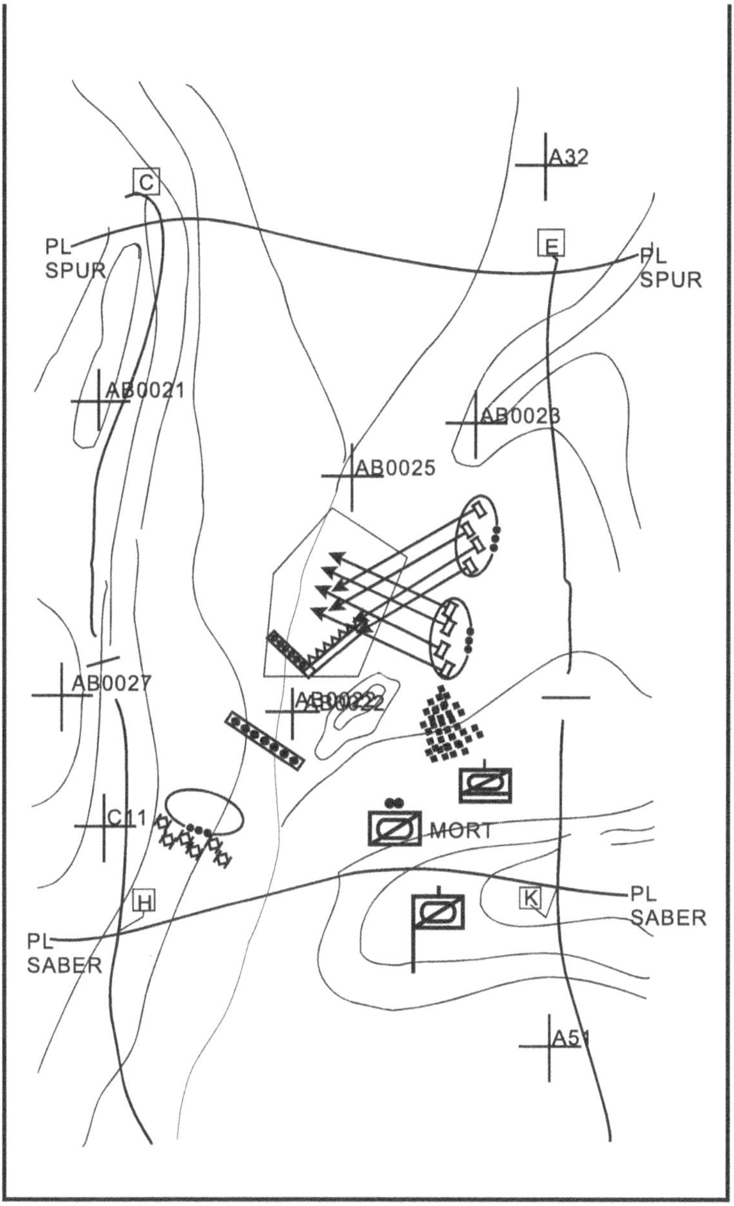

Figure 6-5. Defend in sector (part four).

Section IV. Delay in Troop Sector

Delay is a continuous series of defensive actions over successive positions in depth that trades the enemy space for time while retaining freedom of action. It is an economy-of-force operation that buys time to permit something else to happen at a more critical place on the battlefield.

CRITICAL TASKS

The critical tasks for delay include all the tasks associated with defend in sector as well as—

- Preserve freedom to maneuver.
- Cause the enemy to deploy from march or prebattle formation into attack formation as the troop moves to the rear.

TECHNIQUES

Planning and tactics for delay are identical to those for defend in sector, and vary only in their purpose. The flow of a delay resembles a "hit hard, then move" technique. The troop commander and subordinate platoon leaders must be very aware of disengagement criteria. The troop must mass the effects of fires to temporarily stop the enemy advance, then disengage and move to subsequent positions in depth. The troop cannot become decisively engaged. It must maintain a mobility advantage over the enemy. This means taking advantage of terrain, being familiar with high-speed routes of withdrawal, and rehearsing engagements and movements. The commander will probably have to use one or two platoons to assist disengagement of the others.

Chapter 7

Other Tactical Operations

Several combat operations are routinely associated with successfully accomplishing the missions described in Chapters 3 through 6. These operations require special planning and training considerations and techniques because of their complexity. At troop level, these operations are based on standing operating procedures (SOP) to ensure they can be conducted quickly and efficiently.

CONTENTS

Section I. Tactical Road Marches

Troops not in contact with the enemy often travel long distances to position themselves for future operations. A successful tactical road march depends on unit discipline and the ability to execute the plan strictly in accordance with the SOP.

CRITICAL TASKS

Accomplish the following critical tasks when planning, executing, and supervising a tactical road march:

- Establish the readiness condition (REDCON).
- Issue a warning order.
- Reconnoiter the route to the start point.
- Conduct quartering party operations.
- Issue a movement order.
- Execute a tactical road march in accordance with movement order and SOP.
  - Cross and report the start point, checkpoints, and release point on time.
  - Conduct actions at halts.
  - Conduct actions on contact.
- Maintain security throughout movement and during halts.
- Operate a trail party.
- Clear the release point.

TECHNIQUES

The march discipline necessary to execute a road march with routine precision is attained only by strict adherence to SOP. Ensure the tactical road march portion of the troop SOP addresses, as a minimum, the following:

- Order of march.
- March speed.

- Distance between vehicles.
- Actions on contact (air, ground, and indirect fire).
- Actions at halts.
- Security.
- Contingency plans for vehicle breakdowns, breaks in column, and lost vehicles.
- Quartering party.
- Trail party.

The basic considerations in planning any road march are listed below.

- Time available.
- Distance of the move.
- Current situation.
- Availability and condition of routes.
- Size of the unit.
- Types, numbers, and characteristics of vehicles that must move.

The troop will most often move as part of a squadron operation, and the move will be based on the squadron order. The troop commander must, however, be familiar with planning considerations so he can plan and execute an independent troop move, if required.

March Planning Sequence. If time permits, follow this sequence when preparing for a road march:

- Prepare and issue a warning order, giving the troop's movement and any follow-on missions. Give them time to prepare for the operation.
- Analyze the situation to determine if any of the movement factors (order of march, rate of march, or interval) specified in the troop SOP must be altered to meet mission requirements. If the troop must reconnoiter two assembly areas, the two scout platoons might have to be first in the order of march.
- Conduct a map reconnaissance of the route (if assigned) or determine the best available route.
- Organize and dispatch reconnaissance and quartering parties (if occupying an assembly area).
- Prepare detailed movement plans based on mission requirements and reconnaissance information.
- Prepare and issue the march order to the orders group.

Route Reconnaissance. Reconnaissance is essential to movement planning. It reveals accurate, up-to-date information about the route. A route reconnaissance determines travel times, and identifies capacities of underpasses and bridges, locations of fords and bypasses, and critical points and obstacles. This lets the commander plan his movement and avoid areas or situations that could disrupt it.

Conduct the route reconnaissance with one of the scout platoons. Before the reconnaissance, let members know what information is needed and when and where they should submit it. Get them out as early as possible; troop plans are based on what they see and report.

If no other traffic control assets are available or if the troop is moving independent of the squadron, use the scouts from the platoon conducting the reconnaissance as traffic control teams at critical points along the route. Determine through map reconnaissance any built-up areas or key intersections that could disrupt the troop's movement, and position traffic control teams there. Based on their reconnaissance, the scouts may need to refine these positions.

If the movement is being conducted as part of the squadron, one troop will usually reconnoiter the route. The troop must, however, reconnoiter the route from its present location to the start point to determine if the route is suitable and how long it will take to reach the start point.

Quartering Party. A quartering party is used to reconnoiter and prepare a position before the troop main body arrives. The troop quartering party will often move as part of the squadron quartering party. Organize it with a scout platoon leader or the first sergeant as the party leader, with guides from each platoon, and with any additional personnel needed to clear the area.

Before the quartering party leaves, it must know the troop's route, order of march, estimated time of arrival, and any specifics on establishing the assembly area. Before the main body arrives, the quartering party reconnoiters the area, marks routes, and prepares to guide the main body as it enters the new area.

Trail Party. The troop trail party is made up of personnel and equipment (normally the troop trains) to handle emergency vehicle repair and recovery, medical aid and evacuation, and emergency refueling. The trail party moves just forward of the last maneuver platoon in the main body. The troop motor sergeant or first sergeant is in charge.

Main Body. The troop normally moves as a single march unit in column formation when conducting a tactical road march. The organization of the troop and any attached elements should be standardized during movement. Alter this organization to meet specific mission requirements as needed.

The troop commander has no prescribed place in the column. He positions himself where he can best control the movement of the troop. He will usually be well forward in the column, behind the lead platoon, to respond to contingencies while on the move. As a security measure, the troop command post should be positioned farther back in the column to disperse troop command and control.

The troop's column organization must provide adequate security against air and ground threats, while on the move and during halts. See Figure 7-1 for one way to organize a troop march column.

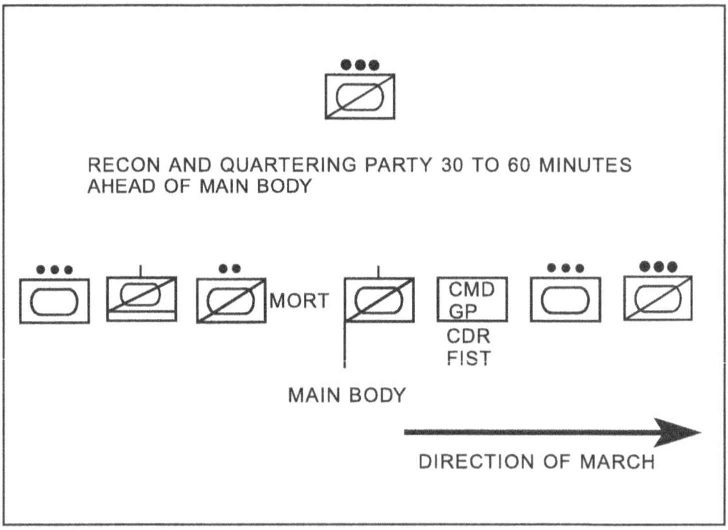

Figure 7-1. Troop column organization.

The 3d platoon, the first sergeant, and the quartering party move ahead of the troop. The troop order of march is 1st platoon, command group, 2d platoon, command post, mortars, trains, and 4th platoon. This order of march provides 360-degree security, disperses the command and control assets of the troop, and provides reconnaissance forward of the main body. Also, vehicle commanders assign sectors of observation to their crews, who search for air and ground threats (see Figure 7-2).

The troop performs the march in open or close column, depending on the situation. Use close column during limited visibility conditions. Vehicles are spaced 25 to 50 meters apart. This method takes advantage of the traffic capacity of the route, but provides little dispersion. Vehicle density is 15 to 30 vehicles per kilometer along the route of march. Use open column to provide greater dispersion and, thus, greater security. The distance between vehicles varies from 50 to 100 meters. Open column is normally used in daylight conditions. Base the troop's march speed on the slowest vehicle in the column. The troop trains will usually limit the rate of march.

Halts. Halts are used to rest personnel, provide personal comfort and relief, facilitate mess operations, refuel vehicles, maintain and inspect equipment, adjust the schedule, and allow other traffic to pass. The squadron march plan or troop SOP will specify the frequency and duration of halts, and will prioritize work to be done. For long movements, plan halts into the troop march table and ensure subordinate platoon leaders understand what actions must occur at the halt. When unscheduled halts occur, find the reason for the halt and let subordinates know how long it will last. Provide for security during halts, and establish OPs to provide early warning of enemy forces during long halts.

Vehicle commanders assign sectors of observation to their personnel so there is 360-degree observation. Each vehicle commander designates an air guard to provide air security, or specific vehicles may be designated air guard vehicles. In that case, the crews concentrate on only vehicle air observation rather than air and ground observation.

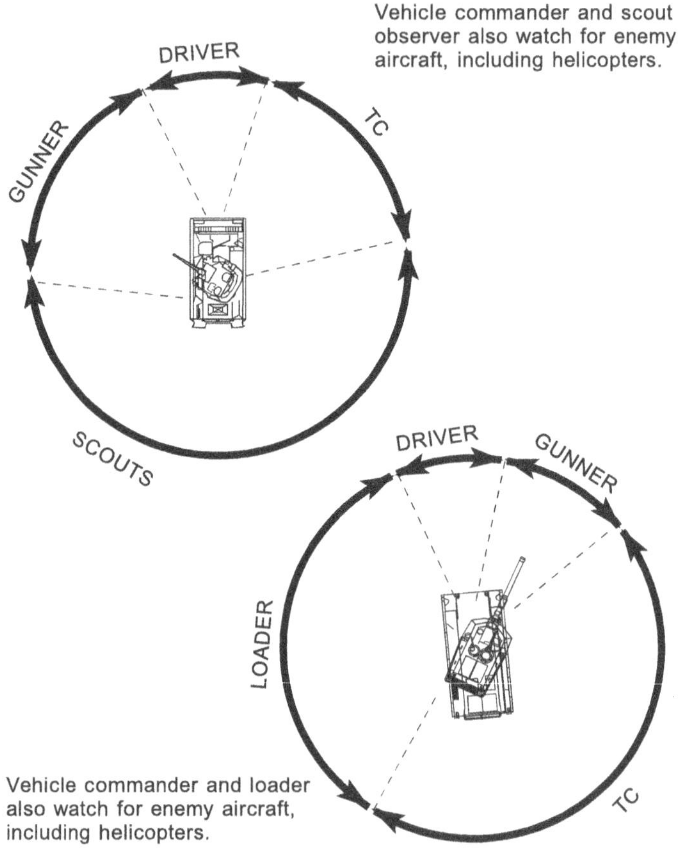

Figure 7-2. Maintain 360-degree security.

Vehicles that become disabled during movement must not obstruct traffic. The crew of the disabled vehicle must move the vehicle off the route, post guides to direct traffic, and find the problem. If the vehicle can be fixed, it rejoins the rear of the column. It does not return to its original position until the column has halted. If the vehicle cannot be readily repaired, the trail party recovers it.

Section II. Assembly Areas

An assembly area serves as a place where the troop gathers to prepare for future operations. In the assembly area, the troop prepares and issues orders, repairs and maintains vehicles, conducts resupply operations, and rests. As a minimum, assembly areas are positioned out of range of enemy light artillery.

TASKS

The following tasks are associated with assembly area operations:

- Plan the occupation of an assembly area.
- Conduct quartering party operations.
- Occupy an assembly area.
- Maintain security.
- Establish communications.
- Prepare for future combat operations.
- Depart an assembly area.

TECHNIQUES

The troop will normally be assigned a specific assembly area location. Within the area available to the troop, conduct a map reconnaissance, and if time is available, a ground reconnaissance. Look for an area that provides overhead concealment. This is extremely important if the troop will remain in the area for any length of time. Select an area that also provides—

- Cover from direct fire.
- Good drainage and a surface that will support troop vehicles.
- Good entrances and exits and an adequate internal road or trail network.
- Space for dispersion of vehicles, personnel, and equipment.

Quartering Party. The troop will often occupy an assembly area at the end of a road march. Before the quartering party leaves the troop's present location, tell them how to organize the assembly area. They must know any special requirements, such as a site for a logistics package (LOGPAC), so they can prepare the position. The quartering party must be organized to provide their own security during this operation. When the quartering party arrives at the forward assembly area, they must—

- Reconnoiter the area. If the area is not suitable, the leader of the quartering party must report immediately and provide a recommendation for another area.
- Organize the area. The leader of the quartering party selects locations for the platoons, mortars, command post, trains, and any attached elements, based on the commander's instructions. He may need to deviate from the commander's guidance to position an element in a suitable location.
- Improve and mark entrances, exits, and internal routes.

- Mark vehicle locations. Platoon representatives in the quartering party select general locations for vehicles in their platoon. Vehicle commanders and the chain of command refine these positions when they arrive.
- Perform guide duties. A platoon representative guides his platoon into position after the platoon clears the release point.

Occupation. Each platoon is guided from the release point into the assembly area by its quartering party member. Color-coded lights can be used to link up guides and lead vehicles (see FKSM 17-97-3). When the troop arrives at the assembly area, all elements move off the route without halting or slowing to keep the route of march clear. Keep this in mind when selecting routes, organizing the order of march for the road march, and allocating space in the assembly area. Once platoons have cleared the route and moved into their areas, they can adjust their positions without slowing the remainder of the troop.

Positioning troop elements in the assembly area is based on the size of the area, the terrain and avenues of approach, the length of time the troop will occupy the area, and any special requirements, such as resupply. Some considerations in allocating space for the troop are shown in Figure 7-3 and in the following list:

- If possible, space vehicles at least 100 meters apart to decrease exposure to enemy observation and fire.
- Keep vehicles in hide positions and establish OPs on terrain that provides good observation of approaches into the assembly areas for early warning of enemy movement. The vehicles can be moved up to a fighting position if necessary.
- Position platoons based on available terrain in the assembly area. Tank platoons may need more space to move vehicles in and out of position; scout platoons may occupy terrain that is more restrictive.

- Position the troop command post in the center of the assembly area. This makes establishing wire communications and issuing orders easier.
- Position troop trains where they have a good road network and space to conduct their maintenance and resupply operations. They may be used to secure the perimeter of the assembly area, but this will reduce their support ability.
- Emplace the mortars in a position from which they can support the troop with indirect fires. Overhead cover will reduce or eliminate their ability to fire.

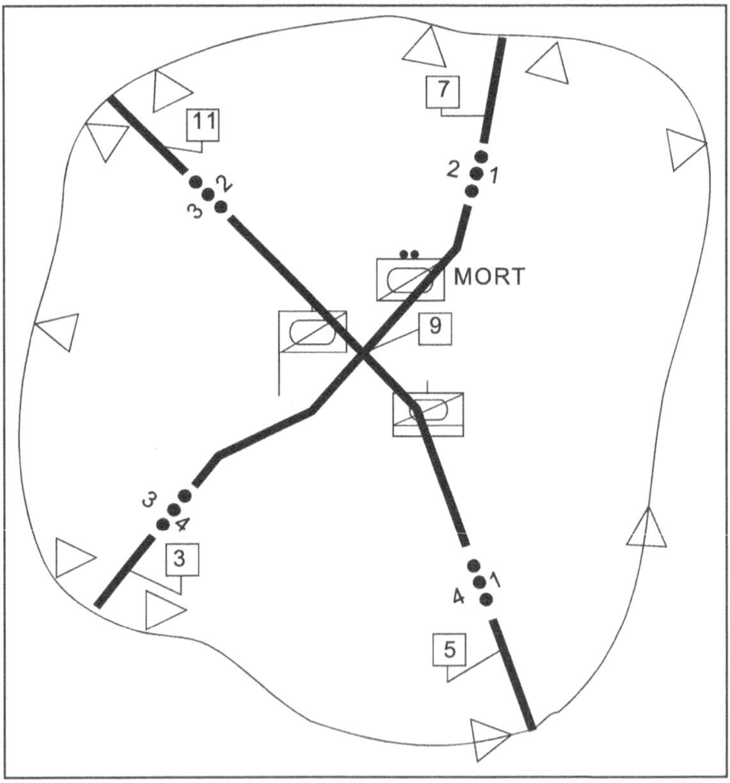

Figure 7-3. Troop assembly area.

Security. Although the assembly area is not a defensive position, the troop must be able to see and defeat enemy ground attacks. The best defense against air attacks is to remain hidden. Post guards at all entrances and exits to stop traffic that tries to enter the area. Establish OPs to observe key terrain features and likely avenues of approach for early warning of enemy approach. Each platoon must provide overlapping observation and fires within its platoon and with the platoons on its flanks. Establish a dismounted patrol plan so platoons make physical contact with their adjacent platoons. Ensure the platoons provide 360-degree coverage of the assembly area. Camouflage vehicles and equipment to prevent enemy detection from the ground and air. Place PEWS (platoon early warning system) in heavily vegetated areas or dead space to provide early warning of enemy movement in and around the position. Emplace NBC alarms upwind and about 150 meters from the troop's positions to provide early warning of an NBC attack. Assign the mortars an azimuth of fire on the most dangerous approach into the assembly area. Give the fire support officer guidance in preparing an indirect-fire plan.

Communications. Messenger and wire are the primary means of communications from the command post to each platoon and to each OP. If practicable, lay wire from the squadron to the troop. Plan on providing a messenger to the squadron command post. Use radio when no other means of communication can be used.

Preparation for Future Operations. Several tasks are routinely accomplished in an assembly area. These tasks are listed in the troop SOP (FKSM 17-97-3) under priority of tasks upon arrival in an assembly area, and include—

- Position vehicles.
- Establish local security.
- Establish OPs.
- Prepare fire plan.

- Establish wire communications.
- Maintain radio watch and man turret weapons.
- Camouflage positions.
- Prepare obstacles/mine plan.
- Select alternate and supplementary positions.
- Reconnoiter routes of withdrawal.
- Perform PMCS.
- Emplace NBC alarms and PEWS.
- Continue to improve positions.
- Conduct logistics resupply (Classes I, III, and V).
- Rest in accordance with REDCON status.

Modify this task listing to accomplish specific tasks (such as conduct rehearsals, test-fire weapons, and conduct inspections) in preparation for future operations. Ensure subordinates know how long the troop will remain in the assembly area and are told of any special requirements. Occupation of the assembly area will often be conducted and supervised by troop NCOs, while the commander and platoon leaders plan for upcoming operations.

Departure. Maintain the appropriate REDCON. Each REDCON level indicates critical tasks and time available to prepare for future operations.

- REDCON 1—be prepared to move immediately.
  - All personnel alert and ready for action.
  - Vehicles loaded and secured, and weapons manned.
  - Vehicle engines running and OPs not manned.

- REDCON 2—be prepared to move in 15 minutes.
  - All personnel alert.
  - OPs and wire pulled in.
- REDCON 3—be prepared to move in half an hour.
  - Fifty percent of each crew/squad stand down for rest, feeding, and maintenance.
  - Remaining 50 percent man vehicles, OPs, weapons, and monitor radios/phones.
- REDCON 4—be prepared to move in one hour.
  - Two men per platoon make dismounted checks of platoon area.
  - One man per vehicle monitors radios/phones and mans turret weapon.

All personnel remain at 100-percent alert until the prioritized work is complete after entering the assembly area. Initiate the appropriate REDCON when the work is finished. As the time for execution of a mission nears, increase the REDCON in accordance with guidance from squadron, assigning REDCON 1 just before the troop must move.

Section III. Relief in Place

A relief in place is an operation in which a unit in combat is replaced by another unit. Responsibilities for the combat mission and the assigned sector, battle position, or zone of the relieved unit are assumed by the relieving unit. A relief in place may be accomplished during offensive and defensive operations and may be conducted during any weather and light conditions.

The primary purpose of a relief in place is to maintain the combat effectiveness of committed units. A relief in place is conducted to replace a committed unit to give it the opportunity to reconstitute, rest, decontaminate, or perform a change in mission.

TASKS

The following tasks are associated with the relief in place:

- Plan and coordinate a relief in place.
- Establish communications.
- Establish liaison.
- Conduct reconnaissance.
- Initiate movement.
- Occupy positions.
- Maintain operations security.

TECHNIQUES

A troop will often perform a relief in place mission as part of a squadron operation to relieve or be relieved by a brigade. The troop will usually conduct a relief in place in conjunction with a battalion. The troop may relieve a battalion and operate in an economy-of-force role or be relieved by a battalion to conduct a change in mission.

To reduce confusion and maintain security, the following factors must be considered when planning a relief in place:

- The time that responsibility for the sector, battle position, or zone is to pass.
- Operations security.
- Deception plans.

- Time, method, and sequence of relief.
- Routes and critical control measures.
- Concept of subsequent missions.
- Plans for additional positions.
- Contingency plans.
- Location of obstacles and transfer of responsibility.
- Transfer of ammunition; wire lines; petroleum, oils, and lubricants (POL); and materiel to the relieving unit.

Communications. At a specified time or upon receipt of an order, the relieving unit minimizes radio traffic or begins to operate at radio-listening silence. The relieving unit command post monitors the command net of the unit to be relieved while continuing to operate on its internal command net.

Reconnaissance and Liaison. The orders group of the relieving unit moves to the command post of the unit being relieved for face-to-face coordination.

In preparation for the relief in place, the relieving unit moves to a forward assembly area under the command of the XO, if necessary.

If the situation permits, the relieving unit conducts a reconnaissance of the area when it completes coordination. The units establish a contact point behind the positions into which they will move; a guide from the relieved unit will meet them there to make any final coordination and guide them into position.

The commanders complete their plans and issue orders to their respective units after the reconnaissance and liaison are complete.

Methods. The relief in place can be conducted in several ways. Use the method appropriate to the troop's situation.

- One unit at a time. This is the slowest method, but the most secure.

- All units simultaneously. This is the quickest method, but the least secure.

- Center units first, followed by flank units.

- Flank units first, followed by relief of the center unit.

- Occupy adjacent or in-depth positions that cover the area of responsibility.

Passage of Command. The troop command post and battalion TOC collocate to facilitate command and control of the operation. The relieving unit departs its location in the prescribed sequence and lines up at the contact points with the guides from the relieved unit. The relieving unit moves into hide positions behind the relieved unit. The relieving unit moves into position after all pertinent information has been exchanged and the relieved unit moves out of position. The relieved unit moves to a rally point or assembly area to link up with the rest of the unit. When each element has completed the relief, the relieving unit assumes responsibility for the area and reports by messenger, wire, or radio to the command post. The relieved unit moves along its designated route to the new location. See Figures 7-4 and 7-5 for illustration of the relief in place.

The higher headquarters will often specify a time the relieving unit assumes responsibility for the sector, battle position, or zone. The commanders conducting the relief coordinate to ensure the relief is conducted on time. The relieving unit normally assumes responsibility when the relieved unit departs its positions. The commanders of the two units are usually collocated and responsibility passed face to face.

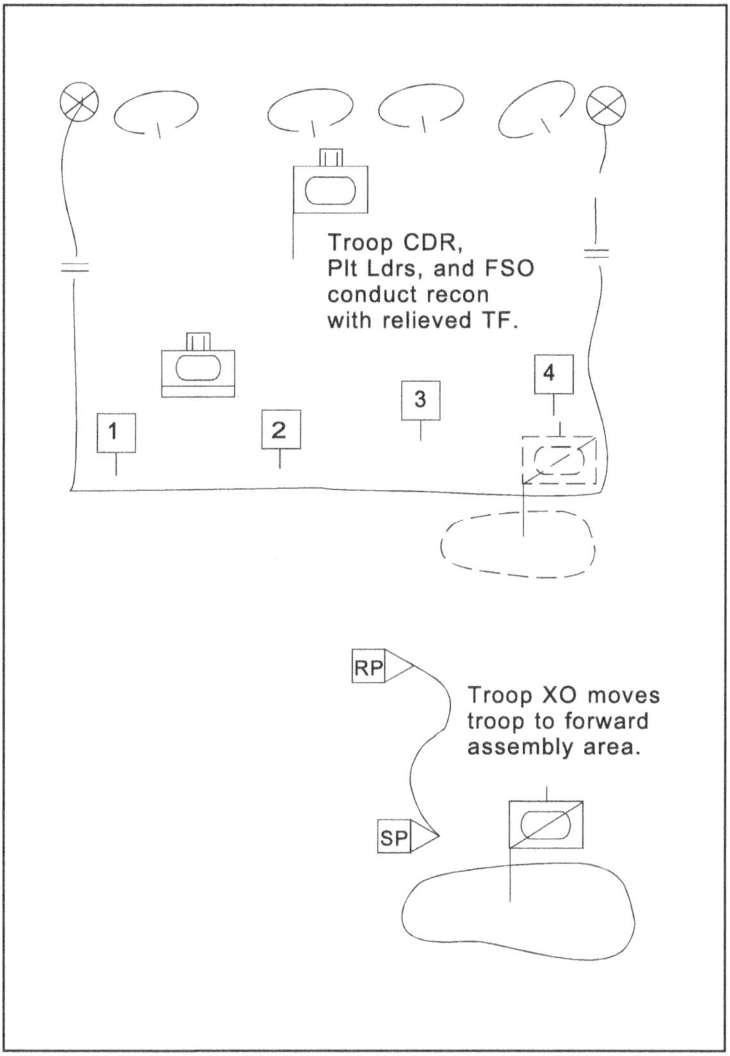

Figure 7-4. Troop prepares to conduct relief in place with task force.

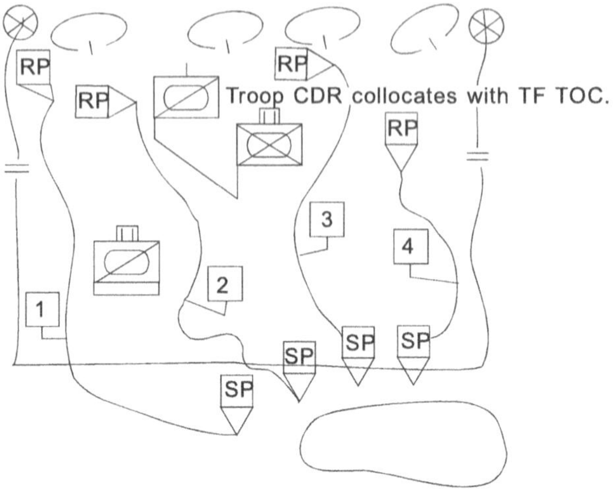

Platoons meet guides at contact points and move into position.

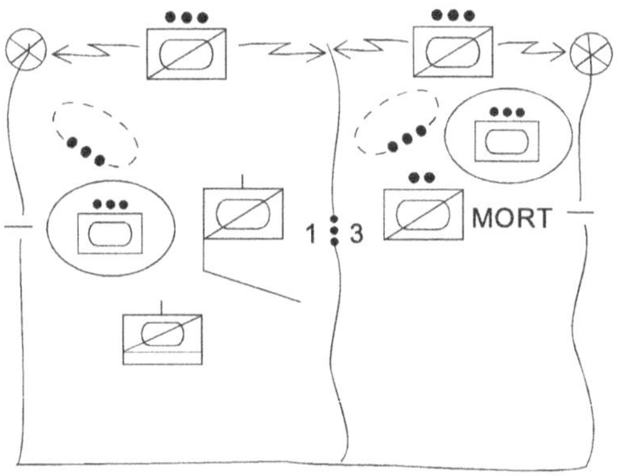

After the relieved unit departs, the troop CDR adjusts his PLT positions if necessary.

Figure 7-5. Troop completes relief in place.

Fire Support Assets. The troop fire support officer coordinates with the squadron fire support officer to ensure there is fire support throughout the operation. The fire support assets of the relieved unit remain in position throughout the relief of the maneuver forces, and are prepared to support both units. Fire support assets of the relieving unit position themselves as quickly as possible to provide additional support.

Enemy Contact. If either unit gains direct-fire contact with an enemy force, it immediately notifies the other unit and the higher headquarters directing the relief. If responsibility has not passed, the relieving unit becomes OPCON to the relieved unit. The relieving unit's mortars fire missions as directed by the relieved commander through his fire support officer. If responsibility has passed, the relieved unit becomes OPCON to the relieving unit. The collocation of the commanders and command posts facilitates rapid coordination and action in this situation.

Section IV. Battle Handover and Passage of Lines

Battle handover is an operation that transfers responsibility for fighting an enemy force from one unit to another in the close-in battle. Conducted by stationary and passing units, battle handover is designed to sustain continuity of the combined arms fight, and to prevent the enemy from getting a "free ride" anywhere on the battlefield as one force picks up the fight from another. It is also designed to preserve the fighting capabilities of both forces as they execute the operation. Battle handover is associated with almost all Army combat operations.

Passage of lines is a tactical event associated with battle handover. It is the controlled movement of one unit through the positions of another stationary unit that does not interfere with either unit's scheme of maneuver. A passage of lines is often used because the combat situation does not permit one unit to bypass another unit's positions. The passing unit must move through the positions of the stationary unit.

Passage of lines is often an integral part of the following operations:

- Deliberate attacks or counterattacks across the FLOT.
- Exploitation and pursuit.
- Route, zone, or area reconnaissance.
- Raids.
- Movement to contact.
- Defend, delay, or withdrawal.
- Defensive or offensive cover.
- Screen or guard.

Although a cavalry troop does not normally perform all of these operations, it may conduct them as part of a larger force.

CRITICAL TASKS

The three players in battle handover and passage of lines are: the *stationary unit*, the *passing unit*, and the *common commander* of both units. Each has critical tasks to perform to achieve smooth and efficient execution of the operation.

Critical Tasks of the Common Commander. The commander exercising command authority over both the stationary unit and the passing unit must accomplish three critical tasks.

- Establish where battle handover will occur by designating a phase line forward of the FEBA as the battle handover line (BHL). The line should be where combat maneuver forces of the stationary unit along the FEBA can effectively overwatch and protect the passing unit as it withdraws behind the FEBA or advances forward of the FEBA or FLOT. The distance forward of the FEBA or FLOT is limited to available fields of fire and the effective range of weapons of the stationary unit.

- Designate contact points just forward of the battle handover line at which stationary and passing units are required to conduct physical coordination (only in defensive operations).
- Ensure the passing unit is provided indirect-fire support while its artillery is displacing during battle handover and passage of lines.

The stationary unit normally recommends battle handover line and contact point locations to the common commander. It remains the responsibility of the common commander to establish the line and contact points. These control measures must be reflected graphically on an overlay and identified in the appropriate operation plan (OPLAN), OPORD, or FRAGO issued to subordinate units. The battle handover line, in effect, establishes a boundary between the ground owned and controlled by the stationary unit commander and the ground controlled by the passing unit commander. The stationary unit commander controls the ground forward of the FEBA up to the battle handover line. He can place security forces, obstacles, and direct and indirect fires into this area to support his scheme of maneuver.

While the battle handover line defines the point at which the battle handover should ideally occur, events normally preclude this from happening. The moving and stationary force commanders and their common commander should understand that the actual handover may occur in a "zone" centered on the battle handover line. Both the moving and stationary force must remain active and responsive enough to perform the battle handover anywhere in this zone, and should not dogmatically "stick to the plan."

Battle handover begins on the order of the common commander. To sustain unity of command in the passing operation, the passing unit is usually placed OPCON to the stationary unit to execute battle handover and passage of lines. OPCON by the stationary unit is limited to those actions necessary to get the passing unit through the stationary unit's area as quickly as possible. In defensive operations, battle handover from the passing unit to the stationary unit is not complete until the combat maneuver forces of the stationary unit along the FEBA have visual contact with the enemy, and the stationary unit commander indicates readiness to assume responsibility for the battle. In offensive operations, battle handover from the stationary unit to the passing unit is not complete until the combat maneuver elements of the passing unit have crossed the FEBA or FLOT, deployed for combat, and maneuvered across the battle handover line.

Passing Unit Critical Tasks. The passing unit must accomplish the following critical tasks to perform battle handover and passage of lines:

- Immediately establish communications with the stationary unit. Enter the command, operations and intelligence (OI), and fire support nets.
- Collocate a command post with the TAC CP or TOC of the stationary unit as soon as possible to enhance communications and unity of effort.
- Continuously report to the stationary unit the location, size, and composition of all enemy forces. Report the enemy's current activity. If the enemy is attacking, report his direction of movement, movement formation, and estimated rate of advance. If he is defending, report his locations, orientation, composition, fire sacks, reserves (if known), obstacle system, or flanks.
- Continuously report to the stationary unit the location, size, and activity of all subordinate elements to include combat support, combat service support, and command and control facilities.
- Given the current disposition of subordinate units, coordinate with the stationary unit to determine

contact points at which each subordinate company-size unit will physically coordinate handover and passage of lines with representatives of the stationary unit. Once contact points are determined, send a FRAGO to each subordinate unit specifying where they will physically coordinate passage with the stationary unit. Confirm recognition signals that must be displayed during passage (defensive operations).

- Once each subordinate unit acknowledges where it must physically coordinate passage, each unit will dispatch representatives to assigned contact points and coordinate passage for its unit. At the contact point, confirm recognition signals and exchange required information (defensive operations).
- Maintain visual contact with all enemy units and delay back to the battle handover line, avoiding decisive engagement (defensive operations).
- Display correct recognition signals and use correct challenge and password as specified in the CEOI during passage.
- Maintain proper weapons orientation.

Stationary Unit Critical Tasks. The stationary unit must accomplish the following critical tasks when ordered to conduct battle handover and passage of lines:

- Establish communications with the passing unit. Coordinate and direct the passing unit to contact points based on current dispositions of the subordinate units (defensive operations).
- Ensure contact points are manned and subordinate commanders have personal communications with their representatives (defensive operations).
- Ensure representatives at the contact points assign each passing unit a passage point into the area of operations and a route that extends from the passage points to the rear boundary or to an assembly area (defensive operations).

- Ensure representatives at the contact points exchange required information with the passing unit as outlined in FKSM 17-97-3.
- If security forces are employed, position them along the battle handover line to observe enemy avenues of approach. Adjust as needed for low visibility conditions (defensive operations).
- If obstacles are emplaced between the FEBA and the battle handover line, ensure routes through the obstacle system are clearly marked and physically controlled by guides, or provide an escort to the passing unit. Ensure that reserve targets on obligated routes are manned by soldiers in direct communication with their commanders.
- Ensure that all routes of withdrawal obligated to the passing unit are unobstructed and facilitate rapid movement to the release point (defensive operations).
- Ensure obligated routes of advance, attack positions, and routes to the battle handover line are unobstructed and permit rapid movement (offensive operations).

GRAPHIC CONTROL MEASURES

Figure 7-6 shows the graphic control measures that support battle handover and rearward passage of lines.

- Battle handover line. This line is established by the common commander in consultation with both commanders. The stationary commander has the major determination in locating the BHL, as his force must be able to overwatch the BHL with direct fires.
- Contact points. These are established on identifiable terrain and normally in the vicinity of the passage lanes. For rearward passage of lines, the contact points are established forward at the BHL. For forward passage of lines, the contact points are established in the stationary unit's rear area rearward of the passage lanes.

- Passage points. The passage point is that point on the passage lane where the moving unit moves through, and responsibility for the battle is passed to the stationary unit. It is usually placed where the passage lane begins.
- Passage lanes. The stationary unit establishes passage lanes to move the passing unit quickly through defending unit positions. This could include passing through gaps in friendly obstacles and moving near or through friendly engagement areas and battle positions. Lanes are restrictive; however, they should ideally be wide enough to allow the passing unit to move in a tactical formation. The passage lane begins at the passage point and ends at the rear of the stationary unit BPs. The passage is considered complete when the moving unit exits the lane.
- Routes. Routes are used to move the passing unit through the stationary unit rear area. The number of routes designated will vary based on METT-T, but as a general rule, multiple lanes/routes should be planned to facilitate rapid passage of moving units and to avoid unnecessary massing of units. The stationary unit may escort or guide the passing unit along the lane/route.
- Assembly area. An assembly area in the rear area of the stationary unit allows the passing unit to conduct hasty reorganization and emergency CSS actions. This assembly area is temporary in nature.
- Infiltration points. Units should plan infiltration points and lanes for personnel unable to complete the passage with the unit. Passing unit liaison officers may remain located with stationary unit CPs to serve as a point of contact for infiltrating personnel/equipment. Personnel who infiltrate must have some way of contacting the stationary unit before crossing into friendly territory.

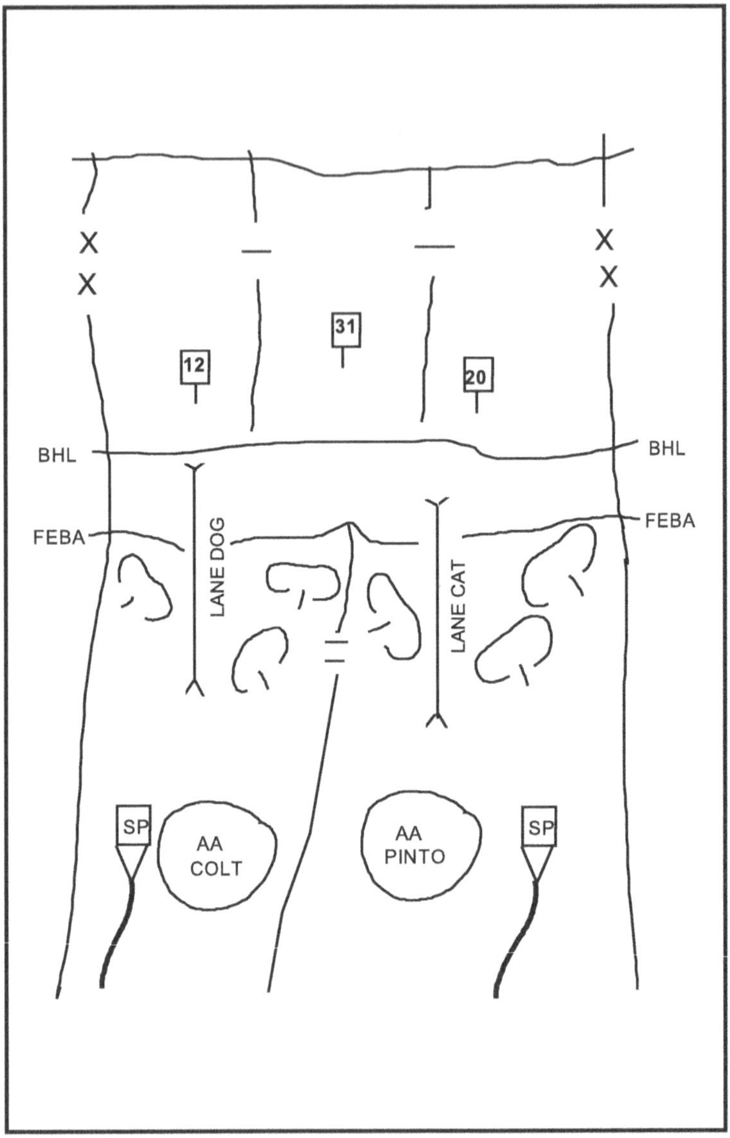

Figure 7-6. Rearward passage of lines.

TECHNIQUES

At troop level, the passage of lines will usually be performed as part of a squadron operation. The passage may be forward, such as to pass through a defending unit to conduct a counterattack, or rearward, such as when a covering force unit withdraws through units in the main battle area.

The troop is particularly vulnerable during a passage of lines. The unit may be concentrated and the fires of the stationary unit may be temporarily masked. Thorough reconnaissance and detailed coordination are critical to ensuring the operation is successful.

A troop will often perform a battle handover and passage of lines through a single battalion while the squadron conducts the operation through a brigade. Ideally, the boundaries will correspond to the battalion boundaries so that coordination for the battle handover is conducted through a single headquarters.

The troop commander has numerous considerations to ensure a successful passage of lines. During reconnaissance, he must confirm—

- The disposition of the stationary force through which the troop will pass.
- The location of contact points where both units are required to make physical contact at a predetermined time.
- The location of passage points on the battle handover line through which friendly forces will pass.
- The location of passage lanes that provide a clear route through a friendly position, and also facilitate a smooth and continuous passage. Areas selected for passage should be unoccupied or on the flanks of units in position. If possible, use multiple routes to reduce vulnerability during the operation.

- The location of an attack position (for forward passage) or assembly area. This position should provide cover and concealment and be located where the passing unit will not interfere with the stationary unit.
- The initial location for combat support and combat service support elements.

Based on the reconnaissance, the troop commander coordinates and plans for—

- Supporting fires. The stationary force supports the passing unit with direct and indirect fires up to the battle handover line. In forward passage, the stationary force supports the passing unit's move through the passage and until it crosses the battle handover line. In a rearward passage, the stationary unit supports the passing unit's move back across the battle handover line and through the passage of lines.

- Time of transfer of responsibility for control of the sector and handover of the enemy.

- Troop density. The passing troop commander should plan for multiple routes of passage to ensure rapid movement and to avoid congestion.

- Traffic control. Guides from the stationary unit pick up passing elements at each contact point and guide them through the position. The passing unit commander tells the stationary unit the type, number, and order of vehicles passing through each contact point.

- Communications. The leaders exchange CEOI information and mutually agreed upon recognition signals.

- CSS. The troop commander must coordinate the evacuation of casualties, PWs, vehicles, and resupply of fuel and ammunition. The stationary unit usually provides emergency service only. The passing unit supports itself.

- Liaison officers. The troop commander should designate a representative to perform the critical duties of a liaison officer. Commanders normally coordinate a forward passage of lines and the XO coordinates a rearward passage. Liaisons are normally located at critical points during the passage. If the commander or XO is not available, a scout platoon leader should perform liaison duties. Ensure he is thoroughly briefed on the situation and follows the checklist in the troop SOP.

Section V. Hasty Water Crossing

A hasty water crossing is the movement across an inland waterway using a crossing means at hand or readily available without significant delay once the waterway is reached. It is preplanned and conducted as a continuation of the operation underway. Although the crossing is termed hasty, detailed planning assures that fire support and crossing means are available on arrival at the water obstacle.

CRITICAL TASKS

The following critical tasks are associated with the successful conduct of a hasty water crossing:

- Plan the water crossing.
- Conduct reconnaissance.
- Establish security.
- Cross the obstacle.
- Provide continuous fire support.
- Continue the mission.

TECHNIQUES

Plan. A troop may perform a hasty water crossing independently or as part of a squadron. A hasty water crossing is performed as an extension of the ongoing operation. It gives the troop the ability to sustain the momentum of an operation, and is often associated with a movement to contact and a zone reconnaissance. The hasty water crossing is characterized by speed, surprise, minimum loss of momentum, and minimum concentration of forces.

The crossing should be preplanned to ensure fire support and crossing means are at the crossing site. The need for a water crossing should be determined in the mission analysis and added to the commander's concept of the operation. This ensures the troop's critical assets are positioned to support the water crossing. Seizing bridges intact before the enemy can destroy them is the quickest and most economical means of crossing, and is used when possible.

Reconnaissance. Scout platoons will be the first to encounter a water obstacle. When this occurs, they report it and reconnoiter it to obtain the following information, which is forwarded to the troop command post:

- Width and depth of the waterway.
- Water velocity.
- Possible entry and exit points and their conditions.
- Enemy situation on the far bank.

Scouts reconnoiter multiple crossing sites to prevent the enemy from discovering the true site and to reduce the concentration of vehicles at the site. If follow-on forces will cross at the same sites, scouts mark the entrances and exits for their use.

The troop commander, with the fire support officer, positions himself forward during the reconnaissance of the water obstacle so he can personally supervise the operation. The fire support officer will support the scouts with indirect and direct fires, if needed.

Security. A crossing can be performed with or without opposition. To avoid enemy fire while crossing, prevent him from knowing where the crossing will be.

Before the main body crosses, the scouts must secure the far bank by fording or swimming across the water obstacle and establishing positions. They should cross on both sides of the site and secure positions on the dominant terrain of the far side to provide early warning of threat activity. After securing the far side of the water obstacle, the scouts establish positions to the flanks of the crossing site to provide local security (see Figure 7-7).

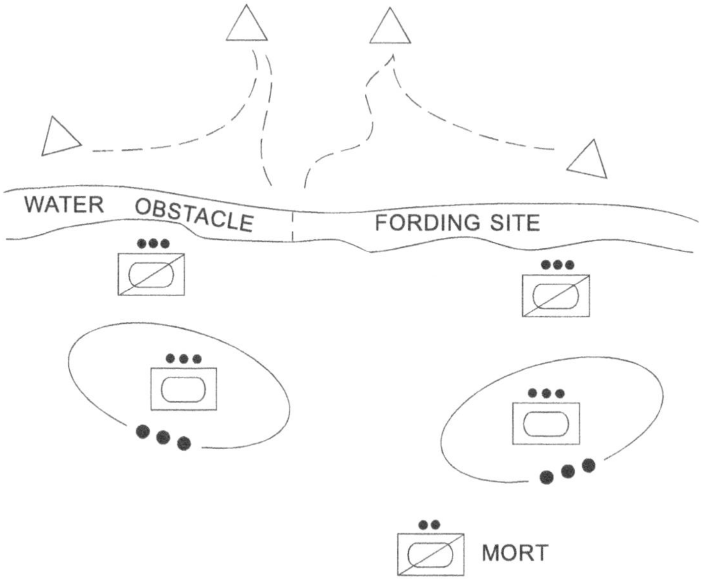

The scouts encounter a water obstacle, find fording sites, and secure the far side. Position the tank platoons and mortars to support with direct and indirect fires.

Figure 7-7. Secure the far bank.

The rest of the troop should not bunch up at the crossing site. Position them a terrain feature short of the water obstacle, so they have quick access to the crossing sites but are not exposed to enemy observation and fires.

Cross the Obstacle. Once the scouts have established positions on the far bank, the troop main body is ready to cross the obstacle. Time the movement to the crossing site so that no vehicles have to pause on the near side, but all can move directly through the crossing site and to their positions on the far side to continue the operation. Once the troop starts the crossing, complete it as quickly as possible (see Figure 7-8).

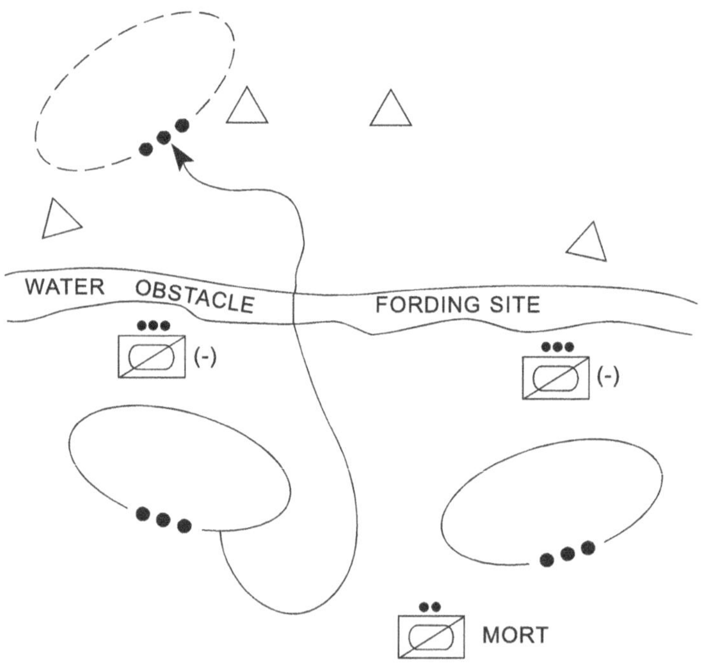

After the far side is secure, rapidly move platoons across the obstacle and continue the mission.

Figure 7-8. Cross the obstacle.

Fire Support. Fire support is used during the hasty water crossing to suppress known and suspected enemy positions at the crossing site. Smoke is employed to the front or flanks to screen the reconnaissance and the crossing. The fire support officer must time the movement of the mortars so they are in position to support the water crossing operation with continuous fire support.

Section VI. In-Stride Breach

When obstacles are encountered, an in-stride breach will maintain the tempo of offensive operations. Treat obstacles with caution and expect them to be covered by enemy observation and fires. The troop's ability to breach obstacles using available assets is important to ensure mission accomplishment. The size of the enemy force covering an obstacle with direct fire will, however, affect the troop's ability to conduct a hasty breach.

TASKS

The following combat tasks are associated with an in-stride breach:

- Detect the obstacle, reconnoiter it, and search for a bypass.
- Suppress all enemy positions with direct and indirect fires.
- Obscure enemy observation with smoke.
- Secure the near side of the obstacle.
- Breach or neutralize the obstacle.
- Move forces across the obstacle.
- Continue the mission.

TECHNIQUES

The enemy will reinforce natural obstacles with man-made ones to slow, disorganize, and canalize the troop. These obstacles may consist of—

- Minefields.
- Log obstacles, such as abatis, log cribs, stumps, and posts.
- Antitank ditches.
- Wire entanglements.

When scouts encounter an obstacle, they must follow the procedures for actions on enemy contact described in Chapter 3.

- Deploy and report.
- Develop the situation.
- Choose a course of action.
- Recommend a course of action.

As they develop the situation, the scouts should rapidly reconnoiter the obstacle to determine if bypass routes exist and whether or not they are covered by enemy fires, if breaches through the obstacle already exist, if the obstacle is part of an occupied defensive position, what the enemy strength is, and where he is located.

To conserve time and manpower, the troop should attempt to bypass all obstacles. Bypassing will not always be possible because available routes around the obstacle may lead into a fire sack. Forcing through an obstacle is the least desirable method of breaching because of the resulting loss of personnel and equipment. A hasty breach may be the only viable course of action.

Before deciding on a course of action, the troop commander should position himself to observe the situation and control the operation, if possible. Once the commander decides that a hasty breach is the only course of action, he must organize the troop into a breaching element and an overwatch element. When executing the hasty breach, the commander should avoid concentrating the troop in any one area, making it vulnerable to enemy fires.

During zone reconnaissance or movement to contact, the troop operates over a 6- to 10-kilometer-wide zone. During the hasty breach, the commander may not be able to use all the platoons in the troop. In this situation, until the breach operation is complete, give the scout platoon not involved in the breach operation a limit of advance so it does not move too far ahead of the rest of the troop. It can provide the troop with an early warning of enemy activity.

The Breaching Element. The breaching element is composed of scouts and engineers attached or in support of the troop. Their mission is to create lanes through the enemy obstacle system to allow the troop to pass.

The Overwatch Element. The overwatch element is the tank/AT platoons, scouts not in the breaching force, and any additional engineers. Its mission is to provide close, continuous overwatching fires in support of the breaching element. After the breaching element has cleared lanes through the obstacle, part of the overwatch element will assault through the lane to defeat any enemy in the area of the breach. The fire support officer coordinates the fires of the mortars and any supporting artillery to suppress enemy fires with HE and to obscure his observation with smoke.

Sequence of Events in a Hasty Breach. After encountering an obstacle and the commander chooses a hasty breach as the course of action, the sequence of events is as follows:

- Scouts.
  - Reconnoiter the near side of the obstacle to determine its front edge and lateral limits, enemy dispositions, if the obstacle is part of a defensive position, and if a partial or complete breach exists.
  - Secure flanks in support of the breaching operation.

- Overwatch element.

  – Occupy overwatch positions and provide direct and indirect suppressive fires on the enemy for elements moving to and through the obstacle.

  – Use smoke to obscure enemy observation of the obstacle (see Figure 7-9).

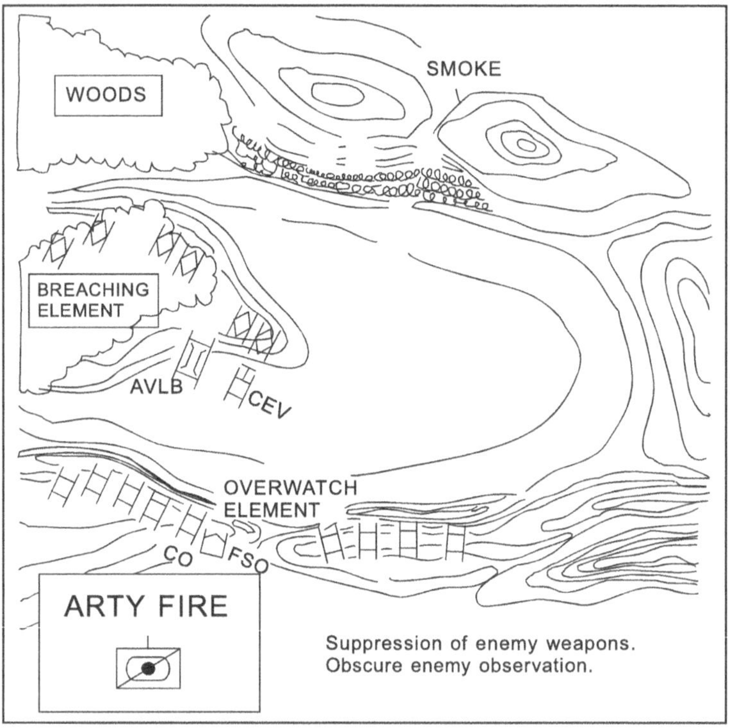

Figure 7-9. Obscure enemy observation.

- Breaching element.
  - Occupy covered and concealed positions to coordinate activities and prepare equipment, demolitions, and routes to the obstacles.
  - After enemy fire has been suppressed, breach the obstacle.
  - Secure fighting positions on the near and far sides of the obstacle (see Figure 7-10).

- Overwatch element.
  - As the breaching element breaches the obstacle, a tank platoon from the overwatch element prepares to attack through the obstacle.
  - Once the breaching element secures positions on the far side of the obstacle, the tank/AT or scout platoon attacks through the obstacle and destroys enemy elements that can fire directly on it. This assault element either continues to advance as the troop lead element or establishes hasty defensive positions while the remainder of the troop passes through the obstacle (see Figure 7-11).
- Troop. Continue the mission.

Neutralizing and Breaching Minefields. Minefields will differ in layout and composition, depending on the availability of mines and the nature of the avenue of approach. When possible, mines should be detonated in place. The objective in breaching the obstacle is to make a safe route to the far side. Multiple routes across the obstacle reduce vulnerability to enemy fires, but may be very time consuming and beyond the capabilities of the troop.

Use the following methods to neutralize and breach minefields:

- **Foot Lanes.** Establish foot lanes through the obstacle when a mounted assault breach is not at first practicable. This lane is normally two meters wide, and can later be widened into vehicle lanes. Foot lanes allow the troop to move forces through the obstacle to secure the far side.

- **The Bangalore Torpedo.** This device will clear foot lanes through mines and wire obstacles. It will clear a path three to four meters wide through wire entanglements and a narrow footpath through a minefield.

- **Vehicle Lanes.** After the first breach is made, foot lanes may be widened to one-way vehicle lanes at least eight meters wide. Vehicle lanes may also be breached separately from foot lanes. Use existing roads when possible. Vehicle-pushed rollers, if available, should be used to proof the lane cleared by line charges.

- **The M173 Rocket-Projected Line Charge.** This is an AT minefield clearing device. When the line charge explodes, it clears a vehicle lane about six meters wide.

Section VII. NBC Defensive Operations

OPERATIONS

NBC defense operations reduce casualties and damage to equipment and materiel, and minimize confusion and interruption of the troop's mission in the event of enemy NBC attacks. These operations are performed concurrently with all combat operations to preserve the fighting strength of the troop.

TASKS

The following combat tasks are associated with the conduct of troop NBC defense operations:

- Prepare for a nuclear attack.
- Respond to the initial effects of a nuclear attack.
- Respond to the residual effects of a nuclear attack.
- Cross a radiologically contaminated area.
- Conduct radiological reconnaissance.
- Perform radiological decontamination.
- Prepare for a chemical agent attack.
- Respond to a chemical agent attack.
- Cross a chemically contaminated area.
- Conduct a chemical reconnaissance.
- Perform hasty decontamination.
- Coordinate for deliberate decontamination of equipment.
- Exchange protective clothing.

ORGANIZATION

The troop's NBC defense system is composed of a trained NBC officer (usually the XO), an NBC NCO, and an enlisted alternate from troop headquarters. Troop soldiers are also designated and trained to operate all assigned NBC equipment and to assist decontamination teams.

The NBC officer supervises troop NBC defense activities and assists the commander in training NBC equipment operators and decontamination teams. The NBC NCO and his alternate directly supervise radiological monitoring, chemical detection, and decontamination operations. During combat operations, the NBC NCO is located in the command post where he—

- Receives, prepares, evaluates, and disseminates information and reports enemy and friendly NBC attacks.
- Supervises employment of detection, monitoring and survey, and decontamination teams.
- Maintains unit radiation exposure status records.
- Assists the troop commander in analyzing guidance from squadron, mission, threat, and weather as they affect NBC operations. Recommends appropriate MOPP level based on this information.

To facilitate operations in an NBC environment, designate the NBC teams from all platoons and sections by vehicle crews so vehicle commanders have responsibility for their crews and the execution of the NBC defense team mission. Each section in the troop has a chemical agent detection team and each platoon has a radiological monitoring and survey team.

Chemical Agent Detection Team. Each troop has twelve chemical agent detection teams. There are three in each scout platoon and one in each tank/AT platoon, one in the mortar section, one in the command post, and one in each combat trains and field trains. Each team has an M256 detector kit, a chemical agent alarm, detector paper, 400 meters of WD-1/TT for remoting the alarm, and enough batteries and supplies for seven days of continuous operations.

Radiological Monitoring and Survey Team. Each troop has seven teams, each equipped with an IM 174/PD dose rate instrument and an IM 93/147 dosimeter. One detection team from each platoon serves as a radiological monitoring and survey team. Also, the mortar section, maintenance section, and troop headquarters each have a radiological monitoring and survey team.

Deliberate Decontamination Team. Deliberate decontamination teams are organized by the vehicle crew. One crew from each scout and tank/AT platoon is designated as a decontamination team. This team is used to support the deliberate decontamination of the troop. One crew from each platoon should receive special training from the squadron chemical section to execute this task.

TECHNIQUES

Proficiency in NBC defense operations is attained only by strict adherence to standardized procedures as outlined in the troop SOP. The troop must be proficient in the following three fundamentals of NBC defense to survive and remain an effective fighting unit:

- Contamination avoidance (before, during, and after an attack).
- Protection (before, during, and after an attack).
- Decontamination (after an attack).

Actions Before the Attack. The best defense against a chemical or nuclear attack is to avoid being detected and targeted. To avoid becoming a lucrative target, implement and strictly enforce the following passive avoidance measures during all operations:

- **Dispersion.** Maintain dispersion between vehicles whether stationary or moving; vehicles bunched together are much easier to find. This is especially important when occupying assembly areas, conducting resupply operations, or when crossing obstacles and waterways.
- **Concealment.** Concealed positions prevent the enemy from observing vehicles and personnel from the ground or air.
- **Camouflage.** When concealment is not available or adequate, use camouflage to hide vehicles and positions.
- **Communications security (COMSEC)/signals security (SIGSEC).** The enemy can easily locate positions with direction-finding equipment if the troop does not adhere to proper communications procedures. Ensure all troop elements properly encode messages sent over unsecured nets.
- **OPSEC.** Take all possible actions to deceive the enemy about troop movement, positioning, intentions, and size. Mask movement by using folds in the terrain. Use hide positions when not moving or engaging the enemy. Enforce noise and light discipline. To deceive the enemy about the number of troop vehicles, start all vehicles simultaneously, and cover all track marks that lead into positions.

Nuclear/Chemical Vulnerability Analysis. Based on the NBC threat and the troop mission, determine the troop's vulnerability and incorporate those factors into the commander's planning.

Take the following actions to reduce the effects of a chemical or nuclear attack:

- **Improve positions.** Use available natural and man-made cover, such as folds in the earth, buildings, and dug-in positions, to protect personnel, equipment, materials, and supplies.
- **Alert personnel.** Inform soldiers to assume the appropriate MOPP level (see Table 7-1), based on the possibility of enemy attack and the MOPP analysis. Inspect unit and individual NBC equipment and prepare for operations.

Table 7-1. Mission-oriented protection posture levels.

MOPP LEVEL	OVERGARMENT	BOOTIES	MASK/HOOD**	GLOVES
0	Stored Nearby*	Stored Nearby*	Carried	Stored Nearby*
MASK ONLY	Carried	Carried	Worn	Carried
1	Worn, open or closed based on temperature	Carried	Carried	Carried
2	Same as MOPP 1	Worn	Carried	Carried
3	Same as MOPP 1	Worn	Worn, hood open or closed based on temperature	Carried
4	Worn, closed	Worn	Worn	Worn

*On combat vehicle or in fighting position.
**Includes M258A1 kit and detector paper.

- **Protect equipment and supplies.** Protect against chemical contamination inside a vehicle since it is very hard to decontaminate. Keep all hatches closed, if possible, and turn on vehicle overpressure system. Cover equipment and supplies stored outdoors or in tank bustles. Nonporous plastic sheets make good covers. Use tarpaulins if plastic is not available.
- **Position alarms.** Activate and emplace automatic alarms to provide the earliest possible warning of an attack. Position the alarms upwind and at least 150 meters away from stationary positions. Keep adjacent alarms about 300 meters apart.
- **Alert NBC personnel and teams.** Ensure NBC personnel and teams are available and prepared to perform their assigned missions.

Actions in Response to an Attack. Every soldier must be proficient at individual NBC basic skills. Masking is conducted immediately in response to these conditions:

- Chemical alarm sounds.
- Positive reading on detector paper.
- Individuals exhibiting symptoms of chemical or biological agent poisoning.

All soldiers mask automatically when an attack is possible, after the initiation of chemical/biological warfare, in response to artillery shells exploding in the area, and when aircraft drop bomblets or spray a mist or fog. In response to these conditions, all soldiers assume MOPP 4 and remain at maximum protection until the absence of contamination is confirmed or the commander has completed the analysis and determined a lower level of MOPP is appropriate.

Actions in response to a chemical or nuclear attack are standardized and included in the troop SOP.

- Initiate immediate masking and sound alarm (visual, audible, or voice).
- Continue mission.

- Assume MOPP 4 as rapidly as possible.
- Treat casualties.
- Send NBC-1 report and update it as more information becomes available.
- Identify the agent.
- Initiate decontamination operations.

These procedures must be rehearsed, and all individuals must adhere to them.

Actions After an Attack. Warn all subordinate elements immediately so they can take appropriate actions. In addition, alert squadron and all adjacent units, giving priority to the unit downwind from the attack. The troop and squadron are extremely vulnerable at the beginning of a chemical attack. Continue the troop mission while preparing for the enemy; he will take advantage of the chemical attack by conducting tactical operations.

In accordance with the troop SOP, prepare and send an NBC-1 report immediately. The personnel observing the attack send the report to the troop XO and NBC NCO in the command post. The XO forwards the report to the squadron TOC. Send the report immediately, and update it as more information is available. Treat any casualties of the attack using self-aid and buddy-aid, and report their status through the troop command post to the squadron S1. Operating in MOPP 4 will, over time, reduce the troop's effectiveness in combat operations. Soldiers become fatigued and lose mental and physical dexterity. Maintaining command and control becomes very difficult during extended MOPP 4 operations.

The troop must determine the type of agent used in the attack and the extent of contamination. Initiate tests with the M256 detector kit to identify the type of agent. The M256 kit detects and identifies field concentrations of nerve, blister, or blood agents. This test will help determine when unmasking may be safe after a chemical attack. Update the NBC-1 report with any information provided by the test. Decontaminate if required, and unmask as soon as possible.

When the troop is contaminated with persistent or semipersistent chemical agents, decontamination is necessary before reducing the MOPP level. Implement the following decontamination techniques:

- Start the skin decontamination within one minute of contamination.
- Apply decontaminating agents to equipment and vehicles to limit the spread of contamination and shorten the duration of the hazard. Use the M258A1 kit for personnel wipe down, and the M11 or M13 decontamination apparatus for vehicle/equipment spray down.

Decontaminate as far forward as possible. After leaving the contaminated area, perform hasty decontamination as soon as the situation permits. Hasty decontamination consists of exchanging MOPP gear and washing vehicles with hot, soapy water or high-pressure water to remove gross contamination. This operation is conducted at platoon or troop level, assisted by the decontamination team and other assets provided by the squadron. Select a site that is uncontaminated, near a water source, and concealed. Hasty decontamination may allow temporary relief from MOPP 4 for eating, drinking, and personal hygiene. Deliberate decontamination is required later when time and additional decontamination assets are available.

Initiate unmasking procedures as soon as possible after a chemical attack, to reduce the performance degradation that occurs while soldiers are in MOPP 4. Unmasking procedures differ based on the availability of detection equipment. When nonpersistent agents are used in the attack and an M256 detector kit is available, use the kit to determine when the agent is no longer present. Instruct two or three individuals to unmask for 5 minutes, remask, and sit in a shady area for 10 minutes. Check the individuals for chemical agent symptoms. If no symptoms appear, it is safe to unmask.

When no detector kit is available, direct two or three soldiers to perform the following procedures. Have them keep their eyes open, take a deep breath, hold it, and break the seal on their mask for 15 seconds. Reseal and clear the mask, and wait 10 minutes in a shady area. Check them for chemical agent symptoms. Have them break the seal and take three or four breaths and reseal the mask. Wait 10 minutes and check them for symptoms. Have them unmask for 5 minutes and remask for 10. Check for symptoms. If no symptoms appear, assume it is safe for all to unmask.

Following a persistent or semipersistent attack, begin unmasking procedures after hasty decontamination is complete and the M256 kit indicates no agent is present.

OPERATIONS IN A NUCLEAR ENVIRONMENT

Radiological Monitoring. This is the detection of radiation and the measurement of dose rate with radiac instruments. Monitoring finds a hazard that would otherwise go unmeasured or undetected. It is conducted by the troop radiological monitoring and survey team, which monitors—

- Total dose using the IM 93/147.
- Dose rate using the IM 174/PD.
- Food, water, and personnel contamination using the AN/PDR-27.

Conduct periodic monitoring when directed, or after nuclear war breaks out. Monitor a specific point in the area at least once an hour. Start continuous monitoring after receiving a fallout warning; when conducting a movement; when a nuclear burst is reported, seen, or heard; after detecting radiation above 1 cGy/hr during periodic monitoring; or when the squadron commander orders. Stop continuous monitoring when the squadron commander orders or, except while on the move, when the dose rate falls below 1 cGy/hr.

Based on NBC-4 monitoring reports, the troop may conduct a radiological survey to determine the degree and extent of contamination. This information is forwarded to squadron in an NBC-4 report and on DA Form 1971-R.

Exposure Control. While operating in a nuclear environment, the troop must control exposure to nuclear radiation. An operational exposure guide is a method of determining the maximum radiation dosage to which platoons can be exposed and still accomplish a mission. Determine radiation exposure by the cumulative dose or the platoon's radiation history.

Platoons record and report their total dose radiation daily, per SOP, to the troop command post where the NBC NCO compiles the data. The dose is determined by averaging the readings from both platoon dosimeters and rounding to the nearest 10. The NBC NCO totals each platoon's average dosage, and coverts it to a radiation status (see Table 7-2). He maintains a record of the radiation exposure status of each platoon and submits a daily radiation exposure status report to the squadron (see Table 7-3).

Table 7-2. Radiation dose categories.

RADIATION DOSE CATEGORIES

THE TABLE INDICATES THAT THE NUMERICAL VALUE OF cGys RECEIVED BY EACH PLATOON IS CONVERTED TO A RADIATION EXPOSURE STATUS (RES) FOR EACH PLATOON. THE CONVERSION FACTORS ARE SHOWN BELOW.

RADIATION STATUS CATEGORY	TOTAL CUMULATIVE DOSE (cGy)
RES-0	0 (NO EXPOSURE)
RES-1	GREATER THAN 0 BUT LESS THAN OR EQUAL TO 70
RES-2	GREATER THAN 70 BUT LESS THAN OR EQUAL TO 150
RES-3	GREATER THAN 150

The regimental and squadron commanders may move units on the battlefield, or relieve a particular unit that has been exposed to a high level of radiation. The squadron commander specifies the degree of exposure that units will not exceed for each operation. There are three degrees of radiation exposure risk—negligible, moderate, and emergency. Each level of risk is correlated to unit radiation status (see Table 7-4).

Table 7-4. Unit radiation status.

RADIATION STATUS (RS) CATEGORY	TOTAL PAST CUMULATIVE	EXPOSURE CRITERIA FOR A SINGLE OPERATION (cGy)
RS-0 UNITS	NO EXPOSURE	NEG RISK: 50 MOD RISK: 70 EMERG RISK: 150
RS-1 UNITS	GREATER THAN 0 BUT NOT MORE THAN 70	NEG RISK: 50-PAST DOSE MOD RISK: 70-PAST DOSE EMERG RISK: 150-PAST DOSE
RS-2 UNITS	GREATER THAN 70, NOT MORE THAN 150	ANY FURTHER EXPOSURE IS CONSIDERED EMERG RISK
RS-3 UNITS	GREATER THAN 150	ALL FURTHER EXPOSURE WILL EXCEED EMERG RISK

1. RADIATION EXPOSURE STATUS CATEGORIES ARE BASED ON PREVIOUS EXPOSURE TO RADIATION.

2. RECLASSIFICATION OF UNITS FROM ONE RADIATION EXPOSURE STATUS CATEGORY TO A LESS SERIOUS ONE IS MADE BY THE COMMANDER UPON ADVICE OF THE MEDICAL PLATOON LEADER AFTER AMPLE OBSERVATION OF ACTUAL STATE OF HEALTH OF THE EXPOSED PERSONNEL.

3. ALL EXPOSURES TO RADIATION ARE CONSIDERED TOTAL BODY AND SIMPLY ADDITIVE. NO ALLOWANCE IS MADE FOR BODY RECOVERY FROM RADIATION INJURY.

4. NEGLIGIBLE RISK WILL CAUSE NO CASUALTIES AND A LESS THAN 2.5 PERCENT INCIDENCE OF VOMITING. MODERATE RISK WILL RESULT IN A LESS THAN 5 PERCENT INCIDENCE OF VOMITING AND HEADACHES, BUT WILL CAUSE NO CASUALTIES. EMERGENCY RISK WILL RESULT IN LESS THAN 5 PERCENT CASUALTIES AND A HIGHER PERCENTAGE OF NAUSEA, VOMITING, AND HEADACHES.

Minimize the Effects of Friendly Nuclear and Chemical Attacks. A NUCWARN/CHEMWARN message is normally not sent below squadron level. Squadron alerts the troops using a code word prescribed in the OPORD. Squadron will also issue a FRAGO containing relocation instructions, the level of protection required, and the time of the friendly strike. To assure maximum reaction time, each soldier in the troop must be informed. Each crew immediately prepares its vehicles, equipment, and position for the attack as is appropriate to its location within the attack area. Crews prepare their vehicles by taking the following actions:

- Move if necessary.
- Position vehicle behind best available cover with front of vehicle toward the blast.
- Point gun away from blast.
- Lock brakes.
- Secure loose equipment in vehicle.
- Secure inside the vehicle all exterior components that could be damaged by the blast.
- Close and lock all hatches, to include ballistic shields.
- Wear helmets and protect eyes.
- Turn off and disconnect all radios but one, which is needed to retain communications with squadron.

Complete these tasks within five minutes of receiving notification. After the friendly strike, continue radiological monitoring and report if necessary. Quickly prepare to continue operations.

If the troop is contaminated with fallout, the level of contamination is monitored by the monitor and survey teams. Decontamination is accomplished by brushing or wiping the dust off the equipment. Brush off or shake out MOPP gear and wash down the equipment and vehicles to reduce radiological contamination. Use the AN/PDR-27 to determine decontamination requirements and to determine if supplies of food and water must be destroyed.

Chapter 8

Combat Support

The effective integration of combat support (CS) will spell the difference between success and failure on the battlefield. The commander must know the capabilities and limitations of CS assets, and how to employ them properly.

CONTENTS

Section I. The Fire Support Team

CAPABILITIES AND LIMITATIONS

The FIST is organized, equipped, and trained to provide—

- A fire support advisor/coordinator—the FSO.
- An observation/laser designation capability.
- A communications link to all available fire support.

The FIST has secure FM communications on four radios, and digital communications on two nets through the use of digital message devices (AN/PSG-5 and AN/PSG-2). These capabilities enable the FIST to maintain communications on the following nets (see Figure 8-1):

- Troop command net.
- Troop fire support net.
- Supporting artillery fire net (digital).
- Squadron fire support net.

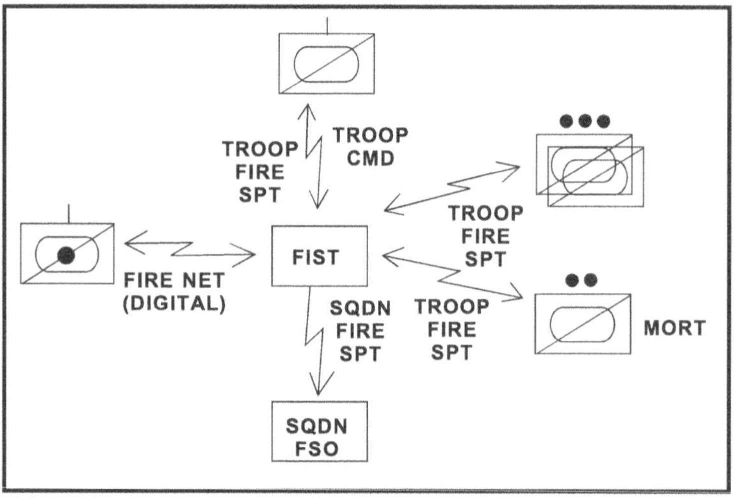

Figure 8-1. Fire support team communication nets.

The FIST is also equipped with a target laser locator designator set to designate targets for terminally guided munitions and to determine target locations accurately at ranges up to 10 kilometers (limited by line of sight). The target designator set is equipped with a thermal sight to use under limited visibility conditions.

The digital message devices allow the FIST to transmit and receive high-speed digital messages from the howitzer battery computer system, the direct support artillery battalion TACFIRE, OH-58D aircraft, and the squadron FSO's variable format message entry devices, among other digital message devices. Digital message devices can quickly transmit standard fire requests, fire adjustments, and plain text messages. The FIST can store nine fire requests on the digital message devices for rapid transmission in support of troop operations.

The FIST's abilities may be constrained by its communications with the supporting field artillery (FA). To coordinate any fire support for the troop, the FIST must be positioned where it can maintain at least a voice communications link to the supporting artillery. The FIST may not be able to see the battlefield and use the laser designator capability in all situations. When the troop is operating in a large sector, the FIST will be able to see only a small portion of the area. The scouts, therefore, must be proficient in requesting indirect-fire support. The scouts' calls for fire will be relayed through the FIST to the appropriate fire support unit.

EMPLOYMENT TECHNIQUES

Planning and Coordination. Fire support planning and coordination begin on receipt of a mission, and continue throughout its planning and execution. The troop FSO should always accompany the commander to mission briefings at squadron, so that he is fully aware of the squadron scheme of maneuver and the availability of fire support assets. He and the squadron FSO can also begin coordinating any specific requirements for fire support, such as preparations, groups, and series.

The troop commander's guidance should include the following to allow the troop FSO to plan the operation:

- What is the purpose of FA and mortar fires?
- How will FA fires support the troop's maneuver?
- What are the attack criteria? Will the commander use FA for immediate suppression, screening, or destruction?
- Are the attack and engagement criteria for mortars different from FA?
- On what and where does the commander want to mass FA and mortar fires?

Plan indirect fires on known or suspected enemy positions; at choke points; on the troop's objective or flanks; or on any other areas that would support the scheme of maneuver. These targets can be fired at a specific time, on call, or when a particular event takes place.

The FSO incorporates his approved fire plan into a target list and forwards it to the squadron FSO. The squadron FSO compiles the lists from all the troop FSOs, eliminates any duplicate targets, forwards the list to the supporting artillery, and informs the troop FSOs of any changes to their target lists.

Once he has completed his plan, the FSO will brief the commander on planned targets, scheduled fires, and fire support coordination measures that will be in effect. The commander will make the final decision on the fire support plan. Ensure it complements the scheme of maneuver and maximizes the available firepower for the troop. Once approved, the fire plan is continuously modified and updated to meet changing situations.

Silence is Consent. Once the FSO has briefed the commander and the commander approves the troop fire support plan, calls for fire are executed during the conduct of the operation by the silence is consent method. This method allows subordinate leaders within the troop to request indirect fires in support of their maneuver within the framework of the commander's guidance and the troop fire support plan. All initial requests for fires will go over the troop command net and the commander will simply acknowledge the request with "ROGER—OUT." The troop FSO will process the call for fire through the appropriate indirect-fire system based on the commander's guidance. All adjustments and corrections for the fires will be communicated on the troop fire direction net directly between the observer and the troop FSO. If the request for fire that is submitted by the subordinate leader is not within the commander's fire support guidance, the troop commander will deny fires or modify his guidance based on the situation.

Execution of the Fire Support Plan. During the operation, the FSO is responsible for—

- Keeping the commander advised of all fire support matters.
- Resolving any fire support conflicts that may arise during the planning and execution of the operation.
- Attacking targets with the most suitable fire support means available to support the scheme of maneuver.

Also, the commander may make the FSO responsible for positioning the troop mortars to best support the operation.

The FSO operates on the four radio nets mentioned previously. The scouts in the troop normally serve as the observers calling for indirect-fire support. The FIST monitors all spot reports on the troop command net and, as necessary, prepares calls for fire. The FSO will allocate fires from mortars and artillery, based on the commander's guidance. The commander may give the FSO additional guidance concerning a fire mission as it is requested (see *Silence is Consent* paragraph). Further information about the fire mission and all other calls for fire are passed on the troop fire support net. Keep the troop command net free for other traffic.

Both the FIST and the mortars monitor the troop fire support net. When scouts forward adjustments or corrections for indirect-fires, they use the troop fire support net. If the FSO forwards the mission to supporting artillery, he sends the message "LIGHTNING UP" on the troop fire support net. This tells the scouts calling for fire and the mortars that the mission is being fired by supporting artillery and that all adjustments and the "end of mission" report must be relayed through the FSO. If the mortars will fire the mission, the FSO sends the message "THUNDER DOWN" on the troop fire support net. This tells the mortar squad to fire the mission, and tells the scouts to send all adjustments and the "end of mission" report to the mortar squad (see Figures 8-2 and 8-3).

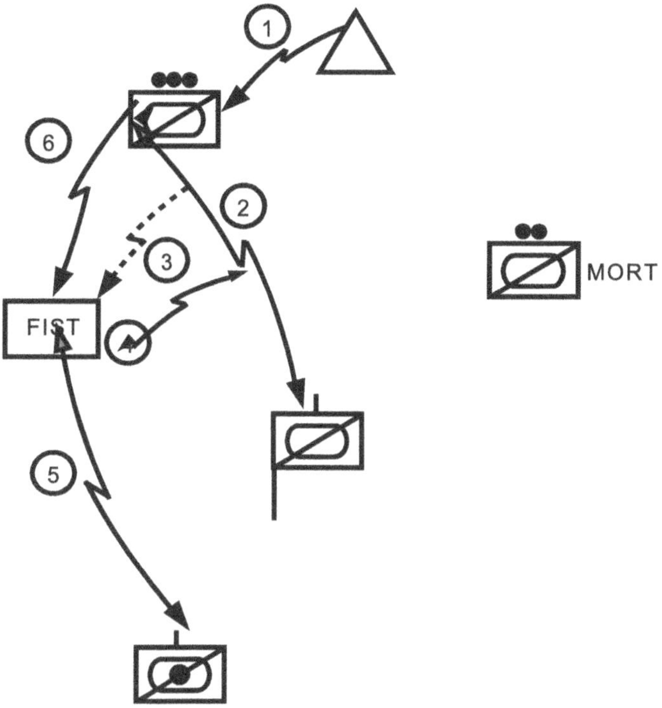

(1) SCOUTS SEND SPOT REPORT TO PLATOON LEADER (PLATOON NET).

(2) PLATOON LEADER FORWARDS THE SPOT REPORT TO THE CP (TROOP COMMAND NET) WITH REQUEST FOR FIRE.

(3) FIST MONITORS SPOT REPORT, RECEIVES PERMISSION TO FIRE THE MISSION, AND DETERMINES WHAT ARTILLERY TO USE.

(4) FIST SENDS MESSAGE "LIGHTNING UP" ON TROOP COMMAND NET. SCOUTS AND MORTARS MONITOR.

(5) FIST SENDS DIGITAL CALL FOR FIRE TO DS ARTILLERY. FIST SENDS SHOT ON TROOP COMMAND NET.

(6) SCOUTS RELAY ALL ADJUSTMENTS TO FIST ON TROOP FSN.

Figure 8-2. Fire request channels (artillery).

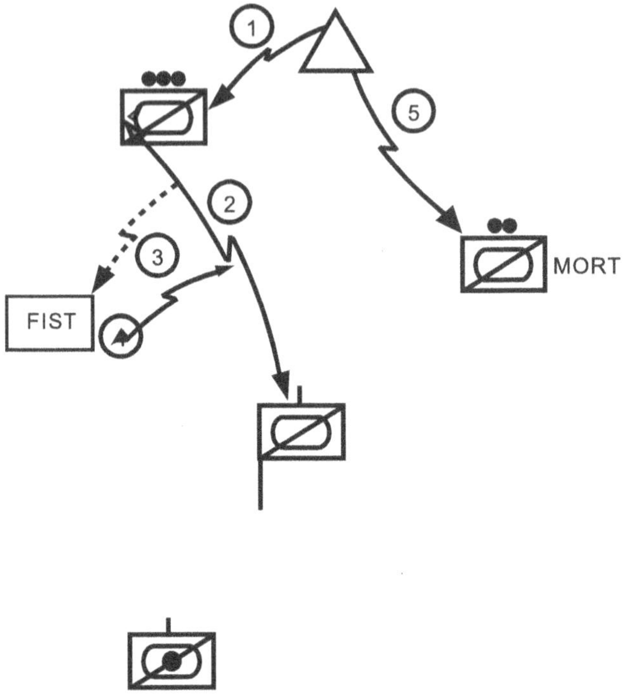

(1) SCOUTS SEND SPOT REPORT TO PLATOON LEADER (PLATOON NET).

(2) PLATOON LEADER FORWARDS THE SPOT REPORT TO THE CP (TROOP COMMAND NET) WITH REQUEST FOR FIRE.

(3) FIST MONITORS SPOT REPORT AND RECEIVES PERMISSION FROM THE TROOP CDR TO SUPPORT WITH INDIRECT FIRES.

(4) FIST SENDS MESSAGE "THUNDER DOWN" ON TROOP COMMAND NET. SCOUTS AND MORTARS MONITOR. FIST SENDS SHOT ON TROOP COMMAND NET.

(5) SCOUTS CALL THE MORTARS ON TROOP FSN TO REQUEST INDIRECT FIRES.

Figure 8-3. Fire request channels (mortar).

Section II. Mortar Support

CAPABILITIES AND LIMITATIONS

The mortar section provides a heavy volume of accurate, sustained indirect fires in response to the needs of the troop. They are ideal weapons for attacking targets on reverse slopes, in narrow ravines, built-up areas and in other areas that are difficult to strike with low-angle fires. Mortars are most effective for—

- **Suppression.** Suppression forces the enemy to button up or move to less advantageous positions. High explosive (HE) ammunition with a variable-time fuze or quick fuze does this best.

- **Screening.** Obscuration smoke is placed directly on or just in front of enemy positions to obscure vision. Screening smoke is placed between troop positions and the enemy to conceal the troop's movements. Smoke will degrade the effectiveness of laser range finders, so the tank platoons must adjust their gunnery techniques when engaging the enemy.

- **Illumination.** The use of illumination permits daytime engagements during some limited visibility conditions by improving the effectiveness of image intensification devices. Illumination rounds are used to mark targets for close air support and attack helicopters, and to allow observers to adjust artillery onto a target. Illumination rounds are usually adjusted so they illuminate above or behind the enemy location to silhouette him.

The height of the illumination should be adjusted to burn out about 200 meters above the ground. This gives maximum illumination time, about 90 seconds for one round, and will not degrade the effectiveness of image intensification devices. Because illumination rounds light the battlefield for both sides, it should be used when night-vision devices are not available in sufficient quantities or when ambient light levels are very low.

EMPLOYMENT TECHNIQUES

The troop mortar section is employed as a separate element during operations. It moves independently of the platoons, and provides its own security. The troop commander normally designates positions for the section sergeant and provides guidance for support forward of the scouts. The commander may find that having the FIST position the mortars is easiest and most effective. This technique allows the mortar section sergeant the ability to concentrate on providing fires and computing fire data and the commander to focus on maneuvering the troop, not one section.

The mortars can operate on two nets. If the FIST is responsible for moving the section, both nets may be on the troop fire support net. The FSO will keep them updated on the situation. However, since many spot reports over the troop net become fire missions, the mortars can eavesdrop and provide more responsive fires if they stay abreast of the situation themselves. If the commander gives the mortar section sergeant authority to select positions and to move the section, he must operate on the troop command net to stay abreast of the situation and on the troop fire support net.

Offense. During offensive operations the movement of the mortars will be based on the progress of the scout platoons. While the section is on the move, it must be prepared to provide immediate fires using direct lay, direct alignment, or hip shoot. Plan the moves of the section so it is in position to support the troop at critical times, such as during river crossings and counterattacks, and when clearing defiles.

Defense. Considerations for using mortars in the defense are similar to those for using them in the offense. Plan on prestocking ammunition at subsequent positions to reduce resupply problems during a defensive operation.

Reconnaissance and Security. During reconnaissance and security missions, the troop will often be operating over a large frontage that cannot be completely covered by the section. The commander must decide whether to position the mortars to cover the most dangerous area, or to move them to a position where they can cover a part of different areas and adjust as necessary. Knowing what other artillery assets are available will help in making that decision.

Commander's Guidance. A list of considerations the commander can use to help plan mortar section employment follows:

- Ammunition constraints by type and quantity.
- Priority of fire to a designated platoon.
- Anticipated changes in mortar employment.
- Communications constraints.
- General designation of positions.
- Movement guidance.
- Coordination requirements.
- Resupply.

Section III. Artillery Support

ORGANIZATION AND SUPPORT RELATIONSHIP

The howitzer battery organic to the regimental cavalry squadron consists of eight 155-mm self-propelled howitzers. The squadron will often be supported by an artillery battalion. The artillery battalion usually has three batteries of 155-mm self-propelled howitzers, and provides direct support to the squadron.

CAPABILITIES AND LIMITATIONS

Field artillery provides the troop its main fire support. It gives accurate fires with a wide variety of munitions. Field artillery adds a powerful dimension to troop direct-fire and maneuver capabilities.

- Artillery can—
  - Provide fire support under all weather conditions and terrain types.
  - Shift and mass fires rapidly without the requirement to displace.
  - Support the battle in depth with long-range fires.
  - Provide a variety of conventional shell/fuze combinations.
  - Provide continuous fire support by careful displacement.
  - Be as mobile as the supported unit.

- Artillery has—
  - Limited capability against moving targets.
  - Limited self-defense capability against air and ground attack.
  - Limited capability to destroy point targets without considerable ammunition expenditure. (Copperhead is the exception. It can be used to engage point targets.)
  - Vulnerability to detection by enemy target acquisition systems because of firing signatures.

Field artillery has a wide variety of munitions tailored for the engagement of different types of targets. The ammunition types include—

AMMUNITION	**TARGET**
High explosive	Personnel, field fortifications, vehicles
HC smoke	Obscuration and screening
White phosphorus	Obscuration
Illumination	
Cannon-launched guided projectiles (Copperhead)	Armored vehicles or high-payoff targets requiring precision target engagement

In addition, the troop must become acquainted with the characteristics of improved conventional munitions (ICM) and scatterable mines. ICMs include antipersonnel (APERS) and dual-purpose ICM. The commander must consider the danger to friendly troops in areas in which APERS is fired. Also, the high ICM dud rate makes maneuver hazardous in the area of an ICM field. Scatterable mines are area denial munitions for use against personnel and remote antiarmor mines for use against armored vehicles.

EMPLOYMENT TECHNIQUES

Artillery units normally provide direct support to the squadron, and are positioned by the artillery commander in close coordination with the squadron S3 and FSO. The troop FSO is the troop's link to the supporting artillery (see Section I, The Fire Support Team, and Chapters 1 and 2).

Section IV. Engineer Support

ORGANIZATION AND SUPPORT RELATIONSHIP

A combat engineer company is organic to the heavy and light armored cavalry regiment. Additional engineer assets are often provided from divisional and corps engineer units to support the regiment. Up to a platoon of engineers is frequently attached to or placed in direct support of a cavalry troop for combat operations. A combat engineer platoon from the heavy regimental engineer company is organized for combat as shown in Figure 8-4.

HEAVY ENGINEER CAPABILITIES AND LIMITATIONS

A combat engineer platoon is uniquely equipped and trained to conduct mobility, countermobility, and survivability operations in support of troop operations. The platoon headquarters has two M9 armored combat earthmovers, which are highly mobile, armored, and amphibious (see Figure 8-5).

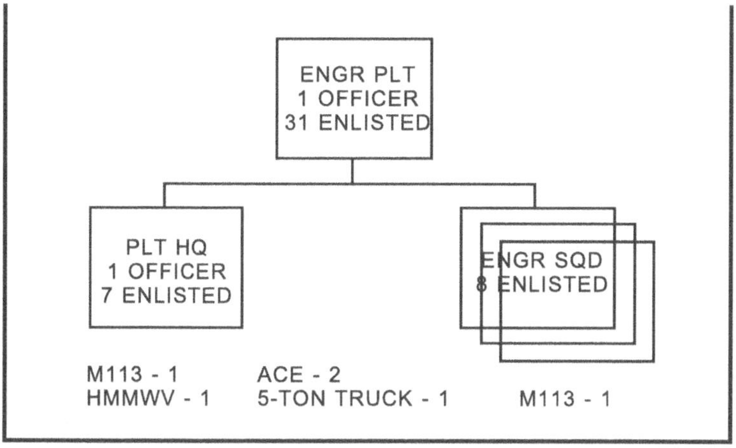

Figure 8-4. Regimental engineer company (heavy) organization.

Figure 8-5. Armored combat earthmover, M9.

In mobility operations, the engineer platoon can provide—

- Obstacle reduction. The engineers can reduce or negate the effects of obstacles to improve the troop's ability to maneuver.

- Route construction. The engineers can construct, improve, and maintain roads, bridges, and fords.

In a countermobility role, the engineers can assist with obstacle construction. The engineers can reinforce terrain by constructing obstacles to delay, canalize, disrupt, and destroy the enemy force to support the scheme of maneuver.

In survivability operations, the engineers can improve positions by constructing berms, dug-in positions, and overhead protection to reduce the effectiveness of enemy weapons.

A cavalry troop is frequently supported by additional engineer assets, such as an armored vehicle launched bridge (AVLB) and combat engineer vehicles (CEV) from the engineer company. The AVLB section of headquarters and headquarters troop of the armored cavalry squadron consists of three AVLBs that could be available during combat operations.

An AVLB is a tank chassis modified to transport, launch, and retrieve a 60-foot class-60 bridge. The bridge is capable of carrying military load class-60 track vehicles across a 17-meter gap and military load class-70 track loads across a 15-meter gap (see Figure 8-6).

Figure 8-6. Armored vehicle launched bridge.

The CEV is an M60A1 tank mounted with a hydraulically operated dozer blade, a 165-mm turret-mounted demolition gun, a retractable boom, and a winch (see Figure 8-7).

Figure 8-7. Combat engineer vehicle.

The small emplacement excavator (SEE) is a small highly mobile truck with a front-end loader and a backhoe (see Figure 8-8).

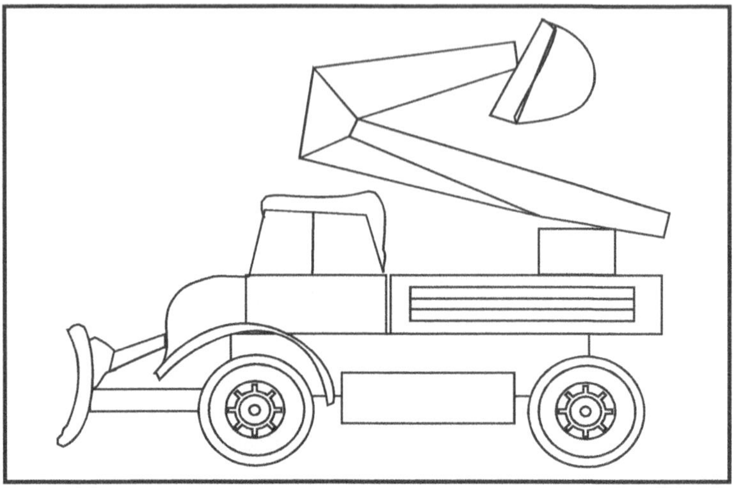

Figure 8-8. Small emplacement excavator.

LIGHT ENGINEER CAPABILITIES AND LIMITATIONS

The engineer platoon from the light regimental engineer company is equipped with five HMMWVs. The platoon can accomplish most of the same tasks that a heavy platoon can, but the time to complete most jobs is greater. The platoon can provide the troop with the following:

- In mobility operations.
  - Obstacle reduction. Engineers can reduce or negate the effects of obstacles to improve the troop's maneuver ability. To improve the platoon's ability, they may be reinforced with an M9 ACE, SEE, MICLIC, and Bangalor torpedo. The engineer company from the light regiment does not provide the regiment (troop) with any bridging capability.
  - Route reconnaissance. Engineers can conduct route reconnaissance.

- In a countermobility role, the engineers are usually left under squadron control to complete priorities of work in accordance with the squadron obstacle plan. The engineers can reinforce terrain by constructing obstacles to delay, canalize, disrupt, and destroy the enemy force to support the scheme of maneuver.

- In survivability operations, the engineers can improve positions by constructing berms, dug-in positions, and overhead protection to reduce the effectiveness of enemy weapons (see Figure 8-9).

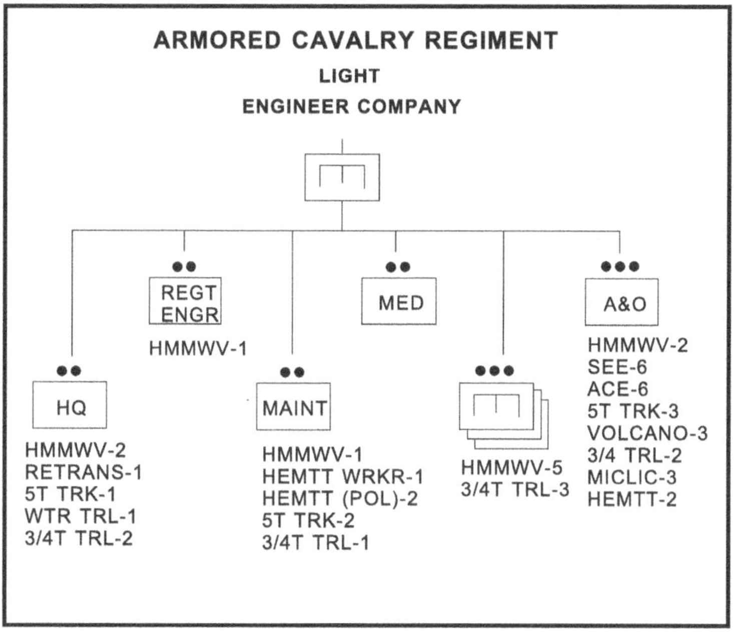

Figure 8-9. Regimental engineer company (light) organization.

EMPLOYMENT TECHNIQUES

The troop has several options in the task organization and employment of engineer assets to support its operation. Normally, the troop will keep the entire platoon under troop control and assign missions just as with a maneuver platoon. The commander can also task organize the engineer platoon, and assign squads to scout platoons to support the operation. Whatever method is chosen, the engineer platoon leader needs to be involved in the troop planning process.

Section V. Ground Surveillance Radar Support

ORGANIZATION AND SUPPORT RELATIONSHIP

A GSR team is organized for combat as shown in Figure 8-10. The team(s) is normally attached or OPCON to a troop during combat operations. GSR is found only in heavy regiments and divisions.

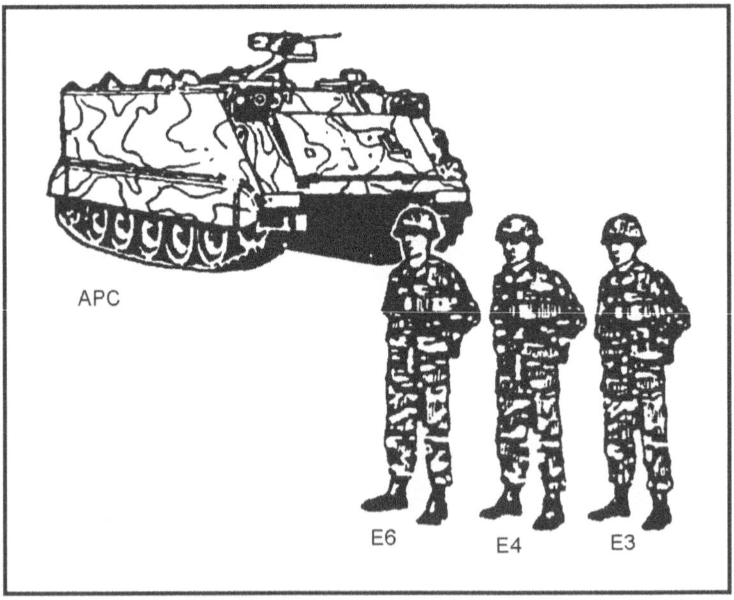

Figure 8-10. Ground surveillance radar team.

CAPABILITIES AND LIMITATIONS

GSRs provide mobile, all-weather battlefield surveillance. GSRs employed in pairs can provide observation from a given vantage point 24 hours a day, detecting targets and providing accurate range and azimuth readings.

The AN/PPS-5 has a line-of-sight range of 10,000 meters against vehicles and 6,000 meters against personnel. A GSR can detect targets through light camouflage, smoke, haze, light snow and rain, darkness, and light foliage. Heavy rain and snow seriously restrict its radar detection capability.

The GSR is generally ineffective against air targets unless the aircraft is flying close to the ground. The GSR is designed to detect targets moving in the presence of a background. It is vulnerable to enemy direction finding and jamming.

EMPLOYMENT TECHNIQUES

The troop will normally employ supporting GSR teams to augment the surveillance capabilities of its scout platoons. The commander should assign each GSR team a specific sector of surveillance and a frequency of coverage, or give the scout platoon leader guidance for employing the GSR team. The enemy can detect radar signals, so the GSR cannot conduct continuous surveillance. The tasks assigned GSR surveillance teams include—

- Searching avenues of approach or possible enemy positions on a time-schedule basis or randomly to determine location, size, composition, and nature of enemy activity.

- Monitoring point targets such as bridges, defiles, and road junctions and reporting quantity, type, and direction of enemy movement through the target.

- Extending the observation capabilities of scouts by enabling them to survey distant points and areas of special interest.
- Vectoring patrols to keep them oriented during times of limited visibility.

Position the GSRs in an area that is free of ground clutter and in one that affords them long-range observation and a wide field of view. Normally, they should be assigned a general area by the troop commander or scout platoon leader and the GSR team leader will select the specific location. The team should be prepared for rapid displacement, and have several alternate positions selected and reconnoitered to avoid enemy suppressive fires.

Offensive Operations. Employ GSRs in the offense to augment reconnaissance and security efforts. Fast-moving operations may preclude the continuous, effective use of GSRs.

In a movement to contact or a zone reconnaissance, employ GSRs on high-speed avenues of approach to the front or flanks of the troop to provide information on enemy movement. A GSR team can provide long-range observation while the scouts conduct a detailed reconnaissance of the local area. Employ GSR teams on the flanks of the troop's movement in pairs so they can bound forward. In this manner, the GSR can provide continuous early warning and keep pace with the operation.

Defensive Operations. In defensive operations, employ GSR teams to augment scouts. Position them to provide long-range observation on expected avenues of approach or to maintain surveillance of flanks.

Position the GSR teams behind the screen line if the terrain affords them long-range observation. Use GSR to assist scouts in maintaining contact with the enemy as the scouts deploy to subsequent screen lines. This also reduces their vulnerability to enemy fires.

Section VI. Air Defense

SUPPORT RELATIONSHIP

An air defense artillery (ADA) battery is organic to the regiment. For combat operations, these assets are normally attached or direct support to the squadron. The squadron commander may provide a slice of ADA to the troop.

Air defense assets are scarce, so the troop cannot plan on dedicated air defense protection. It is the troop's responsibility to protect itself from enemy air attack. The commander must take measures to avoid enemy air attack, to limit the damage if attacked and, if necessary, to fight back.

AIR DEFENSE EMPLOYMENT

Include air defense planning in all tactical operations. Air defense includes passive and active air defense and the employment of ADA assets.

Passive Air Defense

Passive air defense measures are the troop's first line of defense against enemy air attack. They include all measures, other than active defense, taken to minimize the effects of hostile air action. There are two types of passive air defense—attack avoidance and damage limiting measures.

Attack Avoidance. If the enemy pilot cannot find the troop, he cannot attack it. Use concealment, camouflage, obscuration, and deception to hide from the enemy.

Select positions that provide good concealment. When adequate concealment is not available, camouflage the vehicles to blend with their surroundings. Cover all track marks leading into the troop's position. Cover all shiny objects that could reflect light and attract attention. Take all actions necessary to reduce the enemy's ability to find the troop.

Damage Limiting Measures. Damage limiting measures are taken to reduce the effects of enemy air attack. Disperse vehicles while occupying static positions such as assembly areas or when preparing to cross a water obstacle or a breached obstacle. Dispersion reduces the effects of munitions. While on the move, air guards must be alert for enemy air attacks. When an enemy air attack is identified, quickly disperse, go to a concealed position, if possible, and then stop moving. A stationary vehicle is more difficult to see than a moving vehicle, and dispersion will reduce the effect of the air attack.

Use natural or man-made cover to reduce the effects of enemy air attack. Use folds in the earth, depressions, buildings, and sandbagged positions as damage limiting cover. These methods to reduce the effects of air attack must be used scrupulously by troop trains and command posts to reduce their vulnerability on the battlefield.

Active Air Defense

Although passive measures are the first line of defense against air attack, the troop must be prepared to fight back if necessary. The troop gains two things by fighting back. It may kill or drive off the attacker, and the morale and spirit of the troops improve by fighting back.

The decision to fight an air threat is based on the situation and the capabilities of the troop's weapon systems. Generally the troop may defend itself against direct attack, but unless attacked, it does not engage aircraft unless directed to do so by the squadron commander.

Stinger/Avenger Employment

Stingers/Avengers are used most effectively in point defense. However, mobility during tactical operations requires that Stinger crews move with the troop during maneuvering to provide the best possible air defense.

The troop commander must determine the air defense priorities for each phase of the operation and brief the Stinger/Avenger section chief accordingly. In order for the section chief to design an effective troop air defense plan for the commander, the commander must furnish the section chief the mission, objective, routes of march, intended scheme of maneuver, battle formations, and intended response to hostile air activity.

Unit air defense falls under two major categories: air defense of a fixed asset (assembly area or defensive position), and air defense of a mobile asset (convoy or maneuvering troop).

Fixed Asset. The air defense section leader places his crews from 2 to 4 kilometers apart near the troop's position, which provides overlapping and mutually supporting fires for troop defense. He provides command and control to his section, and air defense early warning to the troop unit and his deployed crews. The commander must provide logistical support for the section, indirect protective fires for their defense, and security within the troop defensive perimeter during hours of darkness.

Mobile Asset. While the unit moves, either in a tactical road march or during movement to contact, the air defense crews must move with each element to provide uninterrupted air defense protection.

Convoy operations require the air defense section leader to weight his crews near the front and rear of the troop's column(s), and to spread the remainder of his crews evenly throughout the column(s). While traveling, he provides command and control to his crews and air defense early warning to the troop and his crews.

Section VII. Army Aviation Support

SUPPORT RELATIONSHIP

Army aviation units are seldom OPCON to the troop; however, the troop frequently conducts joint operations with air cavalry or attack helicopter units within the troop area of operations. Air and ground units work together to make an effective team in conducting cavalry operations.

AVIATION TROOP COMMANDER

Air operations are controlled and coordinated by one man, the aviation troop commander. He coordinates directly with the ground troop commander, on the command net, when air operations extend into the troop's area of operations. When time is available, coordinate face-to-face to prepare for an operation. Let him know—

- Situation.
  - Enemy. Ground and ADA units by type and location.
  - Friendly. Locations and FLOT to include supporting artillery.
- Mission.
- Execution.
  - Scheme of maneuver.
  - Fire support.

The aviation troop commander tells the ground troop commander—

- His capabilities.
  - Number and type of aircraft.
  - Armament.
  - Time available.

- Execution. Scheme of maneuver to include approach direction into area.
- Command and signal.
  - Aviation troop commander, frequency, and call sign.
  - Succession of command and designated platoon leader in the troop commander's absence.

The aviation troop commander reports all spot reports and fire requests through the troop command post or FIST while operating in the ground troop's area of operations.

Reconnaissance. The capabilities of air and ground scouts complement each other. Air scouts can quickly reconnoiter a large area while ground scouts are conducting a detailed terrain reconnaissance. Employ the air scouts forward to provide information about areas where the ground scouts need to concentrate their efforts, such as at possible fording sites or route restrictions. The ground scouts can focus their efforts where the air scouts have directed them, instead of conducting a complete and time-consuming reconnaissance. The air scouts can also provide early warning of enemy movement to the front or flanks of the troop.

Security. During screen missions, employ the air assets to the front or flanks of the troop to observe areas between OPs and those areas that are difficult to observe from the ground. The air scouts can place effective indirect fires on the enemy and in maintaining contact with the enemy when the ground scouts move to subsequent screen lines. In offensive and defensive operations, Army aviation will normally work for the squadron to complement the squadron scheme of maneuver.

Section VIII. Close Air Support

CAS sorties are usually allocated to the regiment by the corps commander to meet preplanned requests submitted by the regimental fire support element. CAS for the troop is obtained by submitting an immediate CAS request.

CAS is requested by submitting an immediate request to the squadron air liaison officer. When CAS is on station, the tactical air control party may talk directly to the troop commander to determine enemy strengths, descriptions, and locations, as well as friendly unit locations. A forward air controller (FAC) normally directs the air strike. In the absence of a FAC, the troop FSO is trained to direct air strikes.

The ground troop may be required to mark the target for CAS aircraft. Use the mortar section for this. Fire white phosphorus or illumination rounds set for a ground burst. The FIST can designate targets for aircraft equipped with laser-spot trackers, such as Pave Penny or laser-guided bombs.

CAS capabilities are listed below.

- High speed and long range.
- Versatile weapon and ammunition mix.
- Accurate delivery.
- Excellent air-ground communications in A-10 and A-16 aircraft.
- Can locate and strike moving targets.

CAS limitations include—

- Limited resources.
- Delivery restrictions caused by limited visibility and weather or the proximity of friendly forces.
- Flight restrictions imposed by enemy air defense.
- Delayed response and short time on station.

Chapter 9

Combat Service Support

CONTENTS

Mission accomplishment depends on the troop's ability to obtain, transport, and distribute fighting resources such as fuel, ammunition, replacement personnel, and rations. Dependable ways to treat and evacuate wounded soldiers are crucial to good morale. Evacuation and repair of combat equipment sustain troop combat power and readiness to fight.

Section I. Organization

Troop trains are organized as combat trains and field trains. The combat trains are made up of immediate battlefield service support. The rest of the troop CSS elements are collocated with squadron CSS assets in the squadron field trains and at the unit maintenance collection point (UMCP). The troop CSS team consists of the troop XO, first sergeant, supply sergeant, communications sergeant, maintenance sergeant, and senior troop aidman.

There is no dedicated service support radio net at troop level. All logistical reports and initial requests for logistical support are conducted on the troop command net. Additional coordination is conducted on the platoon radio nets. Routine reports are sent before and after combat operations, and are delivered by messenger when possible. All service support coordination with squadron is conducted on the squadron administration and logistics (A/L) net. The first sergeant operates on this net. The command post does not routinely operate on the A/L net, but uses it when necessary to forward reports and to conduct logistical coordination.

COMBAT TRAINS

The troop combat trains provide CSS for the troop during combat operations. They are organized for combat as shown in Figures 9-1 and 9-2.

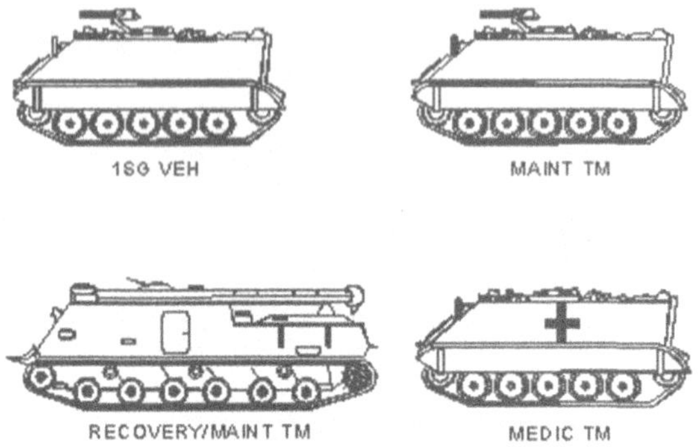

Figure 9-1. Heavy troop combat trains.

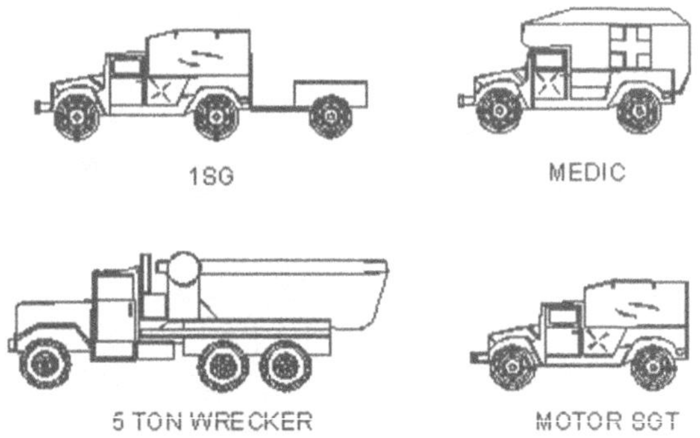

Figure 9-2. Light troop combat trains.

The first sergeant, assisted by the maintenance sergeant, directs movement and employment of the combat trains. He receives logistical reports directly from the platoon sergeants or through the troop command post, coordinates logistical matters with the XO, and assists the XO in ensuring the troop is logistically ready for battle. During operations, the first sergeant directs aidman teams and recovery teams forward as needed, leads the troop LOGPAC forward from the logistics release point (LRP) to resupply the troop and then leads it back to the LRP, and organizes the LOGPAC to conduct troop resupply operations. He also directs the medical evacuation of soldiers who are wounded in action. Additionally, he directs the evacuation of soldiers killed in action, enemy prisoners of war, and disabled vehicles.

The troop combat trains can be consolidated and moved under the control of the first sergeant, or they can be split and moved. If the combat trains are split, the motor sergeant and first sergeant will divide the medics, mechanics, special tools, and high-use PLL items into equal portions and move parallel and behind troop combat elements to provide responsive medical evacuation and battle damage assessment across the troop zone. Recovery assets will continue to move center of zone to facilitate recovery. This technique may be used if the troop zone is wide and split operations are necessary to provide proper support.

Position the combat trains where they can support the troop, but will not be exposed to enemy direct fires. This location is usually from 2 to 5 kilometers to the rear.

Offensive and Reconnaissance Operations.During such operations as a movement to contact or a zone reconnaissance, position the trains in the center of the troop zone and about 2 kilometers, or one terrain feature, behind the tank platoons. Use the established road network, if possible, to move the trains. Bound the trains forward to successive concealed positions, based on the movement of the troop, to keep them in position to support the troop.

Defensive and Security Operations.During defensive and security operations, position the combat trains about 3 kilometers behind the tank platoon positions in the center of the troop sector. If possible, hide the trains in a small built-up area, which provides cover and concealment and a road network to facilitate the movement of the trains. The trains must be ready to move to support the troop or displace to a new location. The XO and first sergeant plan subsequent locations, and as the situation requires, the first sergeant or maintenance sergeant moves the trains to the next location.

Security.Trains must provide their own local security. They do not have the manpower or equipment to clear large areas, so they should move into areas that have already been cleared by scouts. The trains' best defense is to avoid detection. The trains should disperse their vehicles and camouflage the positions by using natural concealment like a wooded area or a small built-up area. A built-up area is best, because there are a limited number of approaches and the thermal signature of the trains is concealed. Enforce strict noise and light discipline. Hide vehicles in barns or garages, if available. The buildings in a built-up area and the trees in a wooded area provide some protection against the effects of indirect fires and air-delivered munitions. Dispersion also limits the damaging effects of these weapons.

Establish OPs around the trains to provide early warning of enemy movement toward the position. Put them where they can cover major avenues of approach. An OP has at least two soldiers, a crew-served weapon (if available), a map, binoculars, a night observation device, and a radio or field phone. The OP must immediately report, by radio or phone, enemy ground and air attacks so the trains can take appropriate actions.

Passive defense is the trains' best defense against air attack. The steps taken to locate them properly, disperse vehicles, camouflage positions, and use the cover and concealment of built-up or wooded areas help prevent the enemy from detecting and attacking, and limit the effects of enemy munitions if attacked.

Air attacks cannot always be avoided. If attacked, take active air defense measures. The first sergeant or senior soldier present must coordinate the small arms fires of the trains for them to be effective against an air threat. Refer to Chapter 8, Section VI, Air Defense, for a complete discussion of passive and active air defense.

FIELD TRAINS

Troop CSS assets not in the combat trains are collocated with the squadron field trains or the UMCP, and sustain the troop's fighting capability by moving rations, ammunition, fuel, repair parts, and replacements from the squadron rear area to the troop on the battlefield. Troop field trains are controlled primarily by the supply sergeant, and consist of the personnel and equipment shown in Figures 9-3 and 9-4. Troop maintenance assets in the field trains may be consolidated under the control of the squadron maintenance officer.

The troop supply sergeant is the troop representative in the squadron field trains. The supply sergeant is responsible for the following tasks:

- Maintaining records of troop clothing and equipment.
- Processing or forwarding requests for all Class I, II, III, IV, V, VI, and VII. He also assists the maintenance section in ordering Class IX repair parts and the medics in resupplying Class VIII (medical).
- Assembling and leading all vehicles in the troop LOGPAC from the field trains to the LRP. If the squadron is conducting resupply operations, the supply sergeant may move under the control and direction of one of the following squadron personnel: squadron support platoon leader or platoon sergeant, S4, HHT commander, or XO.
- Assisting the first sergeant in organizing the LOGPAC site for troop resupply and in evacuating soldiers who are killed in action, enemy prisoners of war, and disabled vehicles; and in medically evacuating wounded and injured soldiers.
- Delivering mail, when available, via the troop LOGPAC.

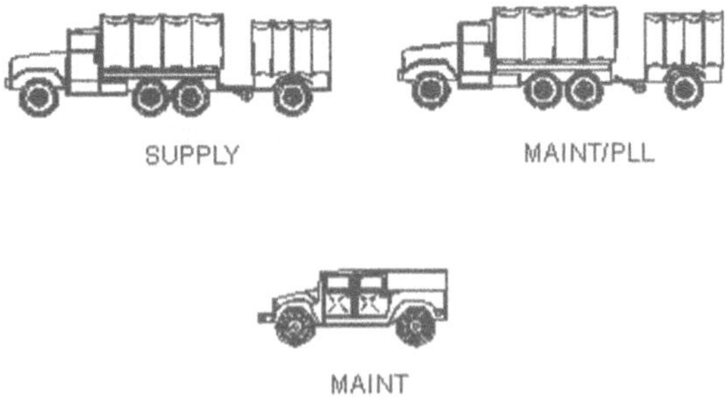

Figure 9-3. Heavy troop field trains.

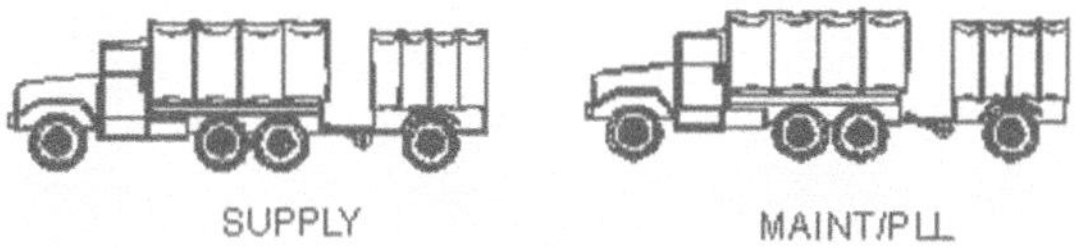

Figure 9-4. Light troop field trains.

At the squadron field trains, the supply sergeant coordinates with the S4 and support platoon leader to ensure the LOGPAC meets the troop's CSS requirements.

The supply sergeant will coordinate with the troop PLL clerk for parts and requisitions and with the squadron personnel and administration center (PAC) section for mail, promotions, awards, Standardization Installation/Division Personnel System (SIDPERS), and other personnel actions. He will coordinate with squadron maintenance and communication for repaired vehicles and equipment and with the squadron food service sergeant for rations. The supply sergeant must frequently check with the HHT command post in the field trains for any additional troop requests reported on the A/L net.

The squadron field trains are responsible for their own local security. The supply sergeant will be tasked to assist the HHT commander in maintaining security of the field trains. He must execute and supervise security operations in accordance with the HHT commander's plan. During the movement through nonsecure areas, the supply sergeant should ensure that all crew-served and individual weapons are manned and ready. He should also ensure that all members of the troop field trains are briefed on immediate action drills for enemy contact or vehicle breakdowns.

If available, the troop should provide the supply sergeant with a radio to enhance communications between the supply sergeant and the first sergeant.

SQUADRON COMBAT SERVICE SUPPORT

In addition to organic troop CSS, one or two medical/evacuation teams are attached to the troop for combat operations from the squadron medical platoon. During resupply operations, the troop normally receives two fuel and two ammunition carriers from the transportation section of the support platoon. These carriers are not usually attached, but are provided to the troop as needed or as part of a standardized LOGPAC. The LOGPAC is discussed in Section II, Logistics.

To ensure responsive CSS for the squadron, the S4 and squadron maintenance officer organize the squadron support into echeloned trains and establish a main supply route (MSR), a UMCP, and an LRP. These trains are made up of combat trains (which provide immediate recovery, maintenance, medical, and emergency resupply support), and field trains (which provide the remainder of the squadron service support and limited direct-support maintenance). The UMCP is established to provide maintenance support for combat operations. It is positioned near or collocated with the squadron combat trains, and is the closest point to which damaged or failed equipment and systems are recovered. The UMCP is the focal point of the squadron maintenance effort. The MSR links troop combat trains to the squadron combat trains or UMCP and the field trains. LRPs are established along the MSR to facilitate the handover of LOGPACs and equipment between troops and squadron. An example of a squadron support plan is shown in Figure 9-5.

Section II. Logistics

SUPPLY OPERATIONS

The supply sergeant is responsible for getting supplies and delivering them to the troop. He delivers small items, but is largely dependent on support platoon assets to deliver bulky or high expenditure items. The commander establishes priorities for delivery, but the demands of combat normally dictate Classes I, III, V, VIII, and IX as most critical.

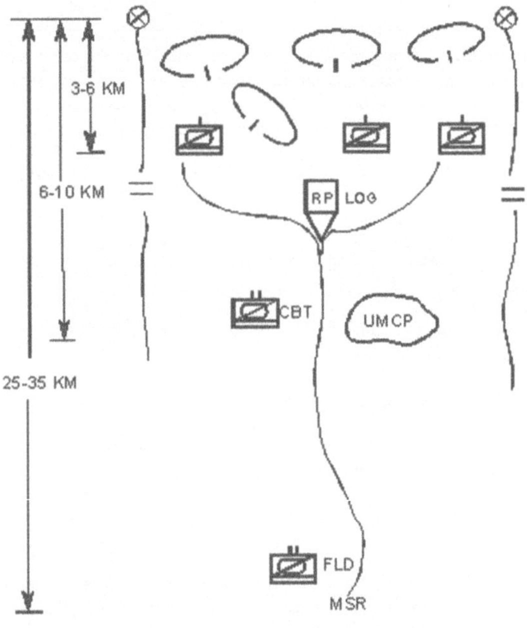

Figure 9-5. Squadron combat service support plan.

Class I (Rations).Meals ready to eat (MRE), C rations, are stocked on board each troop vehicle in a basic load prescribed by SOP (three-day supply). Class I is delivered daily by the supply sergeant as part of the

LOGPAC from the squadron field trains. Hot meals (B or tray-pack rations) are served when possible, but the normal ration cycle is C-C-C during combat operations. This requires resupply of at least 34 cases of MRE per day per troop. Water is a critical item, and must be resupplied daily. Soldiers may require up to 5 gallons of water per soldier per day, or more than 600 gallons for the entire troop. The troop can transport only 400 gallons of water on the water trailer. Soldiers should top off their water cans when possible. The supply sergeant should carry 5-gallon cans of water on his truck to supplement the needs of the troop. When he resupplies the troop, he exchanges the full cans for empty ones.

Class II (Supplies and Equipment).Class II items are requisitioned through the S4 and delivered by the supply sergeant as needed. The supply sergeant also maintains the following supplies, as a minimum, in his troop supply:

- One set of NBC overgarments and filter per soldier.
- One case of BA-30s.
- One mile of WD-1 wire.
- Additional batteries as necessary for night-vision devices and other equipment.
- Five sets of TA-50, to include personal clothing of various sizes, stored in duffel bags.
- Fifteen body bags.
- Map cases, grease pencils, permanent markers, etc.

Class III (POL).Class III is delivered by support platoon assets both as part of the troop LOGPAC and on an as-needed basis. Platoon sergeants report their Class III status to the command post daily, when status is requested, or when any Class III item is below a certain percentage, usually 59 percent.

Each troop vehicle carries a small basic load of packaged products as prescribed by SOP, and the fuel carriers normally have only small amounts of packaged products on board. Requests for large quantities of packaged products must be requested through the S4.

Class IV (Construction Materials). Class IV supplies are requested through the CTCP (combat trains command post) and delivered by the supply sergeant with the LOGPAC. Class IV supplies needed at troop level may include concertina wire, sandbags, and lumber.

Class V (Ammunition).The troop deploys with a basic load of Class V as prescribed by the unit for each type of vehicle. Ammunition status is reported by platoon sergeants to the command post daily, upon completion of enemy contact, or as needed. Normal resupply of Class V is delivered by the support platoon vehicles with the LOGPAC. Emergency resupply is requested through the first sergeant to the S4, and is delivered from the combat trains if it is available. Emergency resupply of Class V will normally consist of ammunition for major weapon systems, such as tank main guns, TOW missiles, and 25-mm guns in the heavy troop; and TOW missiles and caliber .50, 40-mm grenades, 7.62-mm, and individual weapons ammunition in the light troop.

Class VI (Personal Demand Items).Class VI items (such as soap, toothpaste, and cigarettes) are requested through the S4 by the first sergeant. They are usually picked up by the supply sergeant in the field trains and delivered as part of the LOGPAC.

Class VII (Major End Items).Class VII items, such as vehicles and night-vision devices, are automatically requested based on equipment shortages and battle losses. Items are delivered to the S4, who notifies the troop of the availability of the equipment. The equipment is delivered with a LOGPAC or brought forward immediately to an LRP where the first sergeant meets it and delivers it to the troop.

Class VIII (Medical Supplies).Class VIII is provided by the squadron medical platoon. The troop senior medic requests supplies through the squadron aid station. During intense combat, supplies may be pushed forward. They are then delivered to the combat trains where the troop medic picks them up from the squadron aid station, or are brought forward on ambulances.

Class IX (Repair Parts).The troop PLL clerk, who is in the field trains or the UMCP, works under the control of the squadron maintenance officer and the squadron maintenance technician. The PLL clerk requisitions, receives, and maintains records for all Class IX items. All troop PLLs are consolidated under the squadron maintenance officer during combat operations, and one or more PLL trucks remain in the UMCP at all times to provide a supply of Class IX for forward area support. When the troop maintenance sergeant needs a repair part, the request is sent through the squadron maintenance officer. The squadron maintenance officer determines if the part is available through the PLL, and fills the request if possible. If the part is not in the PLL, he then requests it through the squadron maintenance team in the field trains. They have the troop PLL clerk requisition the part for immediate use or replenish the part used from the PLL. The troop PLL clerk submits the requisition through squadron maintenance to the supporting maintenance unit. The part is delivered through squadron maintenance to the troop when the requisition is filled. The PLL clerk updates the records to reflect the requisition.

Maps.Maps are requested through the troop command post to the S4. As maps are available, the supply sergeant picks them up in the field trains and delivers them to the troop as part of a LOGPAC.

RESUPPLY OPERATIONS

Resupply of combat resources is accomplished using standardized procedures to rearm, refuel, and refit the troop as fast as possible to sustain its combat potential. There are two types of resupply operations—routine and supplementary. Methods of resupply are tailgate issue and service station.

Routine Resupply

Routine resupply operations include daily resupply of Classes I, III, V, and IX, and mail and other items needed by the troop. Routine resupply takes place when the troop is not in heavy contact, and may be conducted in an assembly area or behind troop positions when the troop is deployed in sector or zone. Class III is required more than once a day when the vehicles are in continuous operation.

Routine resupply is conducted using the LOGPAC from the field trains and the troop combat trains. The LOGPAC is organized in the field trains. Its composition is based on the troop's needs as reported to the S4, on requisitions, and on the availability of supplies. The troop supply sergeant reports to the S1 and S4 sections, squadron mess section, and squadron maintenance section to ensure all available supplies are picked up and loaded on trucks for troop resupply. The S4, support platoon leader, and troop supply sergeants assemble the LOGPACs for each troop by adding the troop supply trucks to the troop slice of Class III and V trucks from the support platoon. The support platoon leader leads the LOGPACs to the LRP, where the first sergeants meet them. Each first sergeant leads his LOGPAC to the troop resupply site. When it arrives, members of the troop combat trains guide the LOGPAC vehicles into position. Once the LOGPAC is established, the first sergeant reports to the command post that he is set. The troop commander or the XO coordinates resupply operations, and ensures all platoons and sections are resupplied. Resupply is conducted by one of two methods—tailgate issue or service station.

Tailgate Issue Method.The tailgate method is used in static positions such as assembly areas. Class III and V supply vehicles and other bulk-issue vehicles move from vehicle to vehicle to conduct resupply. The rest of the service support vehicles are centrally located in the troop area. Little or no movement is required by the combat vehicles. Personnel move to a centralized location to receive supplies, Class I, and mail. This method provides 360-degree security throughout the resupply operation; however, it is very time-consuming and requires an adequate road network for the wheeled supply vehicles to reach each vehicle. An example of this method is shown in Figure 9-6.

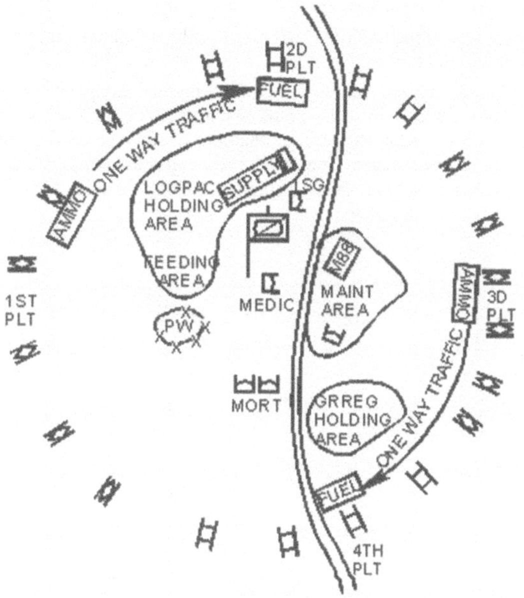

Figure 9-6. Tailgate issue method in a troop assembly area.

The following takes place during tailgate resupply:

- Combat vehicles remain in place. POL and ammunition trucks travel in a clockwise direction around the assembly area to each vehicle position, in turn, to conduct resupply.
- Crewmen rotate through the feeding area and pick up supplies, water, and mail.
- The first sergeant and platoon sergeants arrange for pickup of those killed in action (KIA) and their personal effects. The KIA are brought to a holding area near the medical aid station, but kept out of view.
- Ambulances pick up, treat, and evacuate seriously wounded soldiers. Other wounded soldiers are

carried to the ambulance or walk to it for emergency medical treatment.

- Prisoners are kept together and guarded. As soon as possible, they are moved to the squadron trains on a returning supply vehicle.
- The troop armorer, the radio repairman, and the organizational mechanics repair known problems and spot-check other vehicles.
- Vehicles needing maintenance are brought to the maintenance area.
- The first sergeant and platoon sergeants closely monitor the resupply operation.
- Empty LOGPAC vehicles are moved to a holding area, where they are loaded with KIA, prisoners of war (PW) and inoperative equipment.
- The supply sergeant moves the LOGPAC back to an LRP to link up with the support platoon leader and return to the field trains.

Service Station Method.Service station resupply is used during tactical operations. It is most effective when the troop is positioned in a zone or sector no more than 3 to 5 kilometers wide, such as is found in a defend in sector or defend from a troop BP mission. Platoons or sections are resupplied at the LOGPAC while the rest of the troop stays in position. The first sergeant sites the LOGPAC as shown in Figure 9-7. LOGPAC security is provided by soldiers from the combat trains who are not involved in the resupply, and by platoon vehicles that have completed or are awaiting resupply.

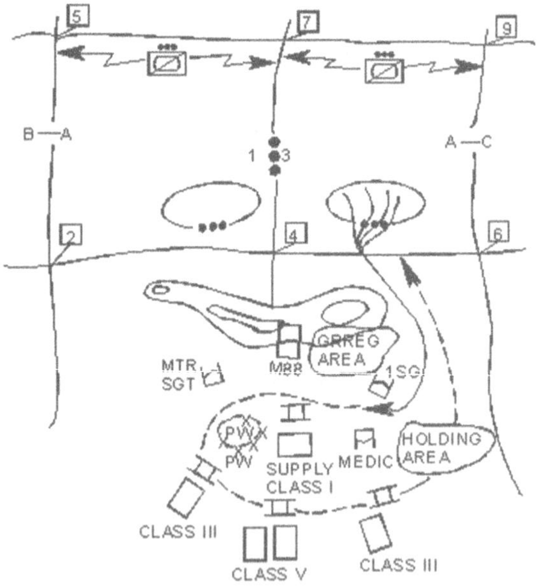

Figure 9-7. Service station method.

The following takes place during service station resupply:

- Vehicles of the first platoon or section enter the LOGPAC at a designated location and follow one-way traffic flow.
- The vehicle carrying the KIA in body bags moves to the holding area, out of view of the troop, and leaves the remains and personal effects.
- Soldiers requiring medical attention are delivered to medics and treated or prepared for evacuation.
- The platoon sergeant supervises the operation and coordinates face-to-face with the first sergeant for any special requirements.
- The platoon leader dismounts his vehicle and uses the first sergeant's vehicle and radio to contact the troop commander for orders and situation reports.
- Crews requiring unit-level maintenance remain in the maintenance holding area.
- Vehicles rotate through stations.
- Unit-level mechanics, the armorer, and the radio repairman repair known problems and spot-check other vehicles.
- Crews rotate to the supply truck to pick up mail, supplies, and Class I.
- The platoon leader and platoon sergeant conduct precombat inspection.
- When the platoon or section has completed resupply, it moves to its designated position.
- The rest of the platoons rotate individually through the LOGPAC for resupply.

At the end of the resupply operation, the troop LOGPAC returns to the LRP where it links up with the support platoon leader and returns to the field trains. In the field trains, the supply sergeant returns the fuel and ammunition carriers to the support platoon, returns ration-serving equipment to the mess team, delivers the KIA to the graves registration (GRREG) collection point, requests any additional supplies from the squadron supply sergeant, and returns to his position in the field trains.

Supplementary Resupply

In the heavy troop, daily routine resupply will not sustain the combat power of the M1/M3 fleet when operating without a break for 8 to 10 hours. The tanks will consume so much fuel that they will have to supplement routine resupply of Class III one or two times daily. The troop may also need to resupply Class V. Supplementary resupply can be conducted in a couple of ways depending on the mission and if the troop is in contact with the enemy. Regardless of the technique used, the commander must push the supplies forward to his platoons; the tempo of combat operations cannot be disrupted in order to resupply. Conduct emergency resupply when in heavy contact with the enemy, and prestock resupply when not. Once the supplies are brought forward, either the tailgate issue or service station methods are used.

Supplementary resupply must be planned when continuous operations are expected. Resupply must be requested through the S4 several hours in advance to ensure the service support assets are prepared to support the operation. The troop must quickly resupply once fuel trucks are available, and then return them to the field trains. There are few fuel carriers and drivers available in the squadron and these valuable assets cannot be wasted.

Platoon Prestock.When operating over wide frontages, such as during screen, zone reconnaissance, or movement to contact operations, break the supplies down into platoon packages and push the supplies forward to each platoon position. Position the vehicles behind the platoon and resupply individual vehicles or sections at a time using the service station method. If the situation allows, the resupply vehicles can move through the platoon position and resupply the platoon in place using the tailgate issue method.

Controlling platoon prestock is difficult because resupply vehicles are moving to several locations rather than remaining under the centralized control of the first sergeant. The platoon leader must also provide security for their own prestock (see Figure 9-8).

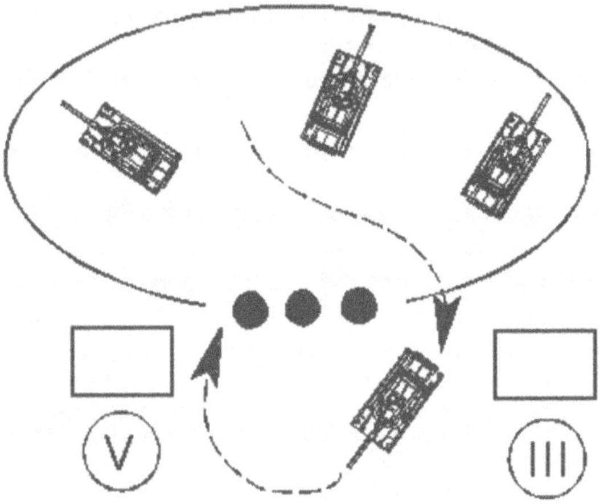

Figure 9-8. Platoon prestock.

Troop Prestock.Troop prestock is used to supplement routine resupply of Classes III and V. Troop prestock is used when the troop is operating in a narrow zone or sector. The resupplies are pushed forward to the troop position using either the service station or tailgate method. Each platoon rotates through the resupply site, using the service station method, while the rest of the troop remains in position; or each vehicle is resupplied in position using the tailgate method (see Figure 9-9).

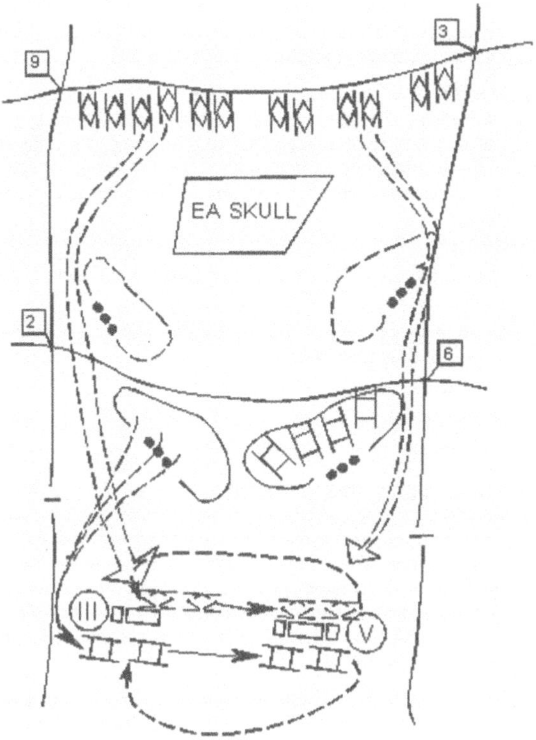

Figure 9-9. Troop prestock.

Variations of these methods can be used to meet the troop's situation. The troop must, however, rehearse these techniques to ensure it can quickly and efficiently conduct resupply operations and continue the mission.

Emergency Resupply

Emergency resupply normally involves only fuel and ammunition and is conducted while in contact with the enemy. The resupply begins at section and platoon level by redistributing ammunition between vehicles to cross-level loads. The platoon sergeant reports his need for emergency resupply to the first sergeant, who relays the request to the S4. The squadron combat trains maintain a small load of Class III and V for these situations. The S4 or support platoon leader coordinates a linkup between the squadron combat trains and the troop first sergeant. The first sergeant meets the resupply trucks and moves back to the troop area. The first sergeant and troop XO choose a resupply point that is just behind the troop position and masked by terrain from enemy direct fire and observation. If fuel is needed, the fuel truck is moved to the resupply point and vehicles or sections go there to refuel.

Resupply Site Selection.The LOGPAC site must be carefully chosen to provide responsive support for the troop, to support the movement of wheeled resupply vehicles, and to limit exposure to enemy fires. The XO selects the general area to be used by the LOGPAC. He knows about the current tactical situation, and can determine what the troop will be doing in the next couple of hours. The first sergeant selects the exact LOGPAC site based on information from the XO and map and ground reconnaissance. He selects a site that provides—

- Cover and concealment.
- Proximity to the platoon positions, from 3 to 8 kilometers behind the FLOT and center of the sector.
- A road or trail network that supports the wheeled resupply vehicles and the heaviest troop vehicles and allows one-way traffic flow to the LOGPAC.
- Enough room to disperse the vehicles.
- Reduction of thermal signature.
- Level enough to allow refueling.

MAINTENANCE AND RECOVERY OPERATIONS

The first sergeant supervises the troop maintenance section in the troop combat trains and the maintenance sergeant runs the section. The maintenance sergeant operates on the troop command net and stands by to provide responsive unit-level maintenance and recovery support for combat vehicles and radio equipment. When needed, squadron maintenance augments troop assets. The troop armorer is responsible for unit-level maintenance and repair of the troop's small arms weapons. Position him in the combat trains where he can provide battlefield support.

Maintenance Operations.Maintenance responsibilities begin at crew level. Operator-level PMCS must be performed before, during, and after operations. During operations, the crew begins the maintenance process when it identifies the fault. Once a problem develops, the vehicle commander determines support requirements, to include self-recovery, assistance from another vehicle, or assistance from troop or squadron maintenance, and takes the following actions:

- Reports the situation to the platoon sergeant, providing condition(s), location, and circumstances.
- Attempts to self-recover, if mired.
- Uses another vehicle to move to a secure location if exposed to enemy observation and fire. Uses smoke to screen the area between the mired vehicle and the enemy.

When repairs are beyond the capability of the crew, the platoon sergeant uses the troop command net to notify the first sergeant of the situation and to request assistance. The crew must maintain radio contact (if the radio is operational) on the platoon net and maintain local security. The maintenance sergeant monitors the radio traffic. He switches his radio to the platoon frequency to coordinate maintenance support and dispatches the appropriate assets to the "down" vehicle. The maintenance team checks whether the problem can be corrected in place within the time criteria (30 minutes in defensive situations and 2 hours in offensive situations). They move the vehicle to a more secure location, if necessary, and fix the vehicle. If repairs will take too long, the maintenance sergeant uses the squadron A/L net to notify the squadron maintenance officer who will coordinate for a handover of the equipment from troop to squadron at the LRP or along the MSR. The squadron maintenance officer needs the following information:

- Identification of the troop.
- Equipment involved.
- Location of equipment.
- Nature of damage.
- Pertinent information about the tactical situation, and any other necessary information.

Based on the information above, the squadron maintenance officer will designate a time and place for equipment handover.

The troop does not have the personnel, equipment, or time to create a holding area in the troop combat trains. If repairs will take more than the allowable time, evacuate the equipment to squadron.

Recovery Operations.When a vehicle cannot be repaired within the allowable time or is damaged beyond repair, the maintenance section must recover it and pull it to the MSR where it is turned over to squadron. Squadron maintenance will pull it to the UMCP or the field trains.

If a vehicle is catastrophic, or not able to be recovered based on the enemy situation, the commander may authorize the destruction of the vehicle. Before using thermite grenades to destroy a vehicle, remove as many valuable items as time permits, such as—

- Classified equipment and documents.
- Communications equipment.
- Weapons.
- Ammunition.

FIELD SERVICES

GRREG is performed by the field services platoon from corps, located in the regimental support area. The initial collection, identification, safeguarding of personal effects, and evacuation of the dead is the troop's responsibility.

When remains are discovered, be careful to preserve all items that may be used for identification. If metal identification tags (dog tags) are on the remains, do not remove them. Secure all personal effects in a bag or poncho and tie it to the remains. Place each casualty in a body bag, poncho, or shelter half and evacuate with the first available means of transportation, such as LOGPAC vehicles or disabled vehicles, to the squadron field trains. The supply sergeant receives all remains, and is responsible for turning them over to the GRREG collection point.

If the tactical and logistical situation makes evacuation impossible, emergency on-site burial is performed. On-site burial requires the permission of the squadron commander. If an on-site burial is performed, do the following:

- Complete two copies of DD Forms 551 and 1077 for each body. These forms are available through the squadron PAC. The first sergeant and supply sergeant should carry them.

- If there are several remains, engineer support may be needed to dig trenches.
- Dig a trench 6-1/2 feet wide and 3-1/2 feet deep. The length is determined by the number of remains.
- Remove one identification tag from each body, and string the tags on a wire in the order in which the remains are buried.
- Place the remains in the grave shoulder to shoulder.
- Bury all personal effects with the remains.
- Mark the ends of each row with a stake. Tag each stake to identify the marker as a grave. Indicate the length of the grave.
- Prepare an overlay that shows the location of the grave site.
- Cover the grave with earth removed from the trench.
- Forward the completed forms, overlay, and identification tags to squadron.

If the remains are contaminated, the grave site must be clearly marked and separated from noncontaminated grave sites. This must also be indicated on the grave-site overlay.

BATH AND LAUNDRY SERVICES

Bath and laundry services are provided by supply and service units from the corps support command. When available, these services are coordinated through the S4.

Section III. Personnel Service Support

PERSONNEL AND ADMINISTRATIVE

All troop administrative and personnel actions are handled by the squadron PAC in the field trains. The troop chain of command is responsible for ensuring that soldiers receive passes, leaves, promotions, awards, mail, legal assistance, financial services, and other personnel and welfare services on a fair and prompt basis. The first sergeant interfaces with PAC daily through the supply sergeant, at the administrative/logistics operations center with the S1/S4, or at the LRP.

Casualties and missing personnel are reported on DA Forms 1155 and 1156. These forms are initiated by the individual who witnessed the incident and are forwarded to the first sergeant. This action is the basis for notification of the next of kin and awarding benefits such as Serviceman's Group Life Insurance (SGLI), so accuracy is a must.

The commander should continuously manage the troop personnel situation to ensure trained personnel are manning key positions, crew shortages are filled with available personnel, and replacement personnel are trained and incorporated into crews. During combat, platoon leaders and platoon sergeants must balance their crews after casualties are evacuated. They must fill key positions on vehicles with the most qualified soldier. If necessary the commander may direct cross-leveling of personnel within the troop.

The troop submits daily personnel reports to the squadron as prescribed by SOP. The personnel reports are important because they allow the squadron commander to make tactical decisions for the employment of the troop. The reports also allow the S1, with the squadron commander's guidance, to properly distribute personnel replacements. Replacement personnel are received by the supply sergeant in the field trains. The supply sergeant in-processes the soldier into the troop, ensures he has the proper equipment, then delivers him to the first sergeant during the troop LOGPAC.

combat health support

First Aid.First aid is the responsibility of all soldiers; they use first aid, self-aid, buddy-aid, and combat lifesavers. All must be trained to take action after a soldier is wounded to keep him breathing, stop the bleeding, prevent shock, and dress the wound until medical personnel are available to treat the soldier. Training must include treatment of NBC casualties and crew evacuation drills to get soldiers out of a vehicle without creating further injury. Selected soldiers from each platoon will be trained in advanced medical skills in the combat lifesaver program.

Treatment.The first sergeant should position troop medical aid and evacuation teams on the battlefield where they can be most responsive. They will usually operate under the control and direction of the first sergeant in the troop combat trains. If the troop zone is pushing 10 kilometers wide, the troop may operate split combat trains with the first sergeant controlling medical evacuation on one side of the zone and the motor sergeant controlling the other half. The medics must know the locations of and routes to each platoon, the troop combat trains, the squadron combat trains, and each collection point.

Combat Stress Control.The psychological effects of combat on soldiers influence their ability to execute their missions. Individual self-control and self-discipline in the face of danger are maintained through unit discipline and firm leadership. The commander and his subordinate leaders are key to the mental toughness of soldiers. Commanders should be visible to soldiers and share their hardships with them. They should talk to their soldiers individually and as a troop to keep them informed. The commander's tone of voice on the radio must indicate that he has the situation under control. He should transmit the successful accomplishments of the troop and tell the soldiers they are doing a good job. Keep details about any casualties in the troop off the radio and ensure that soldiers who are killed in action are covered up and kept out of view of the troop. The chaplain can be a great asset in maintaining troop morale. Do not neglect the mental fitness of soldiers, because uncontrolled fear is contagious and can quickly lead to chaos.

Evacuation.To receive medical assistance, a platoon leader or platoon sergeant calls the first sergeant on the troop command net. The medic team monitors the call and goes to the platoon frequency to make any other necessary coordination. The medic team moves to the reported location to treat the soldier(s). The medics transport as many wounded soldiers as possible in their armored ambulance. Do not transport the dead with the wounded; evacuate them separately. If the medics determine the soldiers are seriously wounded, they either transport them directly to the squadron combat trains, or coordinate to hand them over to squadron at a collection point. Soldiers with slight wounds should not be evacuated unless it is necessary. They should be treated and returned to duty. Secure the seriously wounded soldiers on their litters to prevent any further injury during the evacuation.

As a general rule, do not use combat vehicles to evacuate the wounded; this reduces combat strength. Use combat vehicles only if absolutely necessary, and no other transportation is available.

Weapons and military equipment (except NBC protective equipment) of personnel being evacuated are secured with the platoon sergeant, first sergeant, or supply sergeant. Ensure the soldier keeps his protective mask and overgarments. When the situation permits, all equipment that was secured by the first sergeant or platoon sergeant is turned over to the supply sergeant.

disease and nonbattle injuries

More soldiers are lost in combat to illness and disease than to combat-related wounds. Maintaining the health and fighting fitness of the troop is a leadership responsibility. Personal hygiene, field sanitation, and rest must be incorporated into all troop operations. If these are ignored, the combat effectiveness of the troop will decrease rapidly.

UNIT MINISTRY TEAM

The unit ministry team (UMT), consisting of an appropriate number of chaplains and chaplain assistants, is assigned to provide unit, area, and denominational coverage to all troops in the cavalry squadron. UMT provides the following eight subfunctions of worship opportunities:

- Administration of sacraments, rites, and ordinances.
- Pastoral care and counseling.
- Development and management of the UMT.
- Management of material resources.
- Advising the commander on matters of religion, morale, and morals as affected by religion.
- Ministry in support of soldiers suffering from battle fatigue.
- Development of programs that enhance the total well-being of the soldier.
- Development of activities to enhance unit cohesion.

PERSONAL HYGIENE

The personal hygiene of the troop's soldiers is a leader's responsibility. Rules of hygiene must be observed to ward off disease and improve troop morale. Ensure soldiers wash and change their socks and underwear daily, if possible. All soldiers should shower as time and resources are available. The senior aidman plays a big part in personal hygiene by educating soldiers and by checking for signs of trench foot or immersion foot and frostbite.

Disease can quickly spread through the troop and incapacitate its soldiers. Watch for symptoms and ensure the soldiers' immunizations are current.

REST

Rest is extremely important for both commanders and their soldiers. When the troop is at REDCON 3 or 4, troop leadership should ensure a sleep plan is planned and executed.

FIELD SANITATION

Field sanitation is vital to the prevention of the spread of disease. The medics must assist in this effort by checking troop water supplies to ensure they are potable, and by ensuring mess utensils and equipment are properly cleaned. In static situations, such as when in assembly areas, soldiers will use slit trenches or latrines; at other times soldiers will use cat holes. Slit trenches and cat holes must be covered up after use to prevent the spread of disease.

Section IV. Prisoners of War

CAPTURE AND HANDLING

Maps, military documents, letters, and diaries obtained on the battlefield, as well as PWs, are valuable sources of combat intelligence. Scout platoons usually obtain information while searching the battlefield in the conduct of reconnaissance operations or after contact with the enemy. They should rapidly report this information to the troop commander for evacuation instructions. The scouts may be ordered to take the documents to the troop command post or to rendezvous with the first sergeant at a specific location to turn over the documents. Proper handling and evacuation of captured documents and equipment are important. The materials must be immediately turned over to trained intelligence personnel. In most cases, captured documents lose their value over time.

EVACUATION TECHNIQUES

In most situations, returning supply vehicles, aircraft, or troop headquarters vehicles can be used to evacuate PWs from the troop combat trains to the squadron collection point. The troop is responsible for guarding PWs until they are turned over to the S1. Wounded PWs are treated through normal medical channels, but are kept separate from US soldiers.

When support vehicles are not readily available and the troop must continue its operation, secure the PWs in a holding area such as a basement or compound. Notify the first sergeant and the S1 of the PW location and continue the mission. The first sergeant will pick up the PWs and complete their evacuation to the collection point.

Appendix A

Fratricide Prevention

Fratricide is as old as warfare itself, a complex problem that defies simple solutions. Fratricide can be broadly defined as the employment of friendly weapons and munitions, with the intent to kill the enemy or destroy his equipment or facilities, that results in unforeseen and unintentional death or injury to friendly personnel. This appendix focuses on actions leaders can take with current resources to reduce the risk of fratricide.

CONTENTS

Section I. Magnitude of the Problem

The modern battlefield is more lethal than any in history. The tempo of operations is rapid, and the nonlinear nature of the battlefield creates command and control challenges for all unit leaders.

The accuracy and lethality of modern weapons make it possible to engage and destroy targets at these extended acquisition ranges. At the same time, however, the ability of US forces to acquire targets using thermal imagery and other sophisticated sighting systems exceeds our ability to accurately identify these targets as friend or foe. As a result, friendly elements can be engaged unintentionally and destroyed in a matter of seconds.

Added to this is the problem of battlefield obscuration, which becomes a critical consideration whenever thermal sights are the primary source of target identification. Rain, dust, fog, smoke, and snow degrade identification capability by reducing the intensity and clarity of thermal images.

On the battlefield, positive visual identification cannot be the sole engagement criteria at ranges beyond 1,000 meters. Situational awareness is the key; it must be maintained throughout an operation.

Section II. Risk Identification and Preventive Measures

Reduction of fratricide risk begins during the planning phase of an operation and continues throughout preparation and execution. Risk identification must be conducted at all levels during each phase, and the results clearly communicated up and down the chain of command so risk assessment can begin. This section covers considerations that influence risk identification; it also focuses on measures the platoon leader can implement both to make the identification process more effective and to help prevent friendly fire incidents from occurring. Section III of this appendix covers the risk assessment process; Section IV lists additional fratricide reduction measures and guidelines.

Planning Phase

A plan that is thoroughly developed and understood helps to minimize fratricide risk. The following considerations help indicate the potential for fratricide in a given operation:

- The clarity of the enemy situation.
- The clarity of the friendly situation.
- The clarity of the commander's intent.
- The complexity of the operation.
- The planning time available at each level.

Graphics are a basic tool that commanders at all levels use to clarify their intent, add precision to their concept, and communicate their plan to subordinates. As such, graphics can be a very useful tool in reducing the risk of fratricide. Commanders at all levels must understand the definitions and purpose of operational graphics and the techniques of their employment.

Note. See FM 101-5-1 for the definitions of each type of graphic control measure.

Preparation Phase

The following factors may cause fratricide risks to become evident during rehearsals:

- Number and type of rehearsals.
- Training and proficiency levels of units and individuals.
- The habitual relationships between units conducting the operation.
- The physical readiness (endurance) of the troops conducting the operation.

Backbriefs and rehearsals are primary tools in identifying and reducing fratricide risk during the preparation phase. The following are some considerations for their use:

- Backbriefs ensure subordinates understand the commander's intent. They often highlight areas of confusion, complexity, or planning errors.
- The type of rehearsal conducted determines what types of risks are identified.
- Rehearsals should extend to all levels of command and involve all key players.

Execution Phase

During execution, in-stride risk assessment and reaction are necessary to overcome unforeseen fratricide risk situations. The following are factors to consider when assessing fratricide risks:

- Intervisibility between adjacent units.
- Amount of battlefield obscuration.
- Ability or inability to positively identify targets.
- Similarities and differences in equipment, vehicles, and uniforms among friendly and enemy forces.
- Vehicle density on the battlefield.
- The tempo of the battle.

Maintaining situational awareness at all levels and at all times is another key to fratricide reduction as an operation progresses. Units must develop and employ effective techniques and SOPs to aid leaders and crewmen in this process. These techniques include—

- Monitoring on the next higher net.
- Radio cross-talk between units.
- Accurate position reporting and navigation.
- Training and use/exchange of liaison officers (LO).

Section III. Risk Assessment

Risk assessment must be conducted whenever fratricide risk factors are identified. It must take place at all levels during each phase of operations. As with risk identification, the results of the assessment must be passed on to all levels of the chain of command so that fratricide reduction measures can be developed and implemented. Refer to Section IV for specific reduction measures.

Figure A-1 is a worksheet for evaluating fratricide risk in the context of mission requirements. The worksheet lists six mission-accomplishment factors that affect the risk of fratricide, along with related considerations for each factor. Assess the potential risk in each area as low, medium, or high, and assign a point value to each (one point for low risk, two for medium risk, three for high risk). Add the point values for the overall fratricide assessment score. Use the resulting score only as a guide, however. The final assessment must be based both on observable risk factors like those on the worksheet and on your "feel" for the intangible factors affecting the operation. Note that descriptive terms are listed only in the low- and high-risk columns of the worksheet. The assessment of each factor will determine whether the risk matches one of these extremes or lies somewhere between them as a medium risk.

FACTORS	LOW (1)	MEDIUM (2)	HIGH (3)
1. UNDERSTAND PLAN			
● Commander's Intent	Clear		Foggy
● Complexity	Simple		Complex
● Enemy Situation	Known		Unknown
● Friendly Situation	Clear		Unclear
● ROE	Clear		Unclear

2. ENVIRONMENT ● Intervisibility ● Obscuration ● Battle tempo ● Positive target ID	Favorable Clear Slow 100%		Unfavorable Obscured Fast 0%
3. CONTROL MEASURES ● Command relationships ● Audio ● Visual ● Graphic ● SOPs ● LOs ● Location/Navigation	Organic Loud/Clear Well Seen Standard Standard Proficient Sure		Joint/Combined Jammed Obscured Not understood Not used Untrained Unsure
4. EQUIPMENT (Compared to US) ● Friendly ● Enemy	Similar Different		Different Similar
5. TRAINING ● Individual proficiency ● Unit proficiency ● Rehearsal ● Habitual relationship ● Endurance	MOS Qual Trained Multiple Yes Alert		Untrained Untrained None No Fatigued
6. PLANNING TIME (1/3 - 2/3 Rule) ● Higher HQ ● Own HQ ● Lower HQ	Adequate Adequate Adequate		Inadequate Inadequate Inadequate
OVERALL FRATRICIDE ASSESSMENT	LOW 26-46%*	MEDIUM 42-62%*	HIGH 58-78%*
● Commander may use numbers as the situation dictates. ● Numbers alone may not give accurate fratricide risk.			

Figure A-1. Fratricide risk assessment worksheet.

Section IV. Fratricide Reduction Measures

The following measures are provided as a guide to actions that can reduce fratricide risk. They are not directive in nature, nor are they intended to restrict initiative. Apply them as appropriate based on the specific situation and METT-T factors.

- Identify and assess potential fratricide risks in the estimate of the situation. Express these risks in the OPORD or FRAGO.
- Maintain situational awareness, focusing on such areas as current intelligence; unit locations and dispositions; denial areas (minefields/FASCAM); contaminated areas, such as ICM and NBC; SITREPs; and METT-T factors.
- Ensure positive target identification. Review vehicle/weapon ID cards. Know at what ranges and under what conditions positive identification of friendly vehicles/weapons is possible.
- Establish a command climate that stresses fratricide prevention. Enforce fratricide prevention measures, emphasize the use of doctrinally sound tactics, techniques, and procedures. Ensure constant supervision in the execution of orders and the performance of all tasks and missions to standard.
- Recognize the signs of battlefield stress. Maintain unit cohesion by taking quick, effective action to alleviate it.
- Conduct individual, leader, and collective (unit) training covering fratricide awareness, target identification and recognition, and fire discipline.
- Develop a simple, decisive plan.
- Give complete and concise mission orders.
- Use SOPs that are consistent with doctrine to simplify mission orders. Periodically review and change SOPs as needed.
- Strive for maximum planning time for you and your subordinates.
- Use common language/vocabulary and doctrinally correct standard terminology and control measures, such as fire support coordination line, zone of engagement, and restrictive fire lines.
- Ensure thorough coordination is conducted.
- Plan for and establish effective communications.
- Plan for collocation of command posts whenever it is appropriate to the mission, such as during a passage of lines.
- Designate and employ LOs as appropriate.
- Ensure rules of engagement are clear.
- Include fratricide risk as a key factor in terrain analysis (OCOKA).
- Conduct rehearsals whenever the situation allows time to do so.
- Be in the right place at the right time. Use position location/navigation devices (GPS and POSNAV); know your location and the locations of adjacent units (left, right, leading, and follow-on); and synchronize tactical movement.
- Include discussion of fratricide incidents in after-action reports.

Section V. Fratricide Risk Considerations (OPORD Format)

This section, which parallels the five-paragraph OPORD, contains key factors and considerations in fratricide reduction. This is not a change to the OPORD format; rather, it should be used during OPORD development to ensure fratricide reduction measures are included in the order. It is not a strict guide. The factors and considerations are listed where they would likely appear in the OPORD, but they may warrant evaluation during preparation of other paragraphs.

1. Situation.

a. Enemy forces.

(1) Are there similarities among enemy and friendly equipment and uniforms that could lead to fratricide?
(2) What languages do enemy forces speak? Could these contribute to fratricide risk?

(3) What are the enemy's deception capabilities and his past record of deception activities?

(4) Do you know the locations of enemy forces?

b. Friendly forces.

(1) Among the allied forces, are there differences (or similarities with enemy forces) in language, uniform, and equipment that could increase fratricide risk during combined operations?
(2) Could differences in equipment and uniforms among US armed forces increase fratricide risk during joint operations?

(3) What differences in equipment and uniforms can be stressed to help prevent fratricide?

(4) What is the friendly deception plan?

(5) What are the locations of your unit and adjacent units (left, right, leading, follow-on)?
(6) What are the locations of neutrals and noncombatants?

c. Own forces.

(1) What is the status of training activities? What are the levels of individual, crew, and unit proficiency?

(2) Will fatigue be a factor for friendly forces during the operation? Has an effective sleep plan been developed?

(3) Are friendly forces acclimatized to the area of operations?

(4) What is the age (new, old, or mix) and condition of equipment in friendly units? What is the status of new equipment training?

(5) What are the expected MOPP requirements for the operation?

d. Attachments and detachments.

(1) Do attached elements understand pertinent information regarding enemy and friendly forces?
(2) Are detached elements supplied this pertinent information by their gaining units?

e. Weather.

(1) What are the expected visibility conditions (light data and precipitation) for the operation?
(2) What effect will heat and cold have on soldiers, weapons, and equipment?

f. Terrain.

(1) Do you know the topography and vegetation (such as urban, mountainous, hilly, rolling, flat, desert, swamp/marsh, prairie/steppe, jungle, dense forest, open woods) of the expected area of operations?
(2) Have you evaluated the terrain using the factors of OCOKA?

2. Mission. Is the mission, as well as all associated tasks and purposes, clearly understood?

3. Execution.

a. Task organization.

(1) Has the unit worked under this task organization before?
(2) Are SOPs compatible with the task organization (especially with attached units)?

(3) Are special markings or signals (for example, cats' eyes, chemlites, or panels) needed for positive identification of uniforms and equipment?

(4) What special weapons and/or equipment will be used? Do they look or sound like enemy weapons and/or equipment?

b. Concept of the operation.

(1) Maneuver. Are main and supporting efforts identified to ensure awareness of fratricide risks and prevention measures?

(2) Fires (direct and indirect).

(a) Are priorities of fires identified?

(b) Have target lists been developed?
(c) Has the fire execution matrix/overlay been developed?

(d) Have locations of denial areas (minefields, FASCAM) and contaminated areas (ICM, NBC) been identified?

(e) Are the locations of all supporting fire targets identified in the OPORD/OPLAN overlays?

(f) Are aviation and CAS targets clearly identified? Have signals been established to positively identify these targets for the aircraft? Have airspace coordination areas been developed? Have enemy air defense systems been suppressed?

(g) Has the direct-fire plan been developed and synchronized with the fire support plan?

(h) Have final protective fires been designated?

(i) Have you identified and verified sector limits?

(j) Have executors for each target been assigned and do they understand when and where to shoot? Do the shooters have "eyes on" the target?

(k) Are the observers surveyed in or are they using a map spot? Target location errors can cause big problems.

(l) Do all leaders and executors understand where the fire support coordination measures are and when they go into effect? Rehearsal is the key.

(m) Can the fire support officer hear what targets are being called on the maneuver nets?

(n) Have all targets been rehearsed with the executors and the field artillery battalion?
(o) Does the reinforcing or general support reinforcing field artillery have all the proper graphics and understand where they fit in? Did they attend the rehearsal?

(p) Have restrictions on specific munitions been established and does everyone know where they are planned and emplaced?

(3) Engineer tasks.

(a) Are friendly minefields, including FASCAM and ICM dud-contaminated areas, known?

(b) Are obstacles identified, along with the approximate time needed for reduction/breaching of each?

(4) Tasks to each subordinate unit. Are friendly forces identified, as appropriate, for each subordinate maneuver element?

(5) Tasks to CS/CSS units. Have locations of friendly forces been reported to CS/CSS units?

(6) Coordinating instructions.

(a) Will rehearsals be conducted? Are they necessary? Are direct and indirect fires included?

(b) Is a backbrief necessary?

(c) Are appropriate control measures clearly explained and illustrated in the OPORD and overlays? Have they been disseminated to everyone who has a need to know? What is the plan for using these control measures to synchronize the battle and prevent fratricide?

(d) Have target/vehicle identification drills been practiced?

(e) Do subordinate units know the immediate action, drill, or signal for "CEASE FIRE" or "I AM FRIENDLY" if they come under unknown or friendly fire? Is there a backup action?

(f) Is guidance in handling dud munitions, such as ICM and CBUs, included?

4. Service Support.

a. Are train locations and identification markings known by everyone?

b. Do medical and maintenance personnel know the routes between train units?

5. Command and Signal.

a. Command.

(1) What are the locations of the commander and key staff members?
(2) What is the chain of command and the succession of command?

b. Signal.

(1) Do instructions include backup code words and visual signals for all special and emergency events?

(2) Do instructions cover how to identify friendly forces to aircraft?

(3) Are SOI distributed to all units with a need to know, such as higher, lower, adjacent, leading, and follow-on elements?

Appendix B

Standing Operating Procedures

The intent of this appendix is to provide an outline as an aid for developing a squadron tactical SOP.

I. GENERAL

A. Purpose: This tactical SOP prescribes standard procedures for use during all combat situations and provides a comprehensive reference for conducting operations in a field environment.

B. Conformity: All assigned, attached and OPCON personnel will read and comply with the provisions of this tactical SOP.

II. BATTLE COMMAND

A. Command.

1. Organization.

a. Succession of command.

b. Cues for assuming command.

c. Operation of the command post.

(1) Shifts.

(2) Displacement/set-up/tear-down.

2. Troop-leading procedures.

a. Estimate input (checklist).

b. Precombat inspection (checklist).

c. Backbriefs.

d. Rehearsals.

e. Combat orders.

(1) Formats.

(2) Preparation.

(3) Reproduction.

(4) Dissemination.

3. Coordination with adjacent units (checklist).

4. Liaison with main body elements (checklist).

B. Control.

1. Combat graphics and symbols.

2. Control measures designation (numbering system).

3. Terrain index reference system (TIRS).

4. Operational terms.

5. Vehicle identification marking system.

6. Unit recognition signals.

7. Communication.

a. Net diagrams.

b. Fixed call signs.

c. Brevity codes/cue words.

d. Antijamming actions.

e. Alternate means.

8. Reports.

a. Battle.

b. Intelligence.

c. Logistical.

III. MANEUVER

A. Readiness condition (REDCON).

B. Quartering party.

C. Road marches.

D. Assembly areas.

E. Squadron organization for combat.

F. Formations.

G. Battle plays.

1. Actions on contact.

2. Fix and bypass.

3. Close assault.

4. In-stride breach.

5. Bridges/defile.

6. Formation changes.

7. Passage of lines.

8. Hasty defense.

9. Consolidate on the objective.

10. Ambush.

11. Other plays.

H. Reconnaissance operations.

1. Zone reconnaissance.

a. Graphics.

b. Critical tasks.

2. Area reconnaissance.

a. Graphics.

b. Critical tasks.

3. Route reconnaissance.

a. Graphics.

b. Critical tasks.

I. Security operations.

1. Screen.

a. Graphics.

b. Critical tasks.

2. Area security.

a. Route.

b. Convoy.

J. Offensive operations.

1. Movement to contact.

2. Hasty attack.

3. Offensive graphics.

K. Defensive operations (defensive graphics).

L. Limited visibility operations (checklist).

M. Break in action (checklist).

1. Redistribution of ammunition, personnel, and equipment.

2. Evacuation of casualties and PWs.

3. Redistribution of ammunition under fire.

N. Relief in place.

1. Relief in place graphics.

2. Critical tasks.

O. Air-ground team.

1. Command relationship.

2. Formations/techniques of movement.

IV. MOBILITY AND SURVIVABILITY

A. Mobility.

1. Standard tasks.

2. Standard priorities.

B. Countermobility.

1. Standard tasks.

2. Standard priorities.

3. Engineer target turnover (checklist).

C. Survivability.

1. Standard tasks.

2. Standard priorities.

3. Fighting position construction.

D. NBC defense.

1. NBC team organization and equipment.

2. Unit NBC equipment.

3. Defense against nuclear attack.

4. Defense against chemical/biological attack.

5. Decontamination.

6. NBC reconnaissance.

V. FIRE SUPPORT

A. Fire support request sequence.

B. Fire support planning and execution matrix format.

C. Graphics.

VI. AIR DEFENSE

A. Air defense warning and cue words.

B. Weapons control status and cue words.

VII. INTELLIGENCE/ELECTRONIC WARFARE

A. Intelligence.

1. Standard tasks.

2. Standard priorities.

B. Electronic warfare.

1. Standard tasks.

2. Standard priorities.

3. Countermeasures.

VIII. LOGISTICS

A. Resupply procedures.

1. LOGPAC procedures.

2. Battle loss actions.

B. Combat Health Support.

1. Emergency medical treatment.

2. Medical evacuation.

3. Field hygiene and sanitation.

4. Combat stress control.

5. Routine sick call.

C. Maintenance support.

1. Battle damage assessment and repair.

2. Exchange criteria.

3. Cannibalization criteria.

4. Destruction criteria.

5. Maintenance repair time guidelines.

D. Personnel.

1. Replacements.

2. Accountability.

3. Personnel actions.

E. Combat trains.

1. Layout.

2. Operations.

IX. SAFETY

Appendix C

Operations Order

Operations order format

A sample OPORD is outlined below.

1. Situation.

a. Enemy.

(1) Weather (note effects - include light data).

(a) Light data.
(b) Weather forecast for operation.
(c) Effects of weather and light data on operations.

- Trafficability.
- Visibility.
- Effect on lasers/thermals.
- Effect on air operations.

(2) Terrain.

(a) Obstacles, hills, valleys, road types and conditions, streams, rivers, bridges, towns.
(b) Avenues of approach.

- Size unit supported.
- Start and end point.
- Objective.

(c) Key terrain.

(d) Observation.
(e) Cover and concealment.
(f) Engagement areas.
(g) Effect of terrain on the operation.

(3) Enemy forces.

(a) Identification of enemy forces.

(b) Activity of enemy forces.
(c) Location of enemy units.
(d) Disposition of enemy forces.
(e) Strength of enemy forces.
(f) Composition of enemy forces, to include type of equipment.
(g) Other enemy information critical to the upcoming operation, to include the following:

- Chemical and nuclear capabilities.
- Air defense artillery (ADA).
- Aviation, including helicopters.
- Electronic warfare.

(h) Enemy courses of action.
(i) Probable enemy course of action.

b. Friendly.

(1) Mission of higher headquarters (squadron) and commander's intent.
(2) Mission of adjacent units (left, right, front, rear).
(3) Mission of other organic units in higher headquarters.
(4) Mission of reserves in higher headquarters.
(5) Mission of supporting units who are in direct support/reinforcing (DS/R) to higher headquarters (field artillery, engineer, ADA).
(6) Which element from higher headquarters has priority of fires.
(7) Reinforcing units with a reinforcing/general support (R/GS) role to supporting units.
(8) Close air support (CAS) and number of sorties allocated to higher headquarters.

c. Attachments and detachments to the troop.

2. Mission. The *who, what, when, where,* and *why* for troop/company. State the essential task or tasks to be accomplished by the entire unit, to include on-order missions. Clearly define the troop's objective.

3. Execution.

a. Concept of the Operation. This paragraph explains the commander's intent by stating the *purpose*, *method*, and *endstate* of the operation. The *purpose* tells the "why" of the operation. The *method* tells how the commander visualizes achieving success with respect to the troop as a whole and the utilization of any combat multipliers, in general terms. The *endstate* tells what the final disposition of forces will be and how the endstate will facilitate future operations. For example--

We will conduct the reconnaissance to determine the location, disposition, and composition of enemy main defensive belt in order to facilitate movement through zone, and penetration of the main defensive belt by 52d ID (Mech), allowing 10th Corps to continue offensive operations to the north. We will conduct a forward passage of lines, and perform zone reconnaissance in a troop vee with the engineer platoon conducting a detailed route reconnaissance of Route Cherry behind the scout platoons. Initially movement will be rapid and aggressive until we make contact with the enemy main defensive belt, then transition to a more deliberate, detailed reconnaissance. Indirect fires will be used to provide immediate suppression and assist in the development of the situation by scout platoons. Success is determining the disposition, composition, and location of the enemy main defensive belt then determining the axis and point for the 52d ID (Mech) penetration of the enemy's defense. Our endstate is the troop screening along the enemy main defensive belt ready to pass elements of the 52d ID (Mech) forward for their attack.

(1) Scheme of Maneuver. How major units (maneuver and combat support) will be employed; including movement techniques, bypass criteria (if any), and engagement/disengagement criteria. See the following example.

C Troop moves at 130430 from TAA Viper along Route Dog in troop column. Order of march on Route Dog is Red, Blue, Mortars, FIST, Engineers, Black 6, White, Green, TOC, and combat trains.

C Troop crosses LD at 0500 with Red on left (south), Blue on right (north), Mortars, Engineers, White, Green, TOC, and combat trains moving center of zone (troop vee).

Scout platoons reconnoiter in zone abreast, establishing contact at designated troop internal and flank contact points. Upon enemy contact, develop the situation at platoon level, initially with indirect fire, and report. Destroy squad- and section-sized light-armored and soft-skinned vehicles, within capability. Anything beyond capability, be prepared to assist hasty attack by White and/or Green. Bypass only enemy dismount elements squad-sized or smaller. Bypass all built-up areas larger than 1 kilometer square not on Route Cherry. Upon contact with enemy main defensive belt, conduct dismounted patrols to ascertain enemy disposition, composition, orientation, and location of obstacles.

Mortars cross LD and establish MFP 10 following Blue. Maneuver from MFP 11 to MFP 12 through MFP 25 in sequence. Maintain ability to range at least one-third maximum range forward of lead scouts. Key movement on scouts reporting crossing of phase lines throughout zone.

Engineers cross LD following the FIST and conduct a detailed route reconnaissance of Route Cherry from SP to RP. Maneuver on Route Cherry approximately 500 meters rearward of scouts. Upon identification of enemy's main defensive belt, be prepared to assist scouts with obstacle reconnaissance. White and Green cross LD in platoon columns, White leading. Deploy into platoon wedges forward of LD, terrain permitting. Maneuver in center of troop zone, Green following White by 500 meters. White and Green maneuver approximately one phase line behind Red and Blue through hide positions T-44 through T-52 in sequence. Key movement on scouts reporting crossing of phase lines throughout zone. Scouts will assist in positioning tanks for conduct of hasty attacks by fire or by fire and maneuver upon approval of the troop commander.
TOC and combat trains cross LD in column, following Green. TOC maneuvers center troop zone, establishing positions H-44 through H-48 in sequence. Remains approximately two phase lines rearward of lead scouts. Trains move center troop zone in column, establishing positions T-27 through T-32 in sequence. Remain approximately two phase lines rearward of lead scouts. Key movement on scouts reporting crossing of phase lines throughout zone.

(2) Fires.

(a) Purpose for field artillery and mortar fires. (How will indirect fires support the scheme of maneuver?)

(b) Preparation starting time, duration and description of any fires landing in the area of operations.
(c) Allocation of final protective fires (FPF).
(d) Which element will have priority of fires.
(e) Number of priority targets allocated and who will control them.
(f) Special fires, restrictions, allocation/use of smoke, illumination, or CAS.

(g) Description of scheduled fires (offensive).
(h) Reference to fire support or target annexes.
(3) Obstacles, mines, fortifications.

(a) Priority of engineer effort.

(b) Priority of engineer mission.
(c) Obstacle overlay.
(d) Obstacle list.
(e) Logistical constraints.
(f) On-order missions.
b. Specific tasks to subordinate units. List specific missions in "battle sequence" for each subordinate unit, including attached units. Include movement techniques, flank coordination requirements, other details, and any "be-prepared" missions.

c. Coordinating instructions.
(1) Movement instructions.

(2) Time schedule of events.

- Rehearsal.
- Backbrief times.
- Precombat inspection time.
- First movement time.

(3) Passage of lines.

- Contact points.
- Passage points.
- Lanes and identification.

(4) Mission-oriented protection posture (MOPP) level.

(5) Operational exposure guide (OEG).
(6) Actions on contact.
(7) Actions at danger areas.
(8) Priority intelligence requirements (PIR) and intelligence requirements (IR).
(9) Effective time of attachment or detachment.
(10) Rally points.
(11) Rules of engagement (ROE).
(12) Priority targets for direct-fire weapons.
(13) Uniform and equipment, to include weapon and ammunition.
(14) Any changes regarding battlecarry and battlesight ranges.
(15) Air defense posture and weapons control status.
(16) Any instructions not provided in concept of the operation or specific instructions.

4. Service Support.
a. General.

(1) Organization of trains (squadron and troop).

- Combat.
- Field.

(2) Location of trains (initial and subsequent).

(3) Movement of trains (movement instructions).

b. Materiel and services.
(1) Supply.

- Class I: Time and type.
- Class III: Time, location, and method.
- Class V: Time, amount, and type.
- LOGPAC instructions (include squadron logistics release point [LRP]).

(2) Transportation. Location of main supply route.

(3) Services.

- Mortuary affairs.
- Clothing exchange and bath.
- Maintenance.

c. Medical evacuation and treatment.
(1) Location of squadron aid station.

(2) Displacement of squadron aid station.
(3) Location of regimental clearing station.
(4) Aero-medical evacuation information.
(5) Location of ambulance exchange points.
(6) Handling of contaminated wounded personnel.

d. Personnel.
(1) Prisoner of war (PW) handling and disposition instructions.

(2) PW guard instructions.
(3) Location of PW collection point.
(4) Instructions for interaction with local civil populace (ROE).
(5) Number of expected replacements.
(6) Cross-leveling procedures.
e. Miscellaneous.

5. Command and Signal.

a. Command.

(1) Location of troop commander/troop TOC/squadron command post.
(2) Succession of command.

b. Signal.

(1) SOI index and edition in effect.

- Key frequencies.
- Key call signs.
- Current item number identifier.

(2) KY-57 fill and change over.

(3) Current sign and countersign.
(4) Code words.
(5) Action if jamming or "hot mike" occurs.
(6) Pyrotechnics use.
(7) Periods and conditions of listening silence.

Appendix D

Operations Other Than War

Operations other than war are military activities during peacetime and conflict that do not necessarily involve armed clashes between two organized forces. A cavalry troop will conduct reconnaissance, security, offensive, defensive, and other tactical operations in support of operations other than war. Some examples of operations other than war that a cavalry troop may be involved in are as follows:

PEACETIME CONFLICT
Counterdrug Strikes and raids

Disaster relief Peace enforcement

Civil support Support to insurgency

Peace-building Anti-terrorism

Nation assistance Peacekeeping

Noncombatant evacuation

operations (NEO)

The primary operational constraints during operations other than war are defined by the rules of engagement (ROE). These rules range from the use of force for self-defense only to total impartiality when applying force. The rules of engagement for peacetime will be more restrictive than those for conflict. The troop commander should consider the following in relation to the rules of engagement:

- The degree of force used must only be sufficient to achieve the task at hand and prevent, as far as possible, loss of human life or serious injury.
- Leaders at all levels must ensure that the ROE do not limit the soldiers' ability to defend themselves.
- Leaders must never tape over magazines to keep soldiers from chambering rounds.
- Leaders must ensure all soldiers understand what conditions must be met before chambering rounds.
- The ROE must be realistic, simple, and easy to understand. Develop a single card that clearly outlines the ROE and issue it to all soldiers for reference, keeping in mind that the card in itself is not the answer. Soldiers must know and understand the ROE.
- During peacekeeping, forces have no mandate to prevent violations of peace agreements by the active use of force. (Observe and report only.) To maintain the peace, units may be positioned

between belligerents. Troop commanders must realize that soldiers are being placed at risk. The protection of the troop must be emphasized.

- Peace enforcement missions allow the active use of force. The peacetime ROE resemble the ROE for hostilities (wartime).
- The formulation of ROE should consider the cultural differences of multinational forces.
- Soldiers should be trained in the ROE, using tactical vignettes or simulated events. Train soldiers to avoid unnecessary collateral damage to property.

Actions on contact and the troop's ability to develop the situation will be affected by the rules of engagement for the theater in which the troop is operating. The troop commander should rehearse battle drills based on those rules.

In preparing for operations other than war, a troop may require specialized training. They could benefit from training in the following areas:

- Nature of peacekeeping.
- Regional orientation/culture of belligerents.
- Negotiating skills.
- Mine/booby trap/unexploded ordnance training.
- Checkpoint operations.
- Investigating and reporting.
- Information collection.
- Patrolling.
- Media inter-relationships.
- Establishing a buffer zone.
- Supervising a truce or cease-fire.
- Contributing to maintenance of law and order.
- Demilitarizing cities or geographical areas.
- Border surveillance/security.

Glossary

AAR	after-action review
A&O	assault and obstacle
ACE	armored combat earthmover
ACR	armored cavalry regiment
ACT	air cavalry troop
ADA	air defense artillery
A/L	administrative/logistics
ALO	air liaison officer
ALOC	administrative/logistics operations center
AO	area of operations
APERS	antipersonnel (ammunition)
AT	antitank
APC	armored personnel carrier
arty	artillery
AVLB	armored vehicle launched bridge
BHL	battle handover line
BP	battle position
BUA	built-up area
C2	command and control
cal	caliber
CAS	close air support
cbt	combat
CBU	cluster bomb unit
CEOI	communications-electronics operation instructions
CEV	combat engineer vehicle
CDR	commander
CFV	cavalry fighting vehicle
cGy/hr	centigray(s) per hour
CHEMWARN	chemical warning (message sent in the event of a friendly strike)
COA	course of action
COMSEC	communications security
COT	counter obstacle team
CP	command post
CS	combat support
CSS	combat service support

CTCP	combat trains command post
DA	Department of the Army
DS	direct support
DS/R	direct support/reinforcing
EA	engagement area
engr	engineer
EPW	enemy prisoner of war
evac	evacuate; evacuation
EW	electronic warfare
1SG	first sergeant
FA	field artillery
FASCAM	family of scatterable mines
FAC	forward air controller
FEBA	forward edge of the battle area
FIST	fire support team
FLOT	forward line of own troops
FM	frequency modulation (radio); field manual
FPF	final protective fires
FRAGO	fragmentary order
FS	fire support
FSCL	fire support coordination line
FSE	fire support element
FSO	fire support officer
FSN	fire support net
GLD	ground laser designator
GPS	global positioning system
GRREG	graves registration
GSR	ground surveillance radar
HC	heavy concentration (smoke)
HE	high explosive
HEMTT	heavy expanded mobile tactical truck
HHT	headquarters and headquarters troop
HMMWV	high mobility multipurpose wheeled vehicle
HQ	headquarters
hvy	heavy
ICM	improved conventional munitions

ID	identification
IPB	intelligence preparation of the battlefield
IR	intelligence requirements
KIA	killed in action
km	kilometers
LACR	light armored cavalry regiment
LC	line of contact
LD	line of departure
ldr	leader
LID	light infantry division
LO	liaison officer
LOA	limit of advance
LOGPAC	logistics package
LRP	logistics release point
lt	light
maint	maintenance
MBA	main battle area
MCOO	modified combined obstacle overlay
mech	mechanized
METT-T	mission, enemy, terrain, troops and time available
MFP	mortar firing point
MICLIC	mine clearing line charge
MOPP	mission-oriented protection posture
mort	mortar
MRB	motorized rifle battalion
MRC	motorized rifle company
MRE	meals ready to eat
MRP	motorized rifle platoon
MSR	main supply route
MTOE	modified table(s) of organization and equipment

NAI	named area(s) of interest
NBC	nuclear, biological, chemical
NCO	noncommissioned officer
NCOIC	noncommissioned officer in charge
NEO	noncombatant evacuation operations
NET	new equipment training
NPL	no penetration line
NUCWARN	nuclear warning (message sent in the event of a friendly strike)
obj	objective
OCOKA	observation and fields of fire, cover and concealment, obstacles, key terrain, avenues of approach (considerations in evaluating terrain as part of METT-T)
OEG	operational exposure guidance
OI	operations and intelligence
OIC	officer in charge
OP	observation post
OPCON	operational control
OPLAN	operation plan
OPORD	operations order
OPSEC	operations security
PAC	personnel and administration center
PD	point of departure
PEWS	platoon early warning system
PIR	priority intelligence requirements
PL	phase line
PLL	prescribed load list
plt	platoon
PMCS	preventive maintenance checks and services
POL	petroleum, oils, and lubricants
POSNAV	position navigation
PSG	platoon sergeant
PW	prisoner of war

recon	reconnaissance; reconnoiter
REDCON	readiness condition
RFA	restricted fire area
RFL	restricted fire line
R/GS	reinforcing/general support
ROE	rules of engagement
ROM	refuel on the move
RP	release point
rte	route
S1	adjutant (US Army)
S2	intelligence officer (US Army)
S3	operations and training officer (US Army)
S4	supply officer (US Army)
SCO	squadron commander
sec ldr	section leader
SEE	small emplacement excavator
SGLI	Serviceman's Group Life Insurance
sgt	sergeant
SIDPERS	Standardization Installation/Division Personnel System
SIGSEC	signal security
SITREP	situation report
SOI	signal operation instructions
SOP	standing operating procedure
SP	start point
sqd ldr	squad leader
TAA	tactical assembly area
TAC CP	tactical command post
TACFIRE	tactical fire direction system
TC	tank commander
TIRS	terrain index reference system
TOC	tactical operations center
TOE	table(s) of organization and equipment
TOW	tube-launched, optically-tracked, wire-guided (missile)
TRADOC	Training and Doctrine Command
TRP/trp	target reference point; troop
TTP	tactics, techniques, and procedures
UMCP	unit maintenance collection point

UMT	unit ministry team
US	United States
veh	vehicle
WIA	wounded in action
wrkr	wrecker
XO	executive officer

References

SOURCES USED

These are the sources quoted or paraphrased in this publication.

Army Publications

FM 17-15	Tank Platoon. 7 October 1987.
FM 17-95	Cavalry Operations. 19 September 1991.
FM 17-95-10	Armored Cavalry Regiment and Squadron. 22 September 1993.
FM 17-98	Scout Platoon. 9 September 1994.
FM 71-1	Tank and Mechanized Infantry Company. 22 November 1988.

Command Publications

Command publications cannot be obtained through Armywide resupply channels. Determine availability by contacting the address below.

Commander
USAARMC
ATTN: ATZK-IMO-RS
Fort Knox, KY 40121-5000

FKSM 17-97-3	Cavalry Troop Common SOP. 15 July 1994.

READINGS RECOMMENDED

These readings contain relevant supplemental information.

Army Publications

ARTEP 17-57-10-MTP	Mission Training Plan for the Scout Platoon. 27 December 1988.
ARTEP 17-237-10-MTP	Mission Training Plan for the Tank Platoon. 3 October 1988.
ARTEP 17-487-30-MTP	Mission Training Plan for the Regimental Armored Cavalry Troop. 3 September 1991.

DA Form 1155	Witness Statement on Individual. June 1966.
DA Form 1156	Casualty Feeder Report. June 1966.
DA Form 1971-R	Radiological Data Sheet Monitoring and Point Technique. September 1986.
DA Form 2028	Recommended Changes to Publications and Blank Forms. February 1974.

DD Form 551	Record of Interment. August 1984.
DD Form 1077	Collecting Point Register of Deceased Personnel. July 1984.

FM 1-114	Tactics, Techniques, and Procedures for the Regimental Aviation Squadron. 20 February 1991.
FM 1-116	Tactics, Techniques, and Procedures for the Air Cavalry/Reconnaissance Troop. 20 February 1991.
FM 3-3	Chemical and Biological Contamination Avoidance. 16 November 1992.
FM 3-4	NBC Protection. 20 May 1992.
FM 3-5	NBC Decontamination. 23 July 1992.
FM 3-50	Smoke Operations. 4 December 1990.
FM 3-100	NBC Defense, Chemical Warfare, Smoke, and Flame Operations.

	23 May 1991.
FM 3-101	Chemical Staffs and Units. 22 April 1987.
FM 5-36	Route Reconnaissance and Classification. 10 May 1985.
FM 5-100	Engineer Combat Operations. 22 November 1988.
FM 5-250	Explosives and Demolitions. 15 June 1992.
FM 6-20	Fire Support in the AirLand Battle. 17 May 1988.
FM 7-91	Tactical Employment of Antiarmor Platoons, Companies, and Battalions. 30 September 1987.
FM 7-98	Operations in a Low-Intensity Conflict. 19 October 1992.
FM 8-55	Planning for Health Service Support. 15 February 1985.
FM 10-63-1	Graves Registration Handbook. 17 July 1986.
FM 17-12-1-1	Tank Gunnery (Abrams), Volume I. 19 March 1993.
FM 17-12-1-2	Tank Gunnery (Abrams), Volume II. 19 March 1993.
FM 17-12-8	Light Cavalry Gunnery. 28 October 1994.
FM 17-18	Light Armor Operations. 8 March 1994.
FM 20-32	Mine/Countermine Operations. 30 September 1992.
FM 21-10	Field Hygiene and Sanitation. 22 November 1988.
FM 21-10-1	Unit Field Sanitation Team. 11 October 1989.
FM 21-11	First Aid for Soldiers. 27 October 1988.
FM 22-100	Military Leadership. 31 July 1990.

FM 23-1	Bradley Fighting Vehicle Gunnery. 1 March 1991.
FM 34-3	Intelligence Analysis. 15 March 1990.
FM 44-1	US Army Air Defense Artillery Employment. 9 May 1983.
FM 44-3	Air Defense Artillery Employment: Chaparral/Vulcan/Stinger. 15 June 1984.
FM 44-8	Small Unit Self-Defense Against Air Attack. 30 December 1981.
FM 44-18-1	Stinger Team Operations. 31 December 1984.
FM 71-2	The Tank and Mechanized Infantry Battalion Task Force. 27 September 1988.
FM 100-2-1	Soviet Army Operations and Tactics. 16 July 1984.
FM 100-2-3	The Soviet Army: Troops, Organization, and Equipment. 6 June 1991.
FM 100-5	Operations. 14 June 1993.
FM 101-5-1	Operational Terms and Symbols. 21 October 1985.
TC 6-40A	Field Artillery Automated Cannon Gunnery. 21 April 1989.

Command Publications

Contact the address shown on page References-1 for these command publications.

FKSM 17-15-3	Tank Platoon SOP. February 1989.

FKSM 17-98-3	Scout Platoon SOP. November 1994.
FKSM 17-98-4	Scout Platoon Leader's Notebook.

Standardization Agreements (STANAGs)

To requisition these documents, complete DD Form 1425 and send it to—

Naval Publications and Forms Center
5801 Tabor Avenue
Philadelphia, PA 19120

STANAG 2002	Warning Signs for the Marking of Contaminated or Dangerous Land Areas, Complete Equipment, Supplies and Stores. 2 December 1989.
STANAG 2014	Operation Orders, Warning Orders and Administrative/Logistics Orders. 14 August 1991.
STANAG 2029	Method of Describing Ground Locations, Areas and Boundaries. 10 June 1991.
STANAG 2036	Land Mine Laying, Marking, Recording and Reporting Procedures. 29 April 1988.
STANAG 2044	Procedures for Dealing with Prisoners of War (PW). 12 July 1988.
STANAG 2047	Emergency Alarms of Hazard or Attack (NBC and Air Attack Only). 10 October 1990.
STANAG 2082	Relief of Combat Troops. 24 July 1991.
STANAG 2083	Commanders' Guide on Nuclear Radiation Exposure of Groups.

10 September 1990.

STANAG 2084 Handling and Reporting of Captured Enemy Equipment and Documents. 26 June 1986.

STANAG 2103 Reporting Nuclear Detonations, Biological and Chemical Attacks, and Predicting and Warning of Associated Hazards and Hazard Areas. 10 June 1991.

STANAG 2104 Friendly Nuclear Strike Warning. 12 December 1989.

STANAG 2112 NBC Reconnaissance. 5 August 1986.

STANAG 2113 Denial of a Unit's Military Equipment and Supplies to an Enemy. 11 July 1991.

STANAG 2129 Identification of Land Forces on the Battlefield. 16 May 1989.

STANAG 2889 Marking of Hazardous Areas and Routes Through Them. 8 February 1990.

FIELD MANUAL
NO. 3-22.27

HEADQUARTERS
DEPARTMENTS OF THE ARMY
WASHINGTON, DC, 28 November 2003

MK 19, 40-mm GRENADE MACHINE GUN, MOD 3

CONTENTS

DISTRIBUTION RESTRICTION: Approved for public release; distribution is unlimited.

*This publication supersedes FM 23-27, dated 27 December 1988.

Page

PREFACE

This manual discusses how to train the unit to use the MK 19, 40-mm grenade machine gun, model (MOD) 3, referred to in this manual as the MK 19. This manual highlights mechanical training, weapon capabilities, and gunnery principles, methods, techniques, and standards that apply to the MOD 3. It also includes preliminary gunnery, a gunnery skills test, gunnery tables, and qualification tables. If this information conflicts with an applicable technical manual (TM), revised TM, or TM update, the crew will follow the guidance in the most recently published document.

The primary audiences for this manual are soldiers, trainers, and staff officers. Units can modify the gunnery program to meet local training restraints. Although the gunnery tables are intended for use with live fire, trainers can use the tactical engagement simulator system (TESS), a video disc trainer (VDT), a multipurpose arcade combat simulator (MACS), or other training device, except on qualification tables.

In all cases, units must evaluate their training to ensure that it follows the building-block principle and adheres to sound training policy. Only a crew that is trained and does well in preliminary gunnery exercises is likely to do well in live-fire exercises (LFX) and in combat situations.

The proponent for this publication is the U.S. Army Infantry School. Send comments and recommendations to doctrine@benning.army.mil or on a DA Form 2028 (Recommended Changes to Publications and Blank Forms) directly to Commandant, U.S. Army Infantry School, ATTN: ATSH-ATD, Fort Benning, Georgia 31905-5593.

Unless this manual states otherwise, masculine nouns and pronouns do not refer exclusively to men.

CHAPTER 1

INTRODUCTION

Although the MK 19 is a recent entry into the Army's inventory, development began in 1963. The first version was a hand-cranked, multiple grenade launcher called the MK 18. In 1966 the need for more firepower inspired the development of a self-powered 40-mm machine gun called the MK 19, MOD 0. This model was neither reliable nor safe enough for use as a military weapon system. Product improvements begun in 1971 resulted in the 1972 MOD 1, of which only six were produced. The MOD 1 performed effectively in Navy riverine patrol craft and broader applications for the MK 19 were found. In 1973 the Navy developed the MOD 2, which featured improved reliability, safety, and maintainability. In 1976 a complete redesign resulted in the MK 19, MOD 3, which the Army adopted in 1983. The Army now uses the MK 19 within the tactical environment for defense, retrograde, patrolling, rear area security, urban operations, and special operations.

This chapter provides applications, training strategies, and descriptive, technical, and operational data for the MK 19 (Figure 1-1).

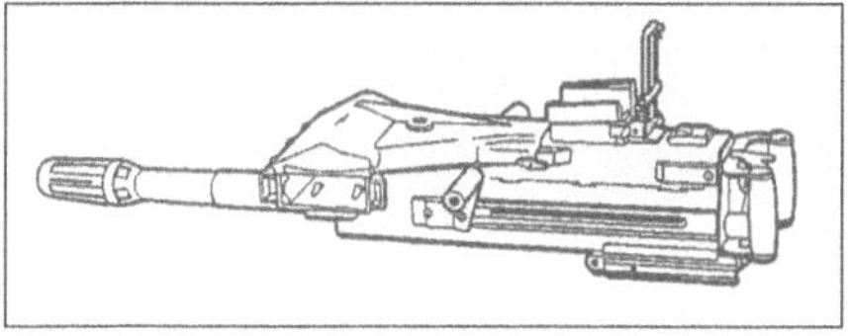

Figure 1-1. MK 19, 40-mm grenade machine gun, MOD 3.

1-1. APPLICATIONS

The MK 19 supports the soldier in both the offense and defense. It gives the unit the capability of laying down a heavy volume of close, accurate, and continuous fire. The MK 19 can also:

- Protect motor movements, assembly areas, and supply trains in a bivouac.
- Defend against hovering rotary aircraft.
- Destroy lightly-armored vehicles.
- Fire on suspected enemy positions.
- Provide high volumes of fire into an engagement area (EA).
- Cover obstacles.
- Provide indirect fires from defilade positions.

1-2. DESCRIPTION

The MK 19 is an air-cooled, blowback-operated machine gun with five major assemblies (Figure 1-2). A disintegrating metallic link belt feeds ammunition through the left side of the weapon. Tables 1-1 and 1-2 provide MK 19 technical and operational data, respectively.

a. **Receiver Assembly.** Holds the barrel and other parts of the gun. Ammunition is fed into the left side of the receiver through the feed throat assembly. The MK 19's barrel will not overheat, even after prolonged firing.

b. **Feed Slide Assembly and Tray.** Holds the rounds in the feeder and indexes the ammunition into position for delinking.

c. **Top Cover Assembly.** Holds the feed slide assembly and tray. It is opened by a latch (left side) for loading or to clean and inspect feeder area. A blade-type front sight is attached to the top cover assembly (Figure 1-3).

d. **Sear Assembly.** Holds the receiver sear. Trigger action releases the sear and allows the bolt to go forward. The safety is attached to the sear assembly.

e. **Bolt and Backplate Assembly.** The bolt fires the round when the sear is depressed by trigger action. The recoil springs drive the bolt forward on the receiver rails. The guide rods hold the springs in position. Trigger and handgrips are located on the backplate assembly.

f. **Feed Throat Assembly.** Allows smooth feeding of 40-mm ammunition. It attaches to the forward left side of the receiver by two sets of spring-loaded retaining pins. Without a feed throat, machine gun stoppages may occur because of twisted or misaligned rounds.

g. **Leaf-Type Rear Sight (with adjustable range plate).** Is marked in 100-meter intervals from 300 to 1,500 meters. The sight is mounted on a spring dovetail base to the receiver assembly (Figure 1-4). Before moving the weapon, the gunner folds the sight forward to a horizontal position. The rear sight can be adjusted for range and windage.

(1) ***Range.*** Different adjustments can be made to the range. Use the rear sight slide release to make *major adjustments* to the range. Use the elevation wheel to make *fine adjustments* to the range.

(2) ***Windage.*** Use the rear sight to adjust for windage. One click equals a 1-mil change. To move the sight to the *right*, turn the windage screw clockwise. To move the sight to the *left*, turn the windage screw counterclockwise.

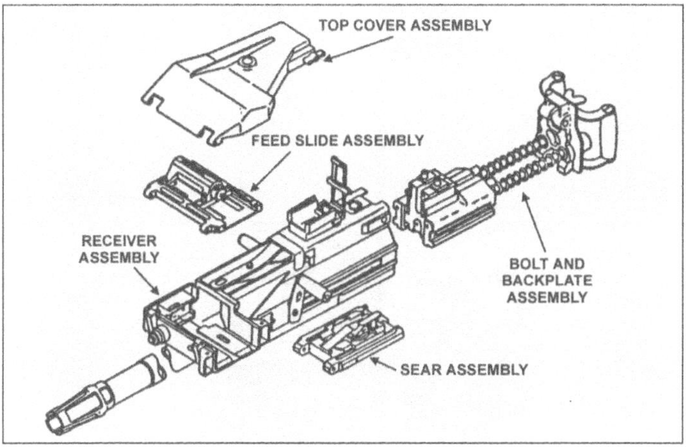

Figure 1-2. Five major assemblies.

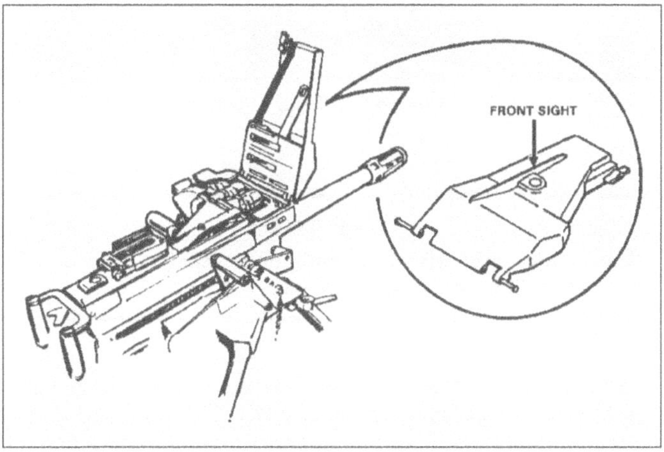

Figure 1-3. Front sight on top cover assembly.

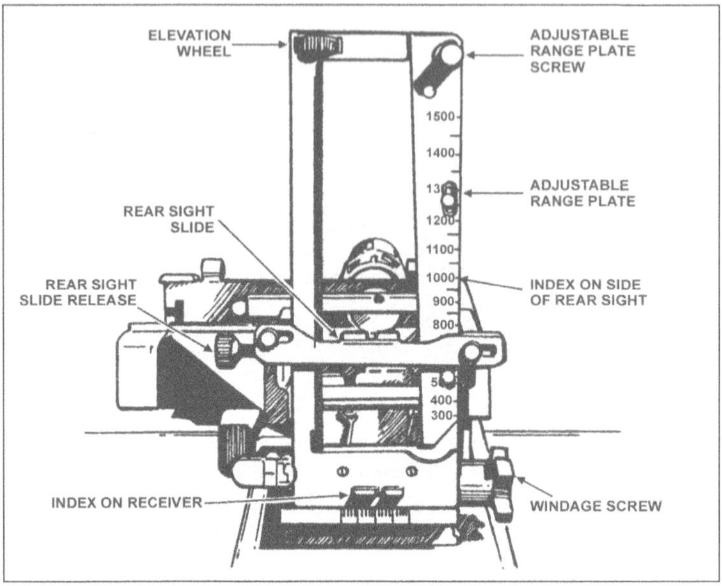

Figure 1-4. Rear sight.

> **WARNING**
>
> **Use only prescribed ammunition. Mixing types of ammunition could result in injury.**

h. The MK 19 uses the following 40-mm cartridges (Figure 1-5, page 1-8).

(1) ***High-Explosive, Dual-Purpose M430 Cartridge.*** The high-explosive, dual-purpose (HEDP) M430 cartridge is the standard round for the MK 19 (Department of Defense Identification Code [DODIC] B542). They are linked with M16A2 links. The HEDP round, the top-curved portion of the projectile, is olive drab with a yellow ogive and yellow markings. It is packed in M548 (48 rounds) or PA120 (32 rounds) ammunition containers. The HEDP, an impact-type round, can penetrate 2 inches of steel armor at 0-degree obliquity and inflict personnel casualties out to 15 meters from impact. It arms within 18 to 30 meters of the gun muzzle and has a point-initiating, base-detonating (PIBD) fuze.

(2) ***High-Explosive M383 Cartridge.*** The high-explosive (HE) M383 cartridge round is olive drab with a yellow ogive and yellow markings. It is packed in a metal ammunition container (48 rounds, linked, in each container). The HE round has a wound radius of 15

meters. It lacks the armor-penetrating ability of the HEDP M430 round. The HE arms between 18 to 36 meters of the gun muzzle fuze.

(3) ***M922 Dummy Cartridges.*** Each MK 19 is issued with one 10-round of inert dummy rounds belt (DODIC B472). M16A2 links join the dummy rounds into a 10-round belt packed in an M2A1 metal box. Trainers use dummy rounds to check weapon function and to train crews.

(4) ***M918 Cartridge.*** The M918 is a training practice cartridge that has the same muzzle velocity of 790 feet per second (fps), signature, and sound as the HE round (DODIC B584).

MK 19 TECHNICAL DATA		
MK 19 (MOD 3):		
Weight without feed throat	77.6	pounds
Weight with feed throat	78.0	pounds
Length	43.1	inches
Width	14.0	inches
Height	8.8	inches
MK 64 (MOD 7) gun cradle:		
Weight	21.0	pounds
Length	17.5	inches
Height	9.5	inches
Tripod (M3) weight:	44.00	pounds
Gun and cradle:		
Weight without feed throat	98.6	pounds
Weight with feed throat	99.0	pounds
Gun, cradle, and tripod:		
Weight without feed throat	142.6	pounds
Weight with feed throat	143.0	pounds

Mounts:	**Ammunition:**
M3 tripod	M430 (HEDP)
M4 pedestal	M383 (HE)
M66 ring	M918 (TP)
HMMWV weapon platform	M922 (dummy)
M113 APC commander's cupola	

NOTE: Keep the feed throat attached to the weapon.

Table 1-1. Technical data.

MK 19 OPERATIONAL DATA		
Maximum range:	2,212	meters
Maximum effective range:	1,500 2,212	meters (point target) meters (area target)
Rates of fire:		
Sustained	40	rounds per minute
Rapid	60	rounds per minute
Cyclic	325 to 375	rounds per minute
Ammunition:		
M430 HEDP	2	inch armor
	15	meter casualty radius
M383 HE	15	meter casualty radius
Service frequency:	50,000	rounds
Elevation, tripod controlled:	100	mils
Depression, tripod controlled:	258	mils
Traverse, tripod controlled:	800	mils (400 left plus 400 right)
Muzzle velocity (average):	798	feet per second
Recoil forces (average):	500	pounds
Angle of automatic fire:	0 to 70	degrees elevation (automatic fire), based on mounting arrangements
Weights:		
Rounds	62	pounds (48 rounds in M548 metal container)
	42	pounds (32 rounds in PA120 metal container)
Planned operating load	400	prescribed by local commanders

Table 1-2. Operational data.

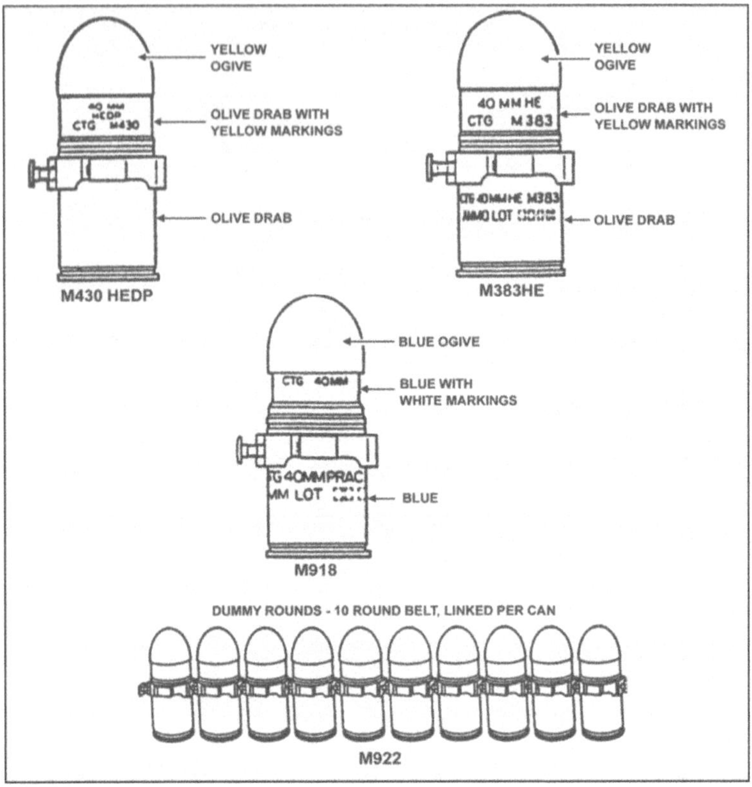

Figure 1-5. 40-mm cartridges.

1-3. TRAINING STRATEGY

A training strategy that integrates resources with the instruction and training of individual and collective skills is necessary to carry out a unit's wartime mission. TRADOC institutions, such as the noncommissioned officers education system (NCOES), and units implement overall training strategies for the MK 19. This multifaceted approach includes specific plans for using resources such as publications, ranges, ammunition, training aids, devices, simulators, and simulations. These specific strategies develop critical soldier skills and training tasks, as well as leader skills needed to support the intended outcome (see Appendix A, Training Strategy, for more details).

CHAPTER 2

OPERATION AND FUNCTION

This chapter includes the cycle of operation; operating precautions; clearing; disassembly and assembly; loading; malfunctions, stoppages, and corrections; operations under adverse conditions; and destruction of the MK 19. Refer to Appendix B, Operations Checklist, for procedures to be used before, during, and after operation of the MK 19 and Appendix C, Stoppages, Malfunctions, Immediate Actions, and Remedial Actions, for procedures to correct stoppages and malfunctions.

2-1. CYCLE OF OPERATION

The MK 19's cycle of operation includes six steps: charging, extracting (delinking), cocking, firing, blowback, and automatic feeding. More than one step may be done at the same time.

a. **Charging.** The charger handles are used to pull the bolt (1) to the rear, aligning the round with the bolt extractors. The rearward movement of the bolt causes the primary drive lever (2) to move to the left, which moves the secondary drive lever (3) to the right. The forked end of the secondary drive lever, which rests on the feed slide pin (4), moves the feed slide (5) to the right. The feed pawls (6) on the feed slide move the linked rounds (7) over one place in the ammunition-feed area of the receiver. The leading round (8) lines up with the bolt extractor (9). See Figure 2-1.

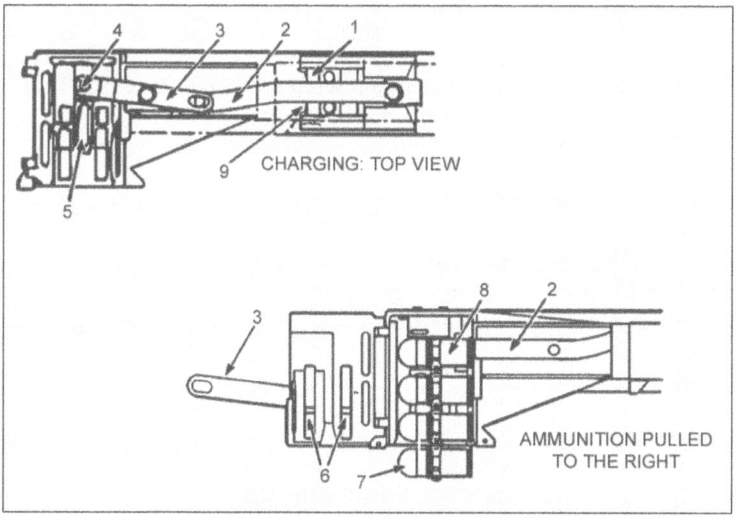

Figure 2-1. Charging cycle.

b. **Extracting (Delinking).** When a round is stripped from the belt, it is extracted or "delinked." This happens, after the MK 19 has been charged once, when the trigger (10) is pressed. The bolt slams forward and the bolt's extractors (11) snap over the rim of the cartridge case. When the MK 19 is charged again, the extractor pulls the leading round to the rear and separates the male and female links. The curved edge of the vertical cam (12) forces the lead round out of the extractors and into the bolt fingers (13). With the bolt completely to the rear, the round lines up with the chamber (14), and is ready to fire. As the original leading round chambers, the next round aligns with the bolt extractors. See Figure 2-2.

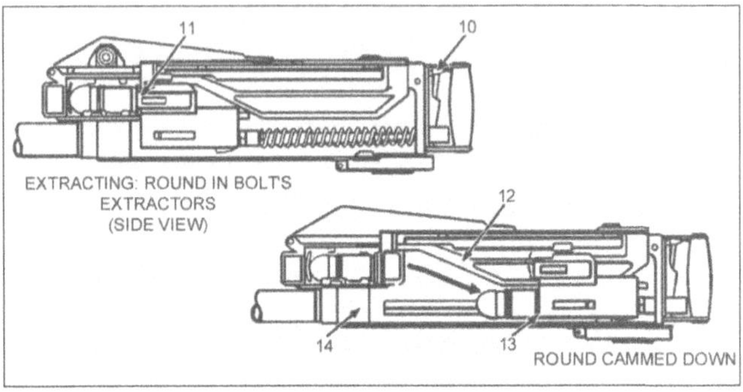

Figure 2-2. Extracting (delinking) round.

c. **Cocking.** The rearward movement of the bolt causes the cocking lever (15) to retract the firing pin (16). When the cocking lever hits the rear end of the left receiver rail slot (17), the cocking lever is forced forward. When the cocking lever retracts the firing pin, the firing pin sear holds the firing pin to the rear (18). See Figure 2-3.

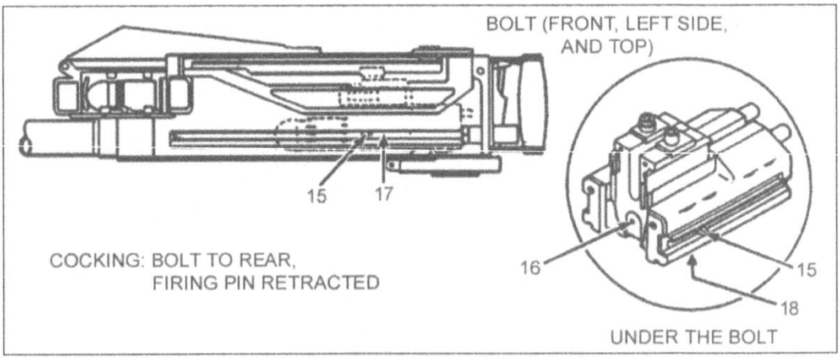

Figure 2-3. Cocking the MK 19.

d. **Firing Cycle.** The safety switch is on F (FIRE). Pressing the trigger (20) depresses the tip of the receiver sear (21). The receiver sear disengages the bolt sear (22), which releases the bolt forward under spring pressure with a round in the bolt fingers. The cocking lever hits the forward end of the left receiver rail slot, forcing the lever to the rear. The bolt sear hits a plate in the bottom of the receiver, which pushes the firing pin sear up to release the firing pin. A combination of the bolt's inertia and pressure from the firing pin spring drive the firing pin forward. The tip of the firing pin detonates the primer. The round is not completely inside the chamber at the moment the weapon is fired. The cartridge case, held by the bolt fingers, protrudes from the chamber (23). The explosion forces the projectile down the bore. See Figure 2-4.

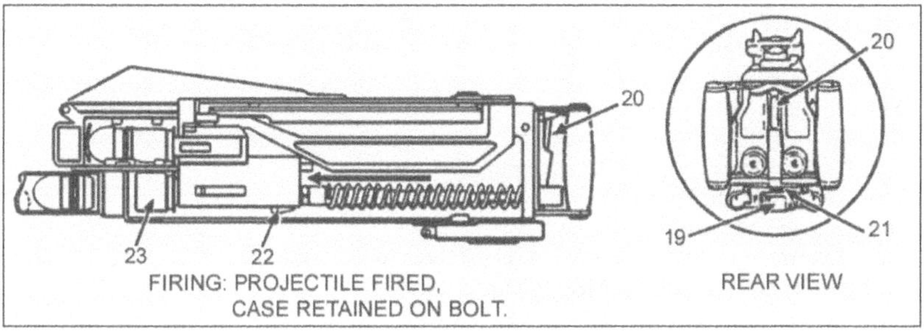

Figure 2-4. Firing cycle.

e. **Blowback and Automatic Feeding Cycle.** The gases from the burning powder force the bolt, with a new round in its extractors, to the rear. During this blowback, several things happen at once. First, the curved rail of the vertical cam delinks and forces down the new round on top of the spent case, which forces the spent case out of the bolt fingers and ejects it out the bottom of the gun. Next, the feed slide assembly pulls the rounds to the right in the receiver ammunition feed area, where a new round is ready to pick up (automatic feed). During the bolt's travel to the rear, the cocking lever is pushed forward, which cocks the firing pin. When the bolt reaches the limit of its rearward travel, the recoil springs (24) are completely compressed. The bolt buffers (25) absorb over-travel, reducing trunnion load (recoil force) at the gun-mount attaching points. The bolt sear will not engage the receiver sear if the trigger is still depressed, and another firing cycle occurs. Release of the trigger causes the bolt sear to engage the receiver sear, which prevents the bolt from going forward, and firing stops. See Figure 2-5.

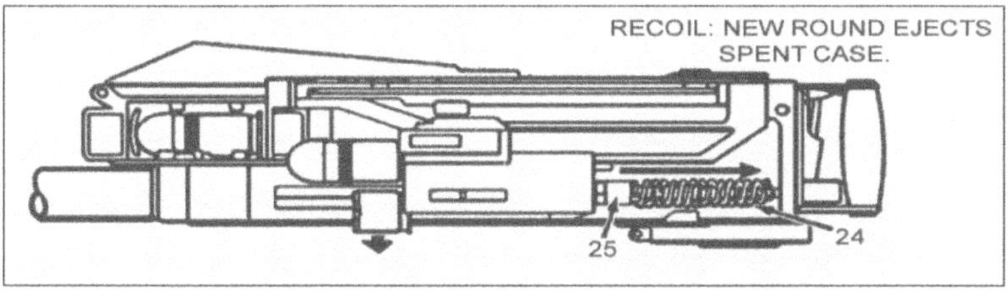

Figure 2-5. Blowback and automatic feed.

2-2. OPERATING PRECAUTIONS

Precautions are taken before, during, and after firing of the MK 19.

a. Before firing the weapon ensure that:

(1) The correct ammunition is on hand.

(2) The ammunition is clean and dent-free.

(3) All ogives on the cartridges are tight.

(4) The line of fire is clear of obstructions.

b. While firing the weapon ensure that:

(1) The top cover is closed.

(2) The muzzle of the MK 19 is pointed downrange at all times.

NOTE: In the event of a runaway gun, lower one of the charging handles.

(3) Use the suggested three to five-round bursts.

c. After firing the weapon:

(1) Unload and clear the MK 19.

(2) Note weapon discrepancies and report them to the armorer.

(3) Clean and lubricate the MK 19 before storage.

2-3. CLEARANCE PROCEDURES

The MK 19 is cleared differently in a firing situation than in a nonfiring situation.

a. **Firing Situation.** In a firing situation, use the following procedures to clear the MK 19:

(1) Move the safety switch to S (SAFE) (Figure 2-6).

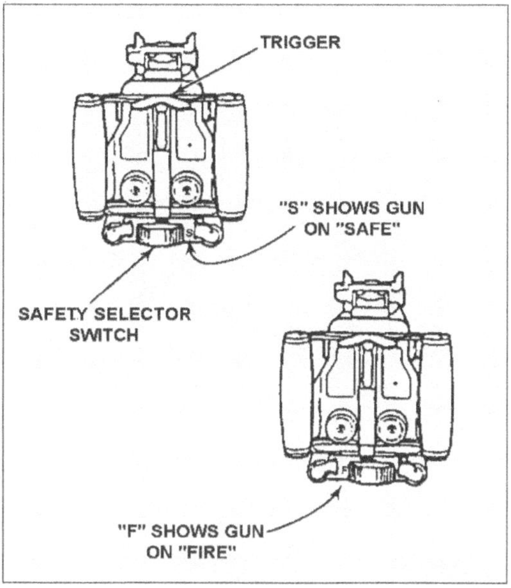

Figure 2-6. S (SAFE) position.

(2) Open the top cover assembly (Figure 2-7). If all the ammunition has NOT been fired, the bolt is to the rear and a round is on the bolt face. If the bolt is forward, lock it to the rear.

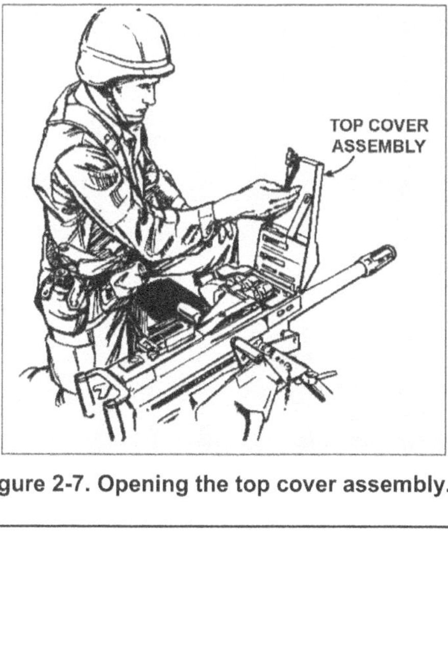

Figure 2-7. Opening the top cover assembly.

(3) Take the ammunition from the feed tray by reaching beneath the feed tray and pressing the primary and secondary positioning pawls. While pressing the position pawls, slide the linked rounds out of the MK 19 through the feed throat (Figure 2-8).

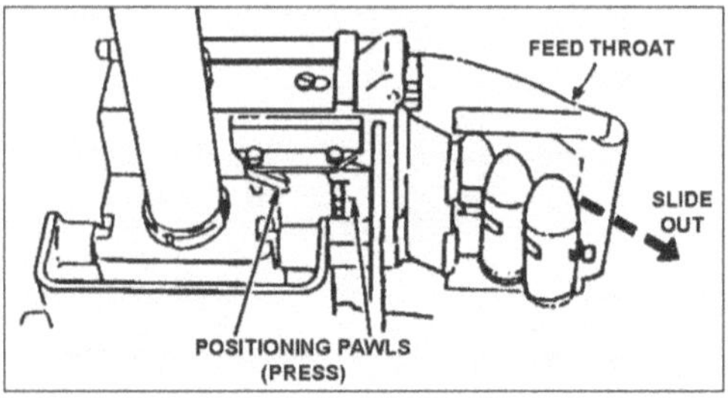

Figure 2-8. Removal of linked rounds from feeder.

(4) Insert a section of the cleaning rod or bayonet through either side of the receiver rail. Place it on top of the live round or cartridge case, as close to the bolt face as possible, and push down. This action forces the round out of the MK 19.

(5) Lower and pull both charger handles to the rear (Figure 2-9).

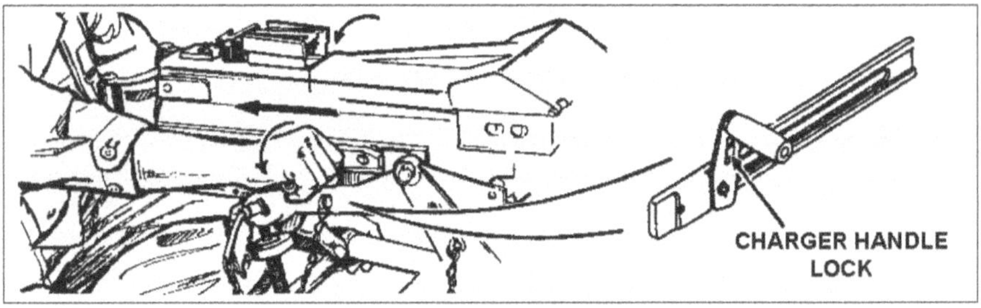

Figure 2-9. Charger handles.

(6) Inspect the chamber and bolt face to ensure that no live rounds are in the weapon.

(7) Place the safety switch on F (FIRE) (Figure 2-6).

(8) Maintain rearward pressure on the charging handle, press the trigger and ease the bolt forward.

(9) Place the safety switch on S (SAFE).

b. **Nonfiring Situation.** In a nonfiring situation, use the following procedures to clear the MK 19:

(1) Place the safety switch on S (SAFE) (Figure 2-6).

(2) Open the top cover assembly.

(3) Lower one or both charger handles.

(4) Pull the charger handle slightly to the rear.

(5) Allow sufficient space between the face of the bolt and the chamber to see both.

(6) Check for live ammunition.

(7) Ride the bolt forward.
(8) Return the charger handle to its original upright position.

2-4. DISASSEMBLY AND ASSEMBLY

Only qualified ordnance personnel should disassemble the MK 19 beyond the steps detailed in this paragraph (Figure 2-10 and Table 2-1).

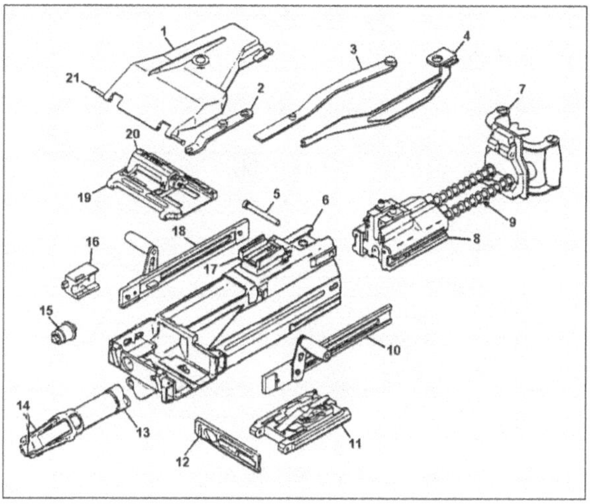

Figure 2-10. MK 19 disassembled.

ITEM	DESCRIPTION
1	Top Cover
2	Secondary Drive Lever
3	Primary Drive Lever
4	Vertical Cam Assembly
5	Backplate Pin
6	Receiver Assembly
7	Control Grip Assembly
8	Bolt
9	Guide Rods and Springs
10	Charger Assembly, Left
11	Sear Assemble
12	Alignment Guide Assembly
13	Barrel
14	Flash Suppressor
15	Ogive Plunger Assembly
16	Round-Positioning Block
17	Rear Sight Assembly
18	Charger Assembly, Right
19	Feed Tray
20	Feed Slide
21	Cover Pins

Table 2-1. Names of the MK 19 major components.

NOTE: High friction areas — the under side of the primary drive lever (3), the vertical cam assembly (4), the bolt (8), the area where the feed slide (20) and feed slide tray (19) touch have been treated to prolong the life of the MK 19 — should be handled carefully during cleaning and handling.

CAUTION

The sear assembly should be removed only in a clean and well-lit area because a small pin could possibly fall out and become lost, which makes the gun inoperable.

a. **Disassembly.** Disassembly includes removal of parts to the extent explained in this chapter. To ensure that parts are not lost and are replaced properly, place them (in the order in which they are taken off) on a clean flat surface. A spent cartridge casing, a section of the cleaning rod, and a cartridge link may be used as removal and replacement tools. To disassemble:

(1) Clear the weapon.

(2) Take out the secondary drive lever by following these procedures:

(a) Raise the top cover assembly and push the secondary drive lever pivot post from the outside of the top cover assembly.

(b) Separate the secondary drive lever from the top cover assembly (Figure 2-11).

(c) Tap lightly on the secondary lever pivot post with any available tool, if necessary.

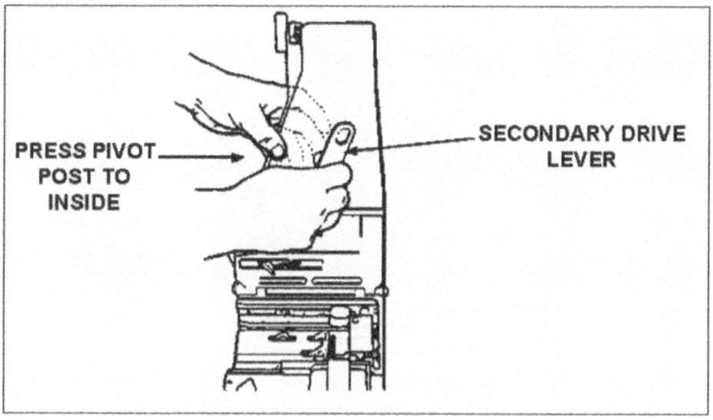

Figure 2-11. Secondary drive lever removal.

(d) Take the secondary drive lever from the slide assembly, and allow the feed slide and tray assembly to close.

(3) Take off the top cover assembly by following these procedures:

(a) Hold the top cover straight up with one hand.

(b) Pull the top cover pins from both sides (Figure 2-12).

(c) Lift the top cover assembly straight up and off.

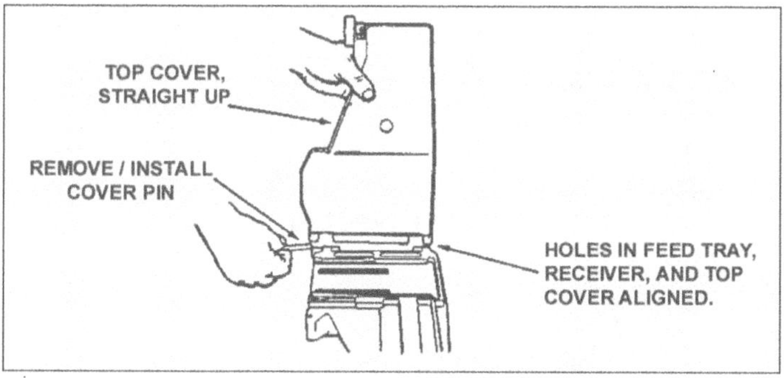

Figure 2-12. Top cover assembly.

(4) Take out the feed slide assembly and separate the feed slide from the feed tray by following these procedures:

(a) Align the tabs on the feed slide with the slots in the feed tray and lift them straight up (Figure 2-13).

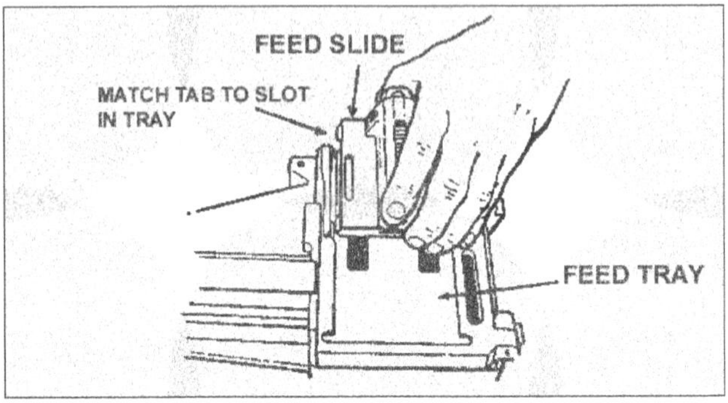

Figure 2-13. Feed slide assembly.

(b) Take out the feed tray by lifting it straight up (Figure 2-14).

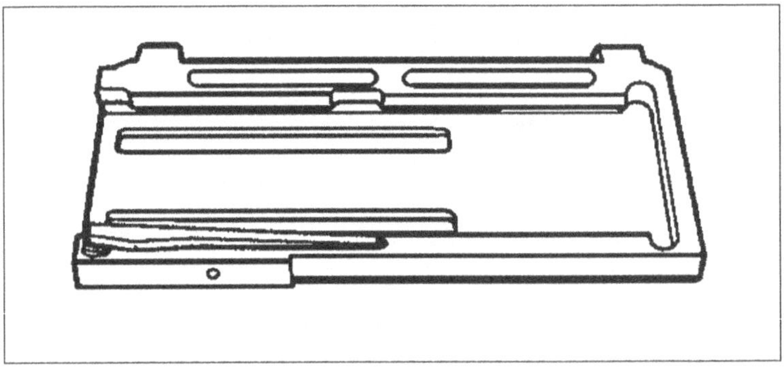

Figure 2-14. Feed tray.

(5) Take off the alignment guide by following these procedures:

(a) Depress the tip of the alignment guide spring with the secondary drive lever or a cleaning rod section.

(b) Slide the alignment guide out of the receiver by pulling the assembly slightly rearward (Figure 2-15).

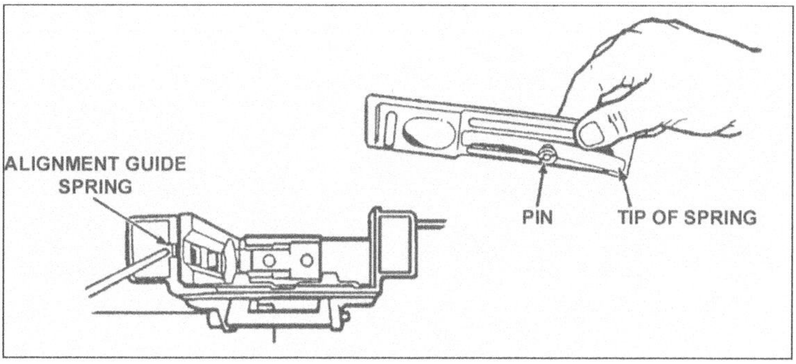

Figure 2-15. Alignment guide.

(6) Take out the ogive plunger by pulling it out through the inside wall of the receiver (Figure 2-16).

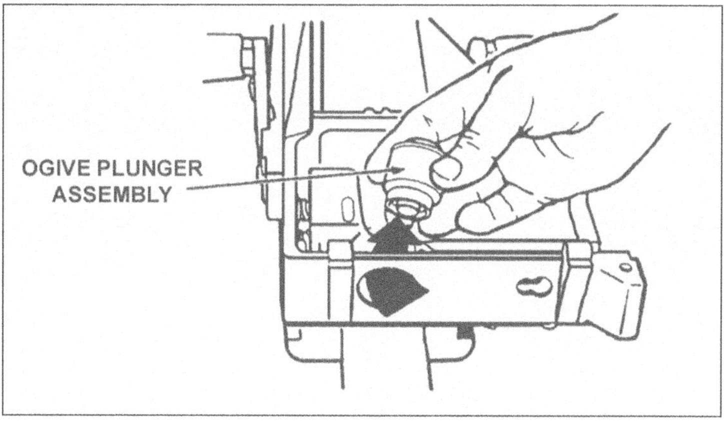

Figure 2-16. Ogive plunger.

(7) Take off the round-positioning block by following these procedures:

(a) Push the round-positioning block into the side of the gun.

(b) Slide the round-positioning block forward.

(c) Release the round-positioning block forward from the keyslots in the receiver wall (Figure 2-17).

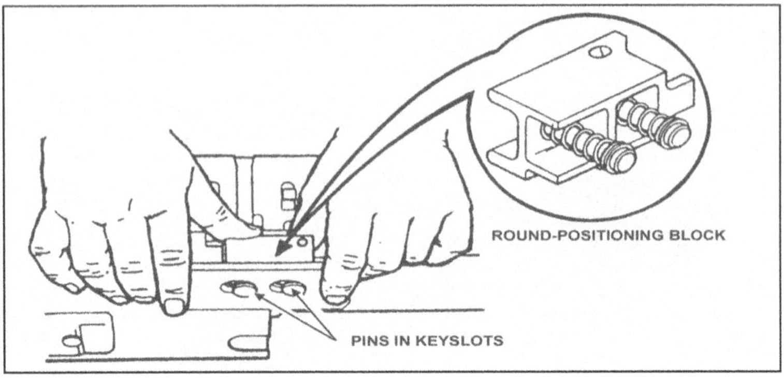

Figure 2-17. Round-positioning block.

(8) Take out the bolt and backplate assembly by following these procedures:

WARNING
The backplate is under extreme pressure when the bolt is in the rear position. To avoid serious injury, ensure the bolt is forward before removing the backplate pin.

(a) Place the safety switch in the F (FIRE) position.

(b) Take out the backplate pin using the rim of a spent cartridge case or metal link.

(c) Pry outward on the pin lip and remove the pin with the fingers (Figure 2-18). If the pin cannot be removed with the fingers, hit the small end of the pin with the secondary drive lever until the pin comes loose.

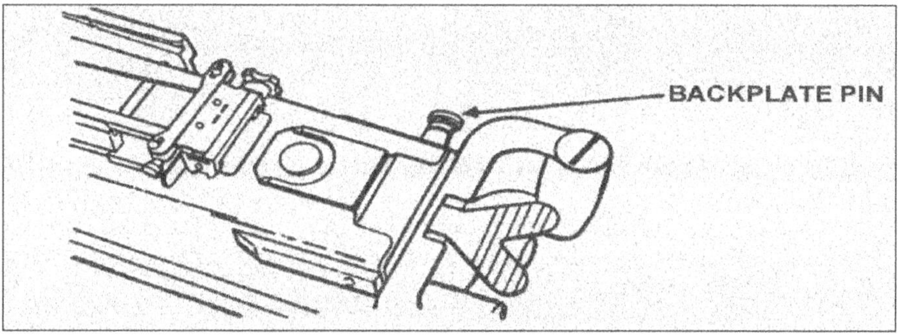

Figure 2-18. Backplate pin.

(d) Grasp the control grips with both hands and lift up slightly to disengage the backplate from the locking lugs in the receiver.

(e) Pull the bolt and backplate assembly to the rear.

(f) Once the bolt clears the sear, catch the bolt in one hand to prevent damage to the backplate assembly (Figure 2-19).

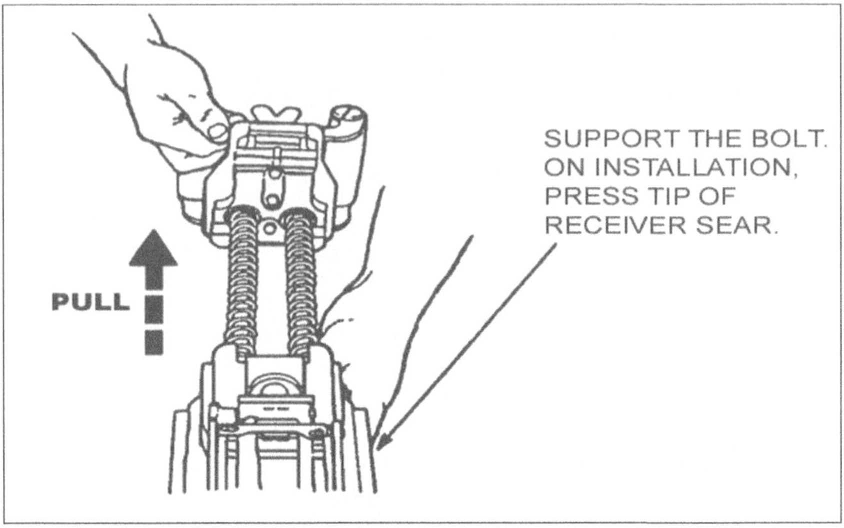

Figure 2-19. Removal of bolt and backplate assembly.

(9) Take off the primary drive lever and vertical cam.

(a) Grasp the primary drive lever and vertical cam with one hand and ensure they do not fall when released.

(b) Reach under the top of the receiver and locate the drive lever lock.

(c) Slide the lock one-quarter of an inch to the rear.

(d) Press down on the primary drive lever pivot post, which releases both the primary drive and vertical cam.

(e) Pull the primary drive lever from the front of the weapon and the vertical cam from the back (Figure 2-20).

CAUTION

Ensure you do not scratch the shiny portion of the vertical cam and primary drive lever. These areas have been specially treated to reduce friction; any scratches or use of abrasives will take off the treatment and increase wear.

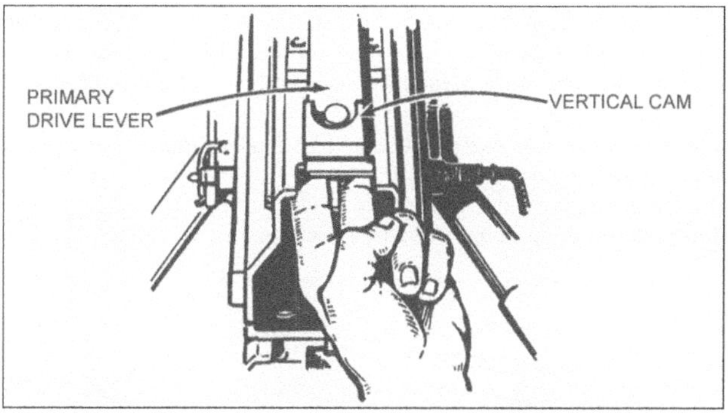

Figure 2-20. Primary drive lever and vertical cam.

(10) Take out the charger assemblies from both sides by following these procedures.

(a) Place the charger assemblies in the upright position.

(b) Using a metal link or spent cartridge case, retract the lock plunger at the base of the charger arm.

(c) Slide the charger housing rearward to disengage the lugs from the keyslots in the receiver.

(d) Lift the charger assembly away from the receiver (Figure 2-21).

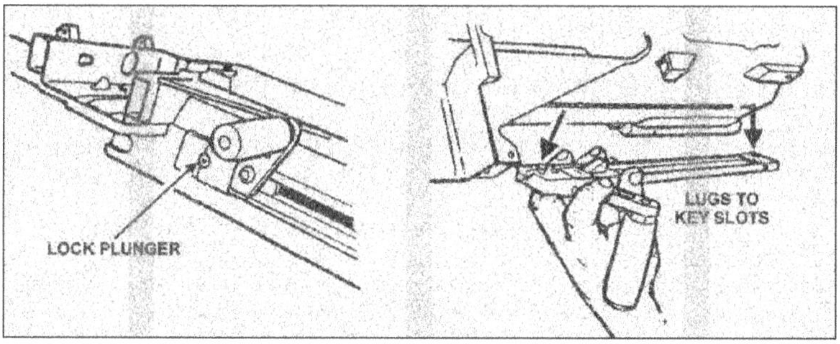

Figure 2-21. Charger assemblies.

(11) Take off the sear assembly by following these procedures:

(a) Turn the MK 19 on its side or upside down.

(b) Move the safety selector switch to F (FIRE), and use the rim of a spent cartridge case to lift up the sear lock plunger.

(c) At the same time, squeeze the sear and rotate the sear assembly (left or right) until you can no longer hold the sear (Figure 2-22).

(d) Release the plunger, and apply downward pressure to the sear assembly. (A spring in the sear will push the sear away from the receiver preventing further turning of the sear.)

(e) With the sear pressed flush to the receiver, continue to turn the sear until you hear a metallic click.

(f) Place the safety selector switch on S (SAFE) and continue to rotate the sear until it is at a 90-degree angle to the receiver.

(g) Lift the sear straight off the receiver.

(h) Place the safety switch in the S (SAFE) position.

Figure 2-22. Sear assembly.

b. **Assembly.** To assemble the gun, replace the groups in the exact reverse order in which they were taken off. Be sure the components are lubricated prior to reassembly.

(1) Install left and right hand charger assemblies by following these procedures:

(a) Turn receiver upright.

(b) Rotate charger handle to the forward straight-up position.

(c) Line up lugs on charger with slots in receiver rail. Insert charger lugs into slots.

(d) Hold charger tightly against rail. Slide charger forward until it locks in place.

(2) Install round-positioning block. Insert pins into the slots on the receiver and slide the block forward until the pins lock into the small sections of the holes.

(3) Install the ogive plunger assembly. Insert ogive plunger as shown in Figure 2-16.

(4) Install alignment guide assembly by following these procedures:

(a) Position the alignment guide assembly so that the pin is lined up with the slot in the feeder wall.

(b) Hold the alignment guide against the front wall and slide the alignment guide into the receiver until it locks or 'clicks.'

(5) Install feed tray and feed slide assembly by following these procedures (Figure 2-23):

(a) Place tray into top of feeder, recessed side up. The tabs on the feed slide should align with the slots on the feed slide tray.

(b) The hinge pinholes on the tray should line up with the hinge pinholes on the receiver.

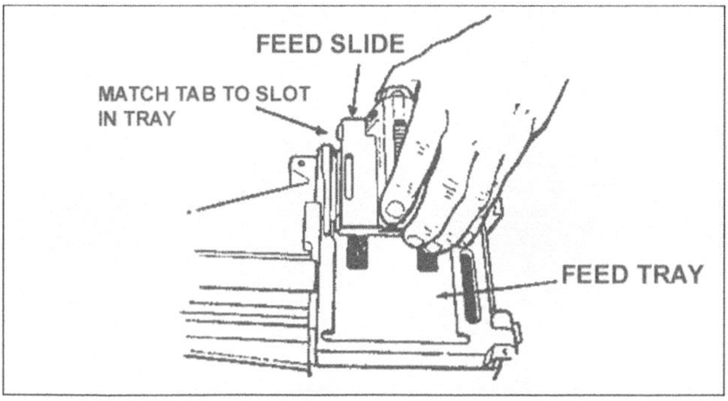

Figure 2-23. Feed slide assembly.

(6) Install top cover assembly by following these procedures:

CAUTION
To avoid breaking the crosspin, be sure it is fully inserted into the receiver before closing the top cover.

(a) Feed tray should be in place resting on the receiver.
(b) Place the top cover on the receiver with the hinge pinholes in line with the receiver lug end feed tray hinge pinholes.
(c) Hold top cover at a right angle to the receiver so that the key slots are aligned. Insert top cover pins on both sides. Ensure the crosspin is fully inserted then rotate the top cover fully open.
(7) Install secondary drive lever by following these procedures (Figure 2-24):

CAUTION
If the secondary drive lever is not properly engaged with the feedslide pin, the gun will not fire.

(a) Rotate the feed slide assembly and tray upward.
(b) Engage forked end of secondary drive lever with the feed slide pin.
(c) Press raised pivot post through hole in top cover.
(d) Press secondary drive lever against top cover until it locks in place.

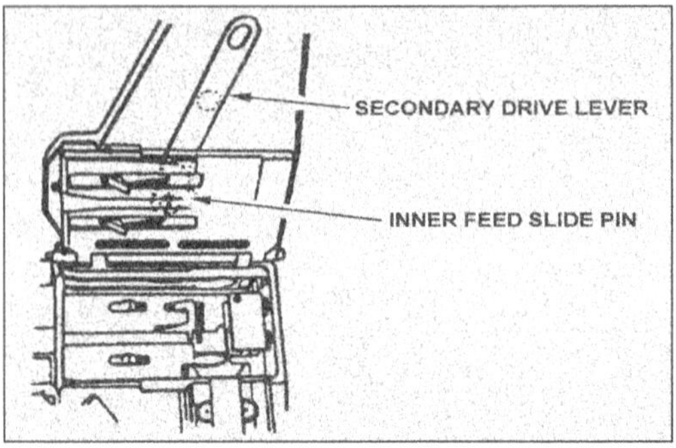

Figure 2-24. Installation of the secondary drive lever.

(8) Install vertical cam assembly by following these procedures:

> **CAUTION**
> Do not damage vertical cam by hitting it on the inside of the receiver.

(a) Slide vertical cam assembly through rear of receiver. Raised portion should slide over the top of the receiver. Drive lever lock should be underneath.

(b) Engage forked end in the notch.

(9) Engage primary drive lever by following these procedures:

(a) Hold vertical cam assembly in place and slide primary drive lever into receiver and rest the front end on the feed area.

(b) Slide drive lever lock rearward and engage pivot post of lever through holes in receiver and vertical cam.

(c) Slide drive lever lock (on the vertical cam just beneath top of receiver) forward.

(10) Install bolt and backplate assembly. With the sear assembly on the gun, assemble bolt and backplate using the following procedure (Figure 2-25):

> **CAUTION**
> Before inserting assembly, put cocking lever in forward position.

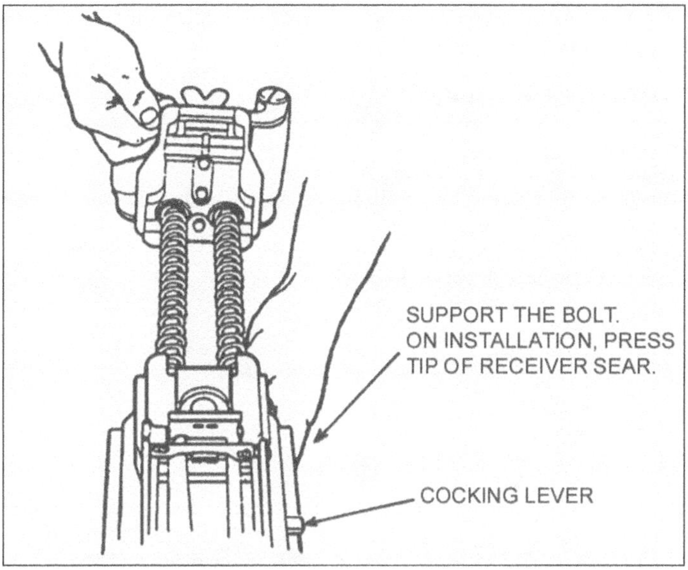

Figure 2-25. Installation of the bolt and backplate assembly.

(a) Place safety in F (FIRE) position.

(b) Ensure that the cocking lever is forward and align the bolt at the rear of the receiver and slide the receiver forward until it stops.

(c) Press down on the bolt release, slide the bolt slightly forward, and remove hands from the bolt release.

(d) Continue to slide the bolt forward until it stops again.

(e) Again, press down on the bolt release, slide the bolt slightly forward, and remove hands from the bolt release.

(f) Continue to slide the bolt forward until it reaches its most forward position.

(g) Seat the bottom of the backplate into the slots on the bottom of the receiver.

(h) Insert backplate pin to lock assembly in place.

(11) Install sear assembly by following these procedures:

(a) Turn receiver over on its top.

(b) Place sear housing on the receiver and line up sear housing assembly at a right angle to the barrel centerline.

(c) Put safety on F (FIRE) position.

(d) Press down and rotate the sear assembly until it locks in place.

(12) Install feed throat assembly by following these procedures:

(a) Squeeze the plungers, and align the pins with the holes in the receiver.

(b) Release plunger to reattach feed throat.

NOTE: Ensure the safety switch is in the F (FIRE) position so the sear can be easily depressed.

2-5. LOADING PROCEDURES

Before loading, the gunner should ensure the MK 19 is on S (SAFE) and the bolt is in the forward position.

a. Before loading, use the following procedures:

(1) Attach the feed throat by squeezing the spring-loaded pins on the feed throat (Figure 2-26).

(2) Insert the feed throat on both sides of the feeder.

(3) Ensure that the feed throat points down.

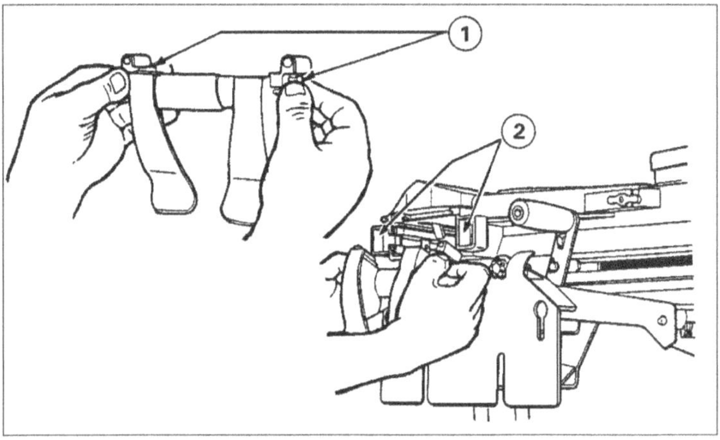

Figure 2-26. Attachment of feed throat.

b. When loading, use the following procedures:

(1) Insert the first round into the feeder (female link first) (Figure 2-27).

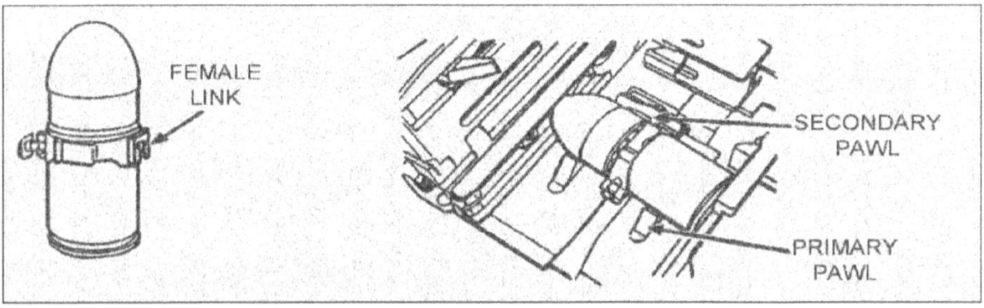

Figure 2-27. Loading of first round.

(2) Push the round across the secondary feed pawl. To move the feed slide to the left, push the secondary drive lever to the right (Figure 2-28).

(3) Close the cover.

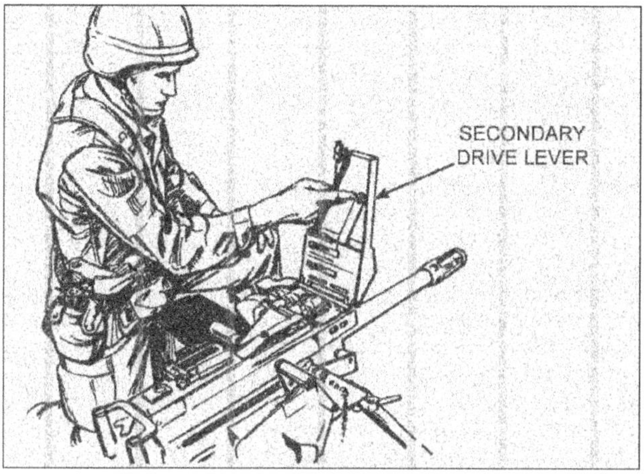

Figure 2-28. Positioning of feed slide assembly.

(4) Grasp the charger handles with the palms down.

(5) Press the charger handle locks in.

(6) Rotate the handles down and pull them sharply to the rear.

(7) Return the charger handles forward to their original upright position, after locking the bolt to the rear.

CAUTION

Failure to completely pull the bolt to the rear may result in the misalignment of the M16A2 links on the round, which causes the round to feed improperly.

(8) Place the safety switch on F (FIRE) and press the trigger. The bolt slams forward and grasps the first round in the bolt extractors.

(9) Grasp, unlock, and turn charger handles down, and lock the bolt to the rear again.

(10) Ensure the safety switch is on S (SAFE).

(11) Return the charger handles to their original upright position. The MK 19 is now ready to fire.

2-6. MALFUNCTIONS, STOPPAGES, AND CORRECTIONS

A malfunction is a failure of the weapon to function properly. Neither defective ammunition nor improper operation of the gun by a crewmember is considered a malfunction of the MK 19. A stoppage is any interruption in the cycle of operation caused by faulty action of the weapon or ammunition. Appendix C, Stoppages, Malfunctions, Immediate Actions, and Remedial Actions, provides inspections, corrective actions, and emergency actions of malfunctions and stoppages. For more information on troubleshooting common malfunctions and stoppages, refer to TM 9-1010-230-10.

2-7. CARE AND CLEANING

The MK 19 requires preventive maintenance checks and services (PMCS) and lubrication before, during, and after firing. It should be checked and cleaned daily when used and not fired, and weekly if not used. For more detailed guidance on the care, cleaning, and PMCS for the MK 19, refer to TM 9-1010-230-10.

a. **Cleaning and Lubrication.** Several types of lubricants can be used on the MK19. High friction areas of the gun have a protective coating and care must be taken to prevent its removal. Listed below are authorized lubricants, and other information about lubricating and cleaning the MK19.

(1) ***Authorized Lubricants.*** These are the authorized lubricants used on the MK 19:

(a) Lubricant, artic weather, automatic weapons (LAW).

(b) Lubricating oil, semifluid, automatic weapons (LSA).

(c) Lubricating oil, semifluid, automatic weapons, temperate (LSAT).

(d) Rifle bore cleaner (RBC). RBC should be used sparingly only in the bore and when necessary.

(e) Grease, molybdenum disulfide (GMD).

(2) ***Use of Lubrications.*** The following is general guidance on the use of lubricants on the MK 19.

(a) Never mix lubricants on the MK19. If you change lubricants or do not know for sure which lubricant is currently applied to the MK19, completely clean off all old materials before you apply a new lubricant.

(b) Never use cleaner lubricant petroleum (CLP) on a MK19. It gums up moving parts instead of lubricating them.

(c) Apply a heavy coat of lubricant and do not wipe it off.

(d) In cold weather, 0 degrees F to -25 degrees F, use LSAT, GMD, or LAW.

(e) In extra cold weather, -25 degrees F and below, use LAW.

(3) ***High Friction Areas.*** The under side of the primary drive lever, the vertical cam assembly, the bolt (where it rides on the rails), and the area where the feed slide and feed slide tray connect have been treated to prolong the life of the MK 19. Special care is required during cleaning and maintenance to prevent damage to the coating. CLP, acids, and abrasives should never be used on the MK 19 because they will remove the protective coating.

b. **General Cleaning and Lubrication Instructions.** The following is general guidance for the cleaning of the MK 19.

(1) Wipe or brush off dirt and grime.

(2) When lubricating the weapon, give extra attention to the feed pawls, the cocking lever rails, the pivot posts on the primary drive lever, and the bolt assembly.

(3) Lubricate each part. Give special care to those hard-to-get spots.
(4) Work in the lubricant by moving the parts.

c. **When to Clean, Lubricate and Inspect the MK 19.** Clean and inspect the gun thoroughly. Report all worn, burred, defective, or missing parts to your armorer or support maintenance. In general, lubricate and clean the MK19 as follows:

(1) Always clean and lubricate the gun after firing.
(2) Clean and lubricate the gun daily if it is used but not fired.
(3) Clean and lubricate the gun weekly if the weapon is stored.

d. **Inspections.** Several parts of the MK 19 should be checked whenever possible. These include:

(1) The vertical cam. Check closely for smoothness.
(2) The gap between the charger assemblies and the receiver.
(3) All parts for wear, cracks, missing items, and abnormal shapes.

2-8. OPERATION UNDER ADVERSE CONDITIONS

The MK 19 is a robust weapon and designed to operate over a wide range of weather conditions but additional care is required when the temperatures are hot or cold, and the conditions are wet, sandy, or icy.

a. During *hot, wet, or salty air conditions* follow these procedures:

(1) Inspect the weapon more frequently for signs of rust.
(2) Keep gun as moisture-free as possible.
(3) Field strip, clean, and lubricate more often to preserve metal and prevent rust.
(4) Use a generous second coat of lubrication for extra protection.

b. During *hot, dry, sandy, or dusty conditions* follow these procedures:

(1) Do not lubricate the entire exposed metal surfaces as this will only collect dust and sand.
(2) Clean the weapon thoroughly and lubricate only the moving components.
(3) Extreme heat dries up lubricant. Clean and lightly lubricate the moving components of the weapon more frequently.

c. During *cold, icy, or snowy conditions* follow these procedures:

(1) Prior to using cold-weather lubricant, be sure to clean off all existing lubrication. From 0 degrees F to –25 degrees F, use either LSAT, GMD, or LAW. At temperatures –25 degrees F and below, use LAW.
(2) If the weapon 'sweats,' dry and lube the parts before taking the weapon outdoors and apply a light second coat to provide protection.
(3) Cover weapon if left outside.
(4) If possible, thoroughly clean, dry, and lubricate the weapon in a warm place.
(5) If the weapon is brought indoors, keep it away from direct heat.
(6) Perform functional checks and lubricate daily to help prevent corrosion.

2-9. DESTRUCTION

Only the commander can direct the destruction of the MK 19. There are several ways to destroy it. Methods include destruction by mechanical means, burning, gunfire, demolition or disposal. Use the same methods to destroy equipment. However, ammunition must be handled differently. The best way to destroy ammunition is to fire it. Otherwise, dispose of it by burial, dumping it in a body of water, or using the demolition method.

WARNING

Ammunition must never be scattered (this may result in a hazard to friendly soldiers), burned (HE and HEDP rounds may explode), or destroyed mechanically.

a. **Mechanical Method.** If possible, use the mechanical method to destroy MK 19s and equipment. There are three ways to do this:

(1) Smash them. Using a sledgehammer or some similar heavy tool, damage the MK 19s so that they cannot be used.

(2) Bend the MK 19s' guide rods over the receivers.

(3) Take off (or out) the same part from each of the MK 19s. If a different part is taken from each gun, new MK 19s may be pieced together.

b. **Burning Method.** To burn equipment or MK 19s, use a fuel source that burns hot enough to take the temper from the metal. For example, use jet propulsion fuel (JP-4), oil, or gas. All friendly soldiers in the area are warned before equipment or MK 19s are bare burned, because a fire will reveal the positions of friendly forces.

c. **Gunfire Method.** This is the least reliable way to destroy MK 19s or equipment. Pile up all of the equipment and MK 19s and use MK 19s, grenades, or rifles to fire on them. They will scatter so check after shooting to see that they are all destroyed.

d. **Demolition Method.** Use an explosive charge such as composition 4 (C4) or any type of HE round. Place the charges so they will completely destroy all the MK 19s and equipment.

e. **Disposal Method.** To dispose of MK 19s and equipment, break them down and either scatter the parts into swamps or marshes, or bury them.

(2) Push the round across the secondary feed pawl. To move the feed slide to the left, push the secondary drive lever to the right (Figure 2-28).

(3) Close the cover.

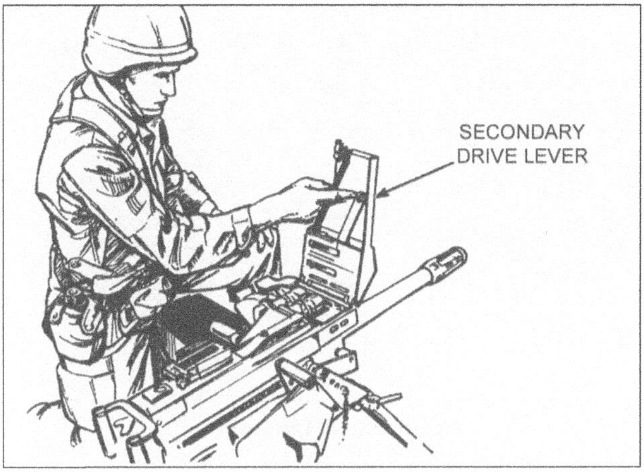

Figure 2-28. Positioning of feed slide assembly.

(4) Grasp the charger handles with the palms down.

(5) Press the charger handle locks in.

(6) Rotate the handles down and pull them sharply to the rear.

(7) Return the charger handles forward to their original upright position, after locking the bolt to the rear.

CAUTION

Failure to completely pull the bolt to the rear may result in the misalignment of the M16A2 links on the round, which causes the round to feed improperly.

(8) Place the safety switch on F (FIRE) and press the trigger. The bolt slams forward and grasps the first round in the bolt extractors.

(9) Grasp, unlock, and turn charger handles down, and lock the bolt to the rear again.

(10) Ensure the safety switch is on S (SAFE).

(11) Return the charger handles to their original upright position. The MK 19 is now ready to fire.

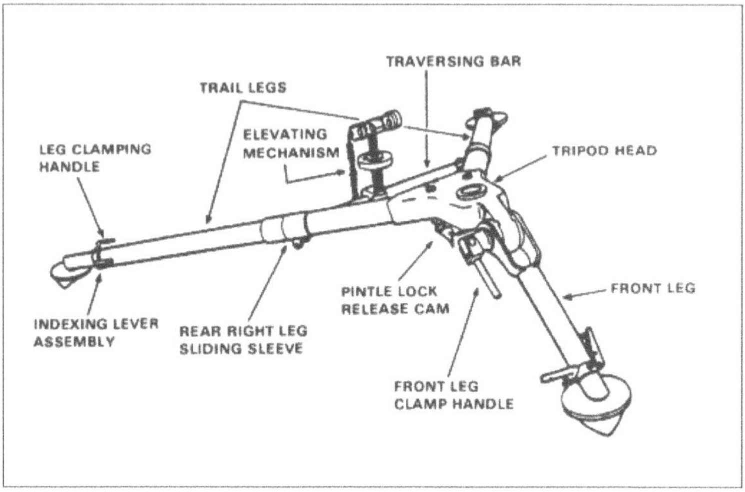

Figure 3-2. M3 tripod.

(1) Unscrew the leg-clamping handle; press down on the indexing lever, and extend the leg to the desired length.

(2) Align the indexing lever stud with one of the holes in the tripod leg extension.

(3) Release pressure on the indexing lever, allowing the stud to fit the desired hole. Tighten the leg-clamping handle.

b. Set the front leg of the tripod by following these procedures:

(1) Turn the front leg clamp handle counterclockwise to loosen the front leg.

(2) Adjust the leg to the desired angle, ensuring the tripod head is level, and tighten the front leg clamp.

c. Secure the tripod legs by following these procedures:

(1) Stamp the metal shoe on each tripod leg into the ground.

(2) Sandbag each leg to stabilize the MK 19 for firing.

d. Mount the M64 gun cradle onto the M3 tripod by following these procedures:

(1) Unlock the tripod pintle lock release cam.

(2) Insert the gun cradle's pintle into the tripod pintle bushing (Figure 3-3).

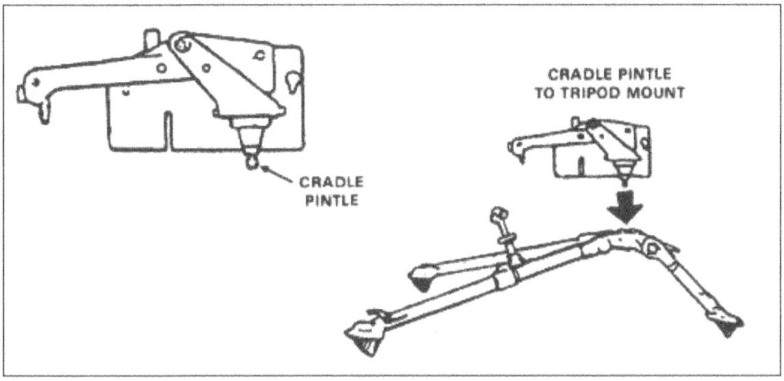

Figure 3-3. Insertion of pintle into pintle bushing.

(3) Lock the pintle lock release cam to secure the gun cradle. Check the gun cradle, by pulling up on it slightly, to ensure that it is seated and locked.

e. Attach the (traverse and elevating) T&E mechanism by following these procedures (Figure 3-4):

(1) Zero the T&E mechanism by zeroing the traversing handwheel, elevating handwheel, and the elevating mechanism sleeve to the lower elevating screw.

(a) To zero the traversing handwheel, hold the T&E mechanism so that the traversing handwheel is on the left when looking at it, then turn the traversing handwheel toward you until it stops. Loosen the locking nut slightly. Align the zero on the scale with the zero on the elevating screw yoke. Hold the scale with the zeros aligned, and tighten the locking nut. Make sure the zeros stay aligned. Turn the traversing handwheel two complete revolutions away from you. If doing this at night, count 50 "clicks" away from you.

(b) To zero the elevating handwheel to the upper elevating screw, align the two zeros. Rotate the elevating handwheel up or down until a zero with a line below it is visible on the upper elevating screw. Position the elevating handwheel so the indicator is pointing at the zero on the handwheel.

(c) Zero the elevating mechanism sleeve to the lower elevating screw. Rotate the elevating mechanism sleeve all the way up; rotate it down until it stops and note the number of complete turns. Rotate the elevating mechanism sleeve up half that number of turns. Position the slide lock lever to face you.

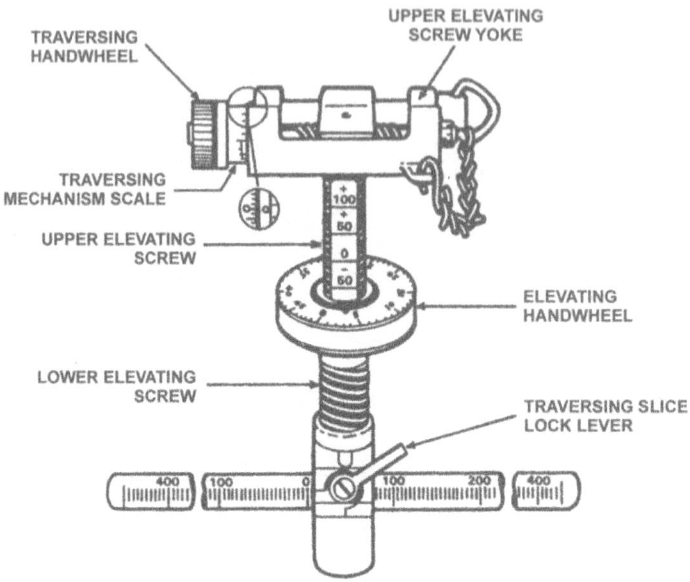

Figure 3-4. T&E mechanism.

(2) Remove the stow pin from the gun cradle (Figure 3-5).

(3) Align holes in the upper elevating screw yoke of the T&E mechanism with the rear holes in the gun cradle (Figure 3-6).

NOTE: The stow pin locks the cradle in a horizontal position, preventing it from depressing or elevating.

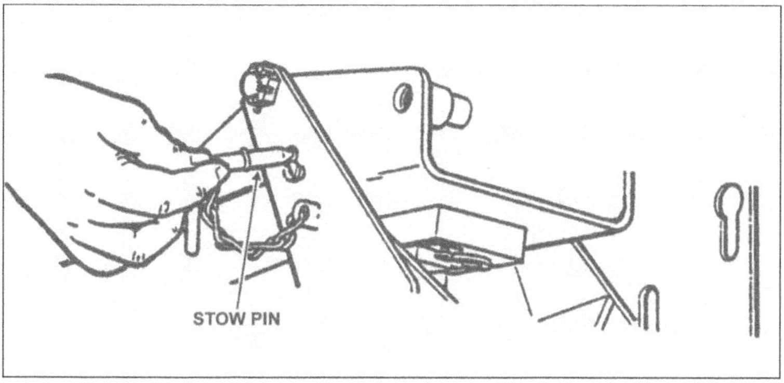

Figure 3-5. Stow pin.

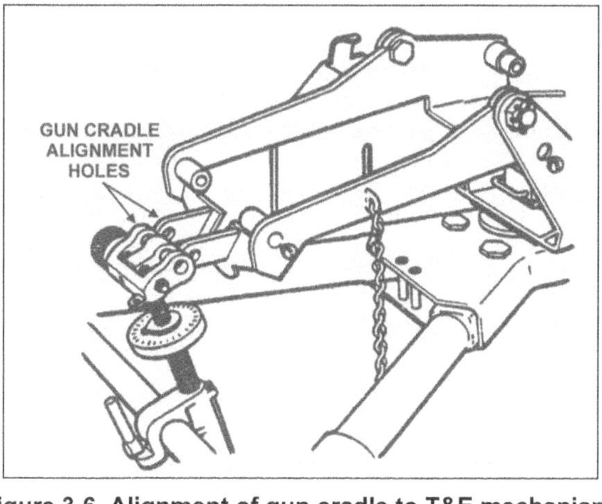

Figure 3-6. Alignment of gun cradle to T&E mechanism.

(4) Lock the elevating sleeve mechanism onto the center of the traversing bar.
(5) Insert the quick-release pin from the right (Figure 3-7).

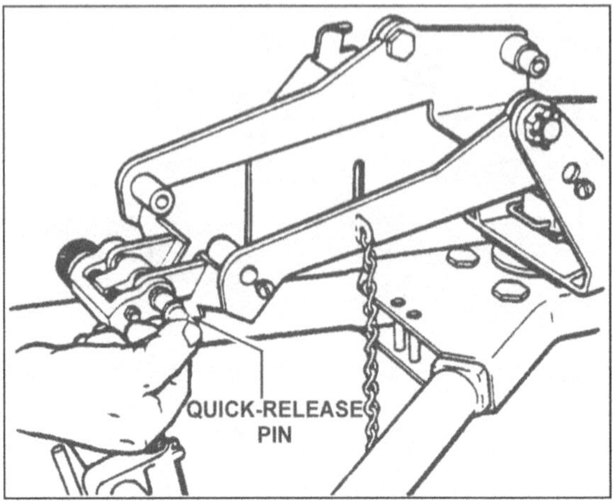

Figure 3-7. Quick-release pin.

f. Mount the MK 19 by following these procedures:

(1) Lift the MK 19 into the gun cradle. One soldier holds the barrel while another holds the control grips. The one holding the grips stands between the trail legs of the tripod; the other straddles the front leg. The soldiers lift the MK 19, barrel first, onto the cradle.

(2) Align the grooves on the receiver with the lugs in the gun cradle, and slide the receiver forward (Figure 3-8).

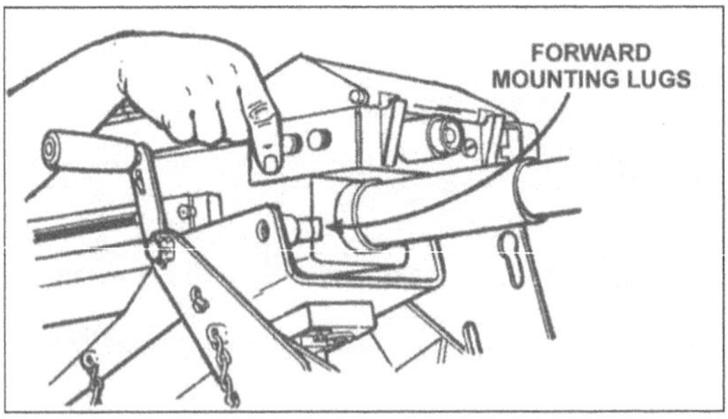

Figure 3-8. Mounting lugs.

(3) Align the sear mounting holes with the gun cradle mounting holes (Figure 3-9). Secure the rear of the receiver by inserting the retaining pin through the cradle and sear assembly and rotate it until it locks in place (Figure 3-10). If a safety clip is attached, use it to secure the retaining pin.

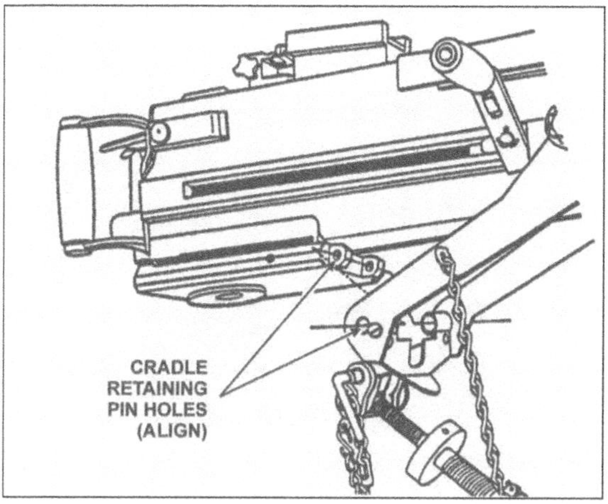

Figure 3-9. Alignment of the sear assembly and pinholes.

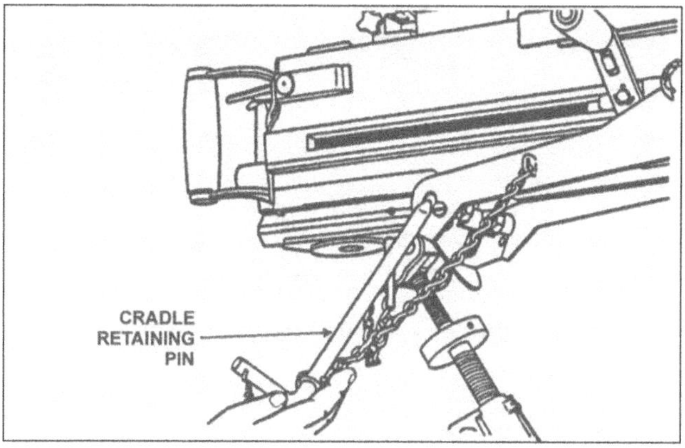

Figure 3-10. Insertion of the cradle retaining pin.

g. Attach the feed throat to the MK 19 by following these procedures:

(1) Squeeze together each set of grip pins (Figure 3-11).

(2) Attach the feed throat to the front left-hand side of the receiver assembly. The pins of the feed throat must line up with the pinholes in the receiver (Figure 3-12).

(3) Relax pressure on the spring-loaded grip pins so they will snap into place (Figure 3-13, page 3-9).

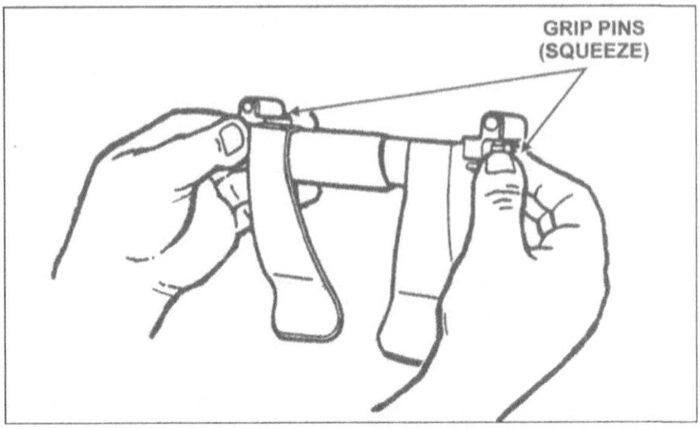

Figure 3-11. Grip pins.

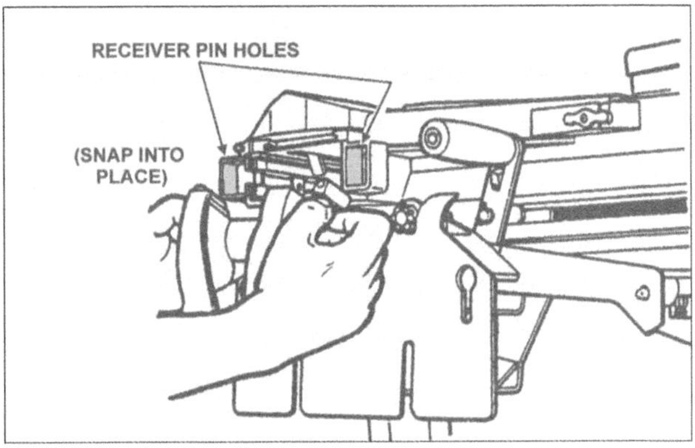

Figure 3-12. Feed throat alignment.

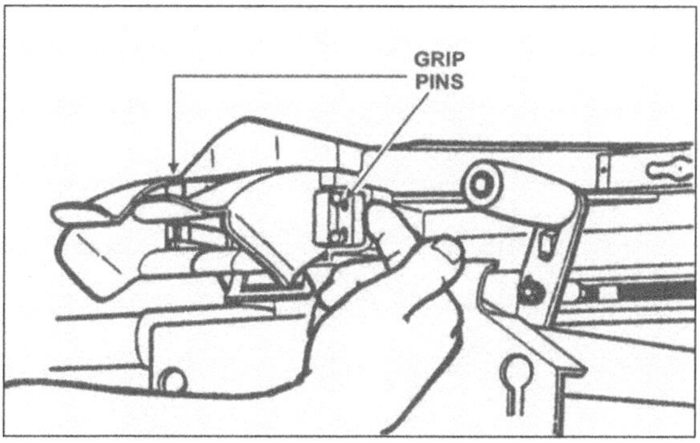

Figure 3-13. Feed throat installation.

NOTE: 1. When the MK 19 is mounted on the M3 tripod or any vehicle, it is called a weapon system.
2. The same procedure is used to mount the MK 19 to the gun cradle regardless of the type of mount used.

3-3. VEHICLE-MOUNT

This section lists the accessories used and outlines the steps to mount the MK 19 onto vehicles. Since mounting procedures for vehicles vary little, this section explains in detail how to mount the MK 19 on the HMMWV and highlights the differences in mounting procedures for other vehicles.

a. **Accessories.** Mounting the MK 19 on vehicles requires several accessories.

(1) ***Gun Cradle.*** The MK 64 gun cradle is used to mount the MK 19 onto any vehicle having a pedestal mount for the M2 machine gun.

(2) ***Pintle Adapter.*** The pintle adapter is used to mount the MK 19 in all vehicular modes (Figure 3-14). Other accessories are used to mount the MK 19 to specific vehicles. The upper end accepts the gun cradle's pintle, which is secured by a quick-release pin. The lower end of the adapter fits the mounting wells on the M4 pedestal, HMMWV weapon platform, M36A2 ring mount with M66 ring, and commander's cupola on the M113 APC.

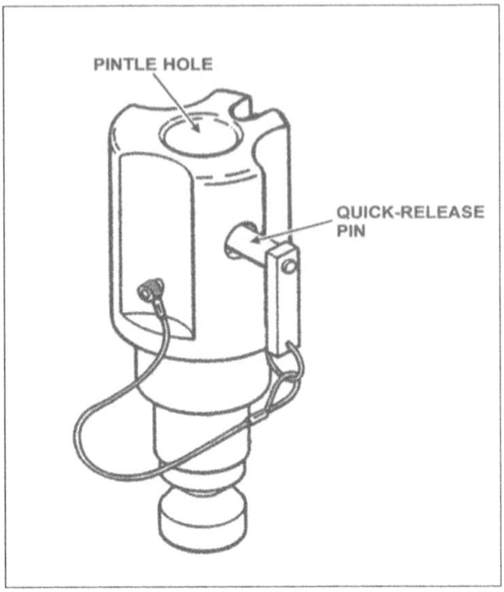

Figure 3-14. Pintle adapter.

(3) ***Train and Elevating Assembly.*** This assembly secures the MK 64 gun cradle to the M4 pedestal or HMMWV weapon platform pedestal and allows directional adjustments. The lower end of the train and elevating assembly is attached to the pedestal by a train lock clamp. The clamp is released or locked in position by a train lock handle. Two positioning clamps are supplied to prevent vertical movement of the train lock clamp on the M4 pedestal. When used on the HMMWV weapon platform pedestal, *only one clamp is needed above the train lock clamp.* The upper end of the train and elevating assembly is a standard caliber .50 T&E mechanism and it is attached to the lower rear holes in the M4 cradle by a retaining pin (Figure 3-15).

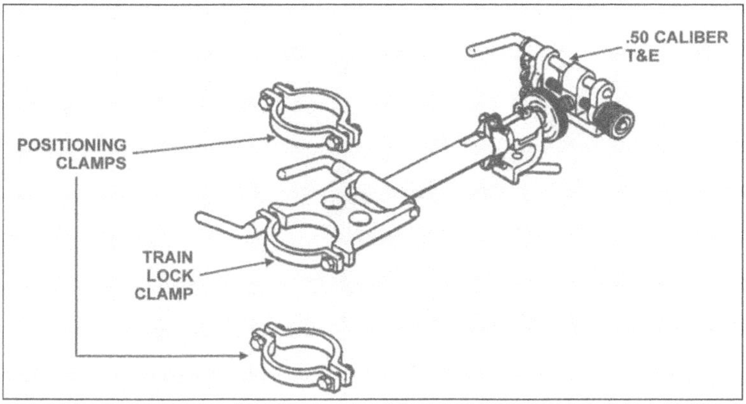

Figure 3-15. Train and elevating assembly.

(4) ***Travel Lock Adapter.*** This adapter is used to hold the gun cradle securely during travel. It is attached to the gun cradle by a retaining pin, which is inserted through the gun cradle train and elevating mechanism mounting holes (Figure 3-16).

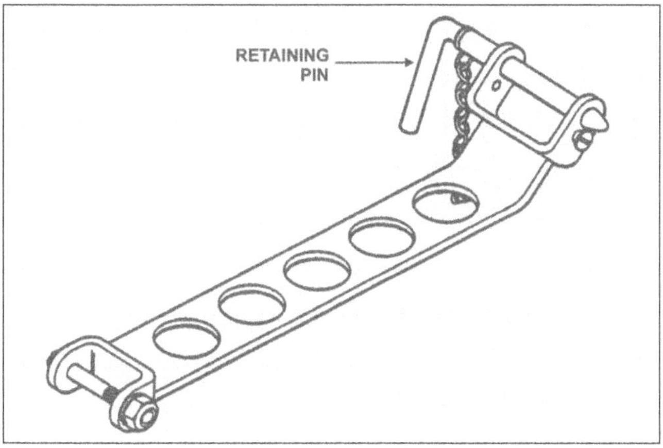

Figure 3-16. Travel lock adapter.

(5) ***Travel Lock Bracket.*** This bracket allows the travel lock adapter to be used with the M4 pedestal mount. It has two halves that are secured around the pedestal with nuts and bolts. The bracket is attached to the travel lock adapter by aligning the lower holes (Figure 3-17).

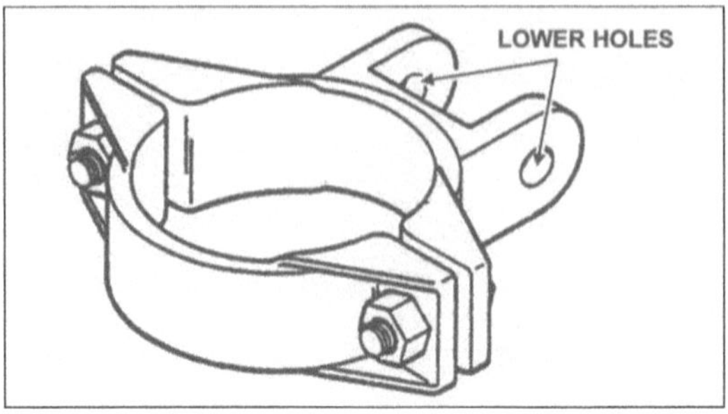

Figure 3-17. Travel lock bracket.

(6) ***Bracket Mounting Assembly.*** This assembly supplies a mount for the M548 metal ammunition container. It has a metal frame that attaches to the gun cradle, and a retaining pin that inserts through the top-inner end of the M548 ammunition container (Figure 3-18).

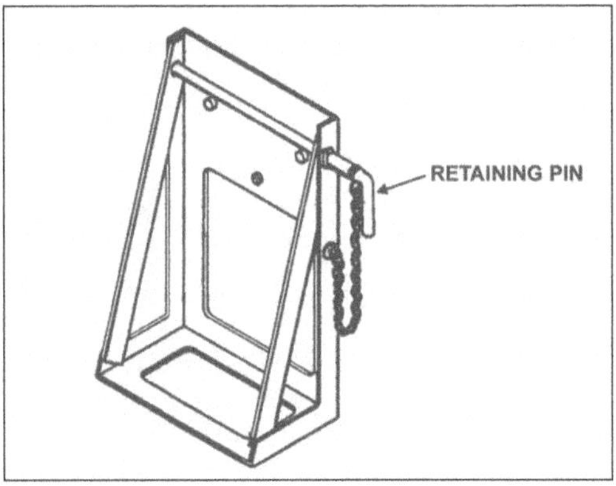

Figure 3-18. Bracket mounting assembly.

(7) ***Empty Case Catch Bag.*** This rubberized canvas bag catches empty cartridge cases as they are ejected. It is held in place by a metal rim, which attaches to the bottom of the cradle under the receiver by two hooks and a hanger (Figure 3-19).

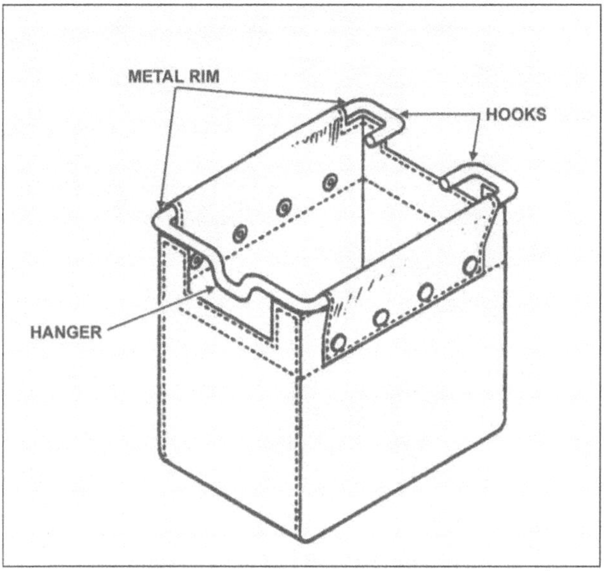

Figure 3-19. Empty case catch bag.

NOTE: The 9/16-, 7/16-, and 3/8-inch open-end wrenches are necessary tools to mount the MK 19 on vehicles.

b. **Vehicle Mounts.** Various vehicle mounts can be used with the MK 19.

(1) ***HMMWV Weapon Platform.*** To mount the MK 19 on the HMMWV weapon platform (Figure 3-20):

(a) Insert the pintle adapter onto the HMMWV pedestal by following these procedures:

(1) Loosen the HMMWV pedestal lock screws by turning counterclockwise until the threaded ends are flushed with the pedestal's inner wall (Figure 3-20).

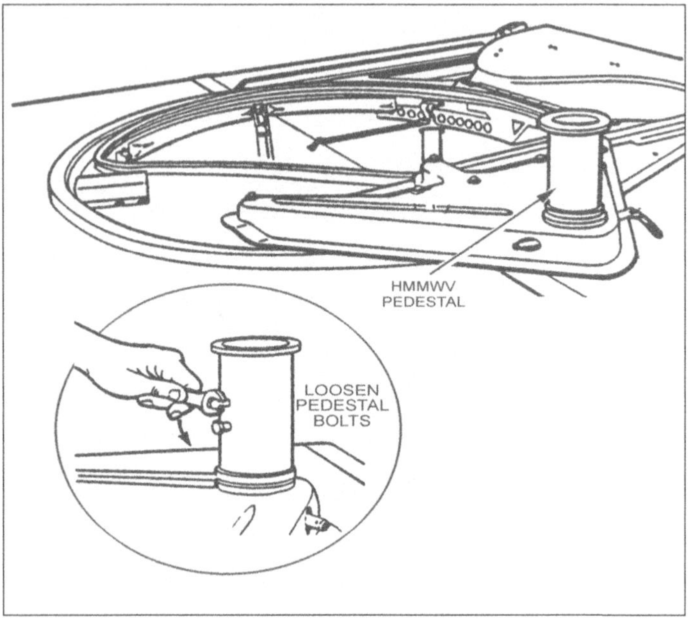

Figure 3-20. HMMWV pedestal.

(2) Insert the pintle adapter assembly by placing the lower end of the pintle adapter into the HMMWV pedestal (Figure 3-21).

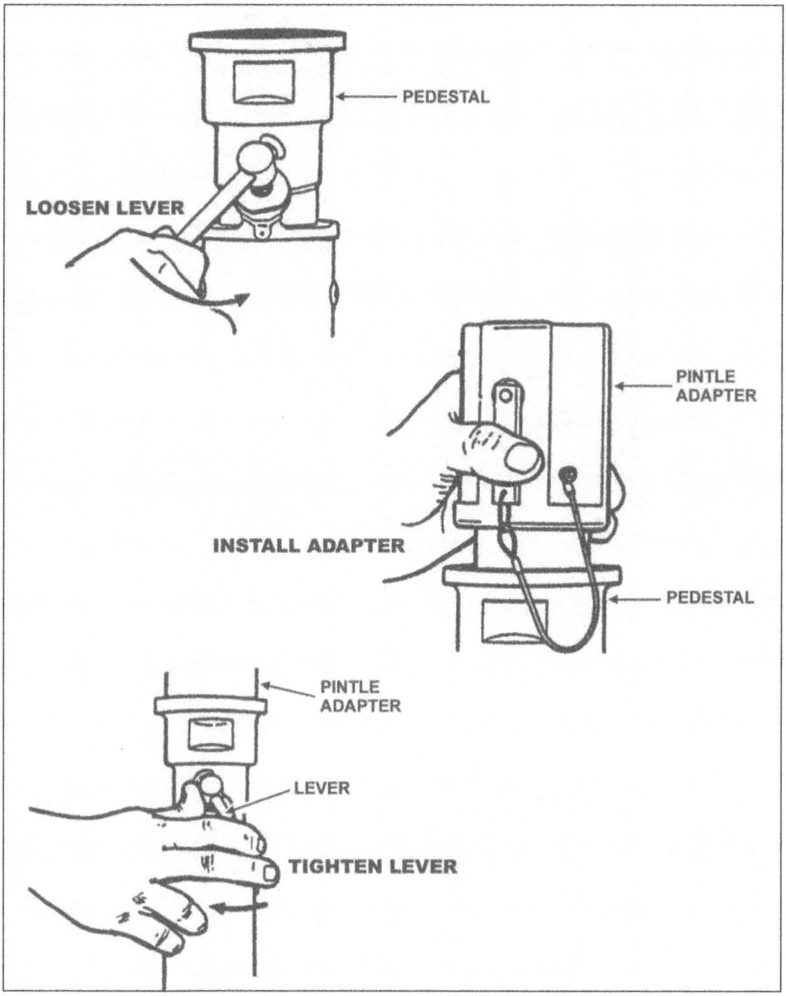

Figure 3-21. Pintle adapter assembly.

(3) Tighten the lock screws with a 3/8-inch open-end wrench. Pull upward on the pintle adapter to ensure it is secured in place.

(4) Remove the pintle adapter quick-release by pressing in on the pin's quick-release button and pulling the pin from the pintle adapter.

(b) Install the carriage and cradle assembly by following these procedures (Figure 3-22):

(1) Insert the front stow pin to prevent relative movement to the carriage and cradle.

(2) Insert the carriage pintle into the top of the pintle adapter assembly.

(3) Press in on the pin's quick-release button and insert the pin.

(4) Pull upward and twist the carriage and cradle assembly. It should lock into the pintle adapter, but it should traverse freely left and right.

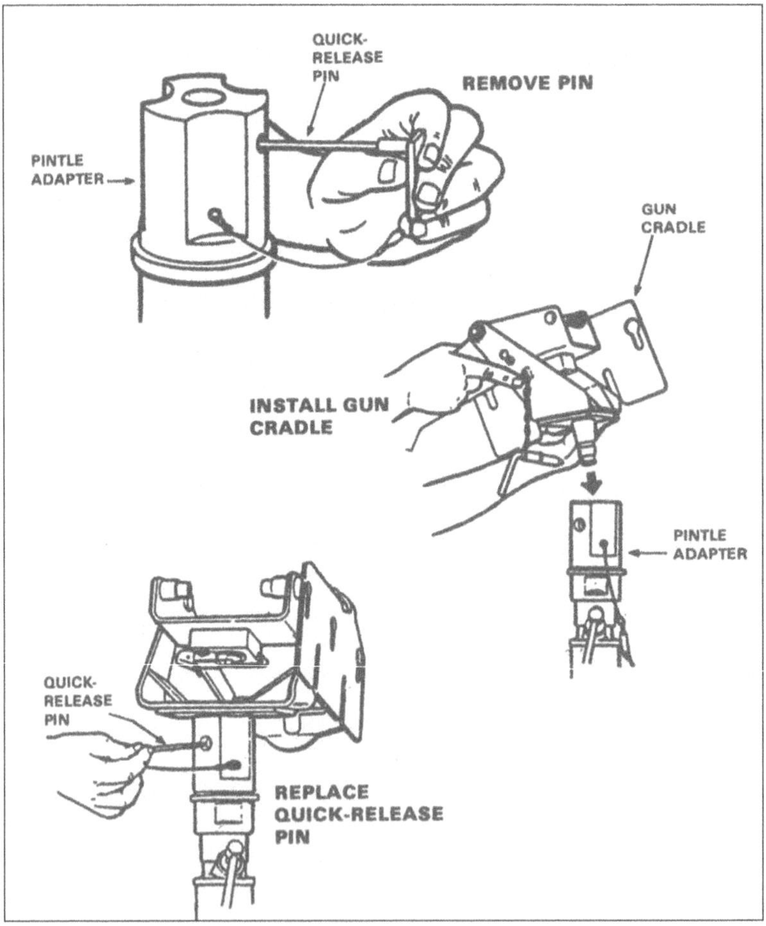

Figure 3-22. Install the carriage and cradle assembly.

(c) Install the T&E mechanism in the train and elevation assembly by following these procedures:

(1) Turn the elevating handwheel to read 250 mils.

(2) Loosen the pivot arm screws, using two 7/16-inch open-end wrenches.

(3) Insert the T&E mechanism all the way into the train and elevation assembly.

NOTE: The T&E mechanism is the upper portion of the training and elevation assembly; it must be installed in the train and elevation assembly *before* the training and elevation assembly can be installed on the vehicle.

(4) Position the traversing lock lever to the rear and the traversing knob to the left by tightening the pivot arm screws, using the two wrenches.

(d) Separate the middle clamp on the train and elevation assembly by following these procedures:

(1) Unscrew the train lock handle counterclockwise to remove.

(2) Using the 9/16-inch open-end wrench, remove the hex head screw on the other side of the middle clamp.

(e) Attach the train and elevation middle clamp to the HMMWV pedestal post by following these procedures (Figure 3-23):

(1) Assemble the middle clamp around the pedestal.

(2) Tighten the clamp by turning the train lock handle clockwise.

(3) Using the 9/16-inch open-end wrench, equally tighten the screw on the other side of the middle clamp.

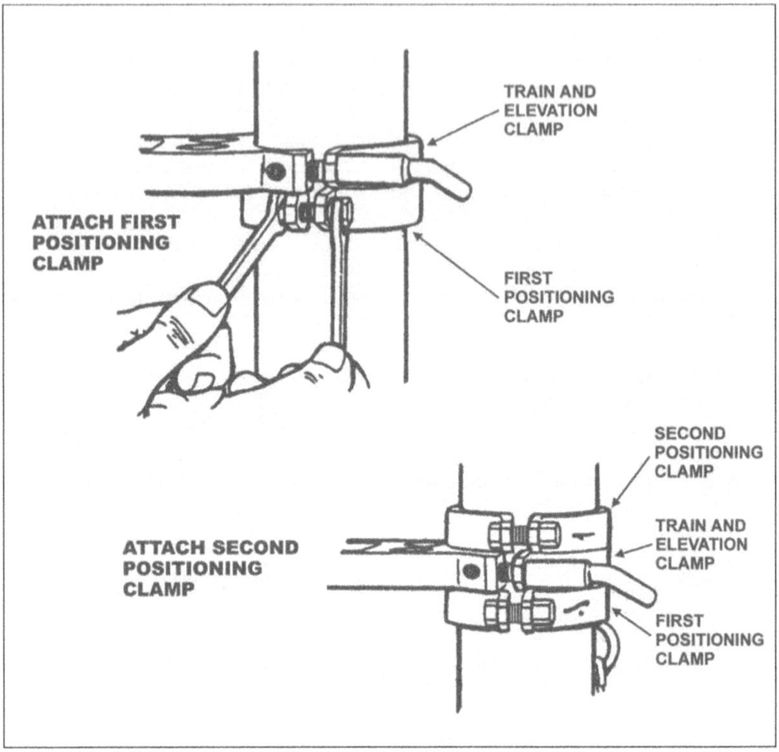

Figure 3-23. Train and elevation middle clamp.

(f) Install a support clamp above the middle clamp. To allow free left and right traverse of the carriage and cradle assembly, loosen the train lock handle. To lock in place, tighten the handle.

NOTE: The second or lower set of clamps is not used in this application since the train lock rests on the built-up base of the HMMWV pedestal. This will also allow for increased elevation. Using two 9/16-inch open-end wrenches, fasten one set of clamps above the middle clamp. Tighten each screw two turns until snug.

(g) Attach the train and elevation assembly to the carriage and cradle assembly by following these procedures:

(1) Pull out the front stow pin to allow the cradle to move.

(2) Pull out the train and elevation assembly's retaining pin.

(3) Ensure the elevating handwheel is set at 250 mils. Position the lock lever to the rear and ensure the traversing knob is on the left.

(4) Turn the traversing knob to center the elevating screw in the yoke.

NOTE: Ensure the elevating handwheel does not get jammed under the lip of the cradle.

(5) Turn the elevating handwheel to align the elevating mechanism holes with the lower rear holes in the cradle.

(6) Insert the retaining pin and rotate to the locked position.

(h) Mount the ammunition can bracket by following these procedures:

(1) Partly unscrew the wing nut on the threaded stud of the bracket mounting assembly. Align the stud with the forward groove in the gun cradle's side plate. Push the bracket mounting assembly up until the heads of the two mounting pins align with the two forward keyholes. Push the heads of the two mounting pins into the keyholes and allow the bracket mounting assembly to slide down. Tighten the wing nut behind the side plate of the cradle (Figure 3-24).

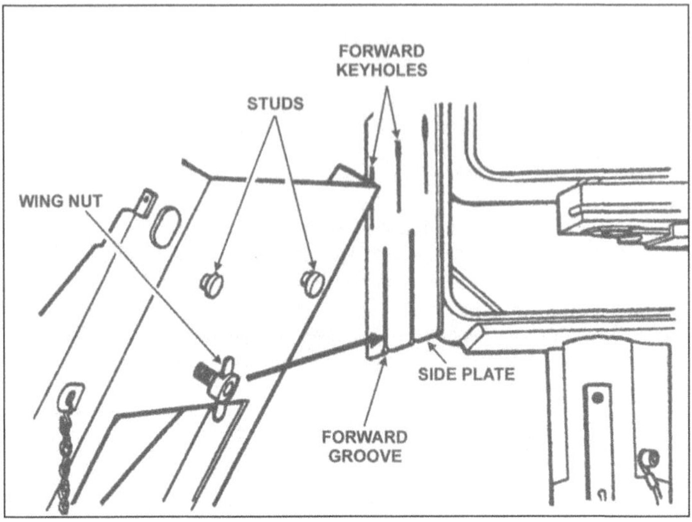

Figure 3-24. Attachment of bracket mounting assembly.

(2) Insert the two hooks on the empty case catch bag through the rear holes in the gun cradle. Engage the single front hanger on the catch bag with the hook on the gun cradle (Figure 3-25).

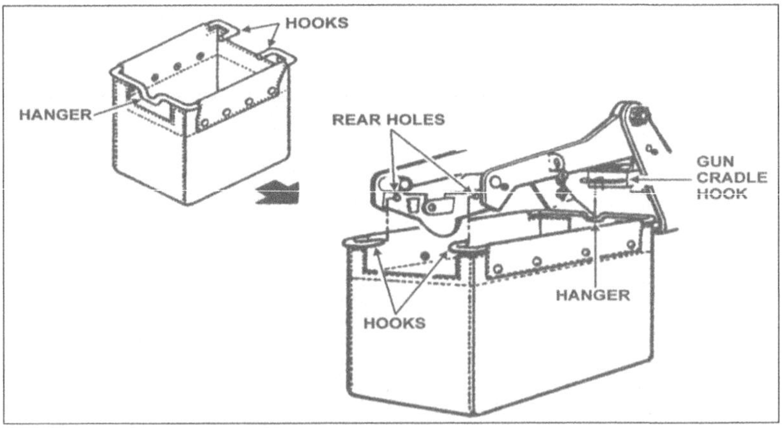

Figure 3-25. Attachment of empty case catch bag.

(2) ***M66 Ring Mount (2 1/2- to 5-ton cargo trucks).*** To mount the MK 19 on the M66 ring mount:

(a) Insert the pintle adapter into the M36A2 ring mount with the M66 ring receptacle.

(b) Install the gun cradle and mount the MK 19.

(c) Attach the bracket mounting assembly and the empty case catch bag.

(3) ***Commander's Cupola (M113 APC).*** To mount the MK 19 on the commander's cupola follow these procedures:

(a) Install the pintle adapter and the gun cradle.

(b) Mount the MK 19 and attach the bracket mounting assembly and the empty case catch bag.

(c) Install the machine gun pintle adapter in the commander's cupola mount by pushing the lock handle down.

(d) Pull the lock handle up, if it does not come up when the pintle is installed.

(e) Insert the front stow pin to prevent relative movement of the carriage and cradle.

(f) Insert the carriage pintle into the top of the pintle adapter assembly.

(g) Press in on the pin's quick-release button and insert the pin.

(h) Pull upward and twist the carriage and cradle assembly. It should be locked into the pintle adapter, but it should traverse freely left and right.

(i) To remove, reverse the steps.

(4) ***The M88 Recovery Vehicle Mount.*** To mount the MK 19 on the M88 recovery vehicle mount follow these procedures:

(a) Loosen the traverse locking screw.

(b) Install the machine gun pintle in the machine gun mount.

(c) Insert the front stow pin to prevent movement of the carriage and cradle. Insert the carriage pintle into the top of the pintle adapter assembly.

(d) Press in on the pin's quick-release button and insert the pin.

(e) Pull upward and twist the carriage and cradle assembly. It should be locked into the pintle adapter, but it should traverse freely left and right.

(f) To remove, reverse the steps.

(5) ***Weather Cover.*** To install the weather cover on the MK 19 follow these procedures:

(a) Unzip the zipper on the weather cover.

(b) Pull the weather cover over the barrel, ammunition can, and rear of the MK 19.

(c) Zip the zipper to secure.

NOTE: The weather cover for the MK 19 is an auxiliary component not common to all MK 64 mounts.

CHAPTER 4

MARKSMANSHIP

This chapter gives guidance for instruction and development of MK 19 gunnery skills. It provides data needed to develop training programs, plans, and lessons used to build the gunner's confidence and skill in firing the MK 19. Marksmanship training ensures that each soldier keeps his combat firing skills at a consistent level with the unit's mission. See Appendix D, Gunnery Tests, for more details concerning diagnostic and intermediate tests.

Section I. PREPARATORY MARKSMANSHIP TRAINING

Preparatory marksmanship training for the MK 19 covers the firing positions, the MK 19 fighting position, use of the T&E mechanism, and manipulation exercises. Thorough instruction during the preparatory training and exercises helps ensure efficient use of time and ammunition during range firing.

4-1. FIRING POSITIONS

Use the basic sitting, standing, and kneeling positions during training and range firing of the MK 19.

a. **Sitting.** When the tripod is used in the low or high position, sit directly behind the gun between the trail legs of the tripod. Extend your legs under the tripod, cross them, or brace your feet on the tripod (Figure 4-1). Place your elbows on the inside of your thighs for support when crossing your legs or bracing your feet on the tripod.

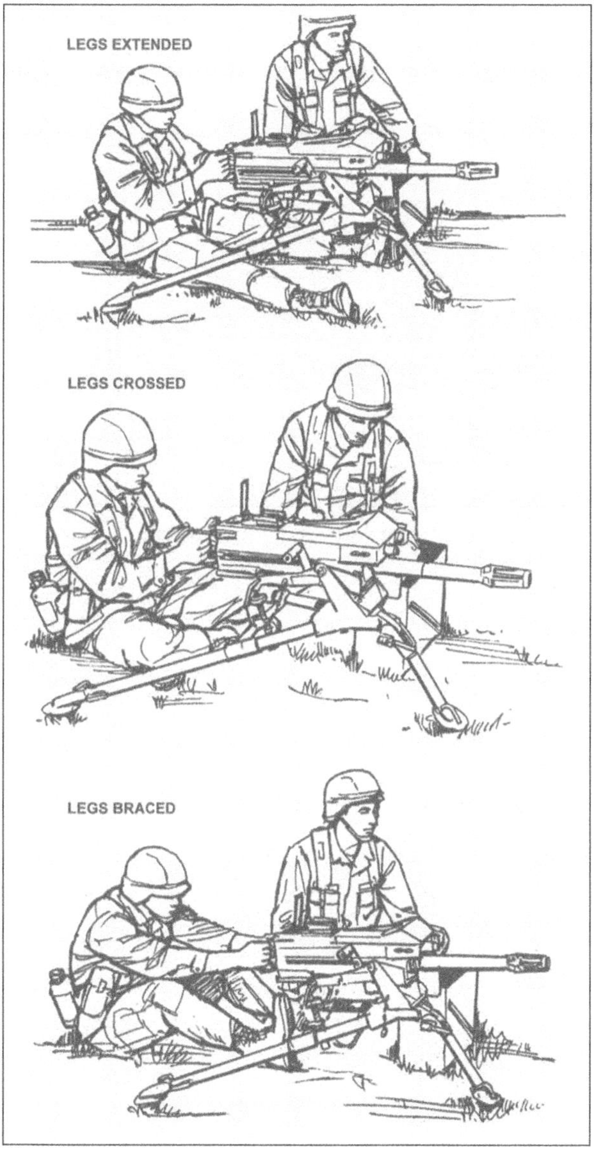

Figure 4-1. Seated firing positions.

b. **Standing.** When the MK 19 is mounted on a vehicle, stand with both of your hands on the control grips and your thumbs resting on the trigger. Keep your elbows against your body, your body forward, and your chest against your hands to brace the gun (Figure 4-2).

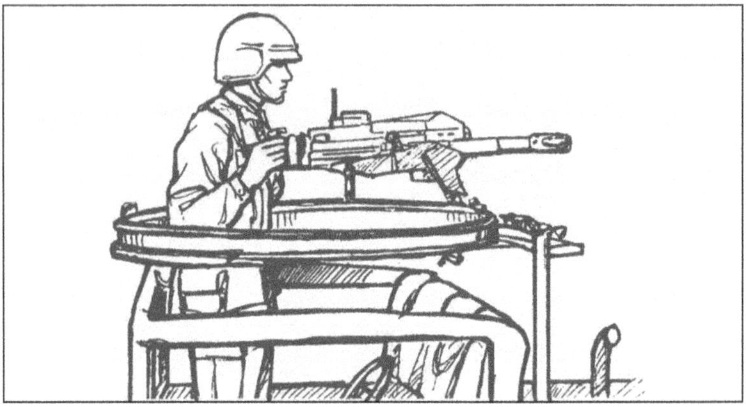

Figure 4-2. Standing position.

c. **Kneeling.** When the MK 19 is in a fighting or hasty tripod-mounted position, kneel and grasp the control grips with your thumbs on the trigger (Figure 4-3).

Figure 4-3. Hasty tripod position.

4-2. FIGHTING POSITION

A fighting position is dug when the unit has time to prepare it or is in a defensive position. Make the fighting position according to Figure 4-4.

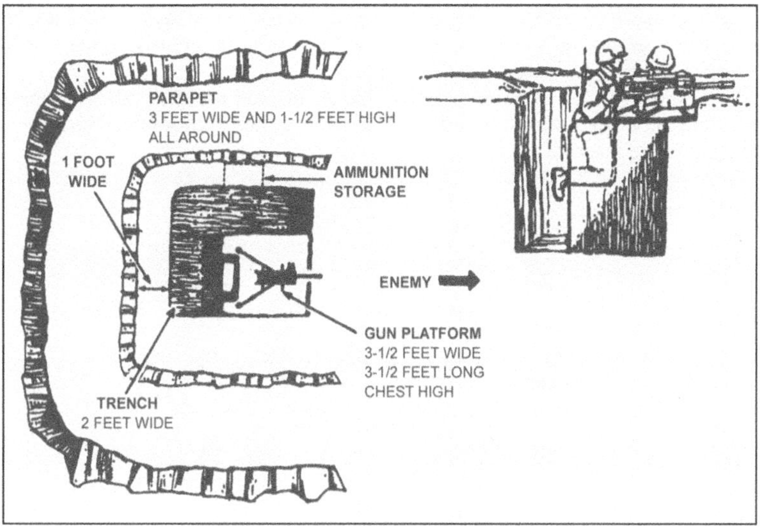

Figure 4-4. Dug-in position.

a. Position the MK 19 so it is oriented on the center of the assigned sector of fire. Mark the position of the tripod legs and trace the outline of the firing platform, which will be dug about 3 inches deep.

b. Dig the position in an "L" shape around the firing platform. The position should be chest-deep and wide enough to allow the MK 19 crew to load, operate, and place effective fire on an assigned sector of fire. The position can be made so that the gunner can fire from the kneeling position by digging the desired depth directly behind the gun.

c. Use the dirt from the hole to build flank parapets 3 feet wide and 1 ½ feet high. Dig the ammunition storage area into the left wall of the "L" and close to the ground.

d. Make sure the wall behind the firing platform is sloped to allow for entrance to and exit from the fighting position.

e. Follow these steps to build overhead cover:

(1) Put flank support logs, 4 to 6 inches in diameter, on top of each other along the entire length of the flank parapets.

(2) Put logs, 4 to 6 inches in diameter, side by side across the support logs as the base for the overhead cover.

(3) Put a waterproof layer over the base logs.

(4) Put 18 inches of dirt on top of the waterproofing material.

(5) Mold and camouflage the cover to blend with the terrain.

f. Camouflage the position with natural materials such as rocks, logs, live bushes, and grass.

4-3. TRAVERSE BAR AND T&E MECHANISM

Use the T&E mechanism to engage preselected target areas at night or during degraded light conditions. Record direction and elevation readings from the traversing bar and T&E mechanism. Record all readings in mils.

a. **Zero the T&E Mechanism.** Before the gunner can use the T&E mechanism to engage targets, he must zero it (Figure 4-5). Refer to section 3-2 for more information on zeroing the T&E mechanism.

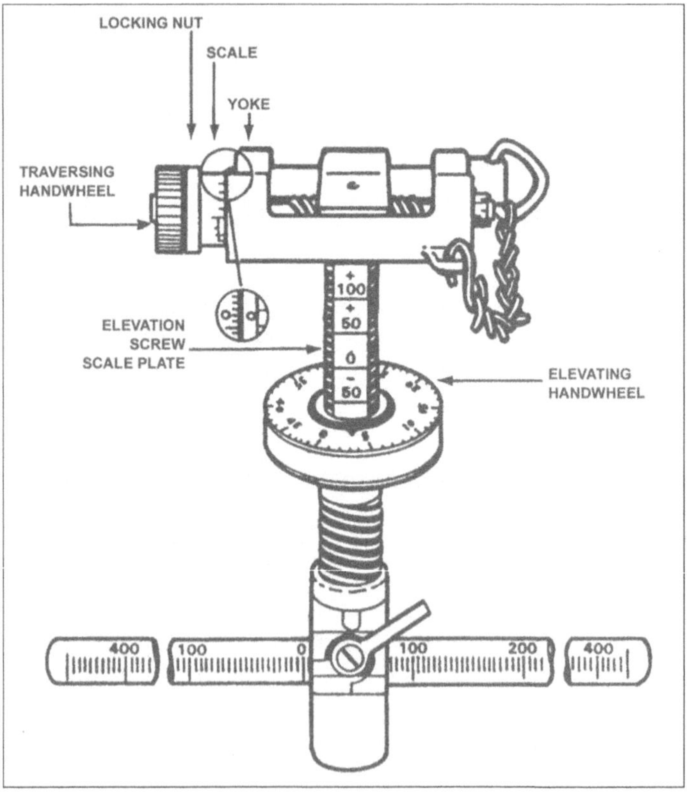

Figure 4-5. Zeroing the T&E mechanism.

b. **Lay the Gun for Direction.** When a sector of fire has been assigned, pick up the rear legs of the tripod and shift the tripod until the muzzle of the weapon points to the center of the sector of fire. Once the MK 19 is laid for direction, firmly stamp in the tripod shoes and place sandbags on the legs. This aids stability and may prevent accidental movement.

c. **Obtain and Record Direction Readings to all Targets Within the Sector of Fire** (Figure 4-6)**.** Loosen the traversing slide lock lever and move the slide along the traversing bar until the MK 19 is laid on the center of a point target or on either flank of a linear target. Lock the traversing bar and read the direction from the scale on the traversing bar. If the left edge of the traversing slide does not fall exactly on a 5-mil tickmark, move the left edge of the traversing bar slide back to the next smaller mil reading. Use the traversing handwheel to complete the initial lay. Obtain a reading to the target by the direction of the MK 19 barrel. If the barrel is moved to the right, record a right reading. Read the number on the traversing bar from the left side of the traversing slide lock. If the barrel is moved to the left and the traversing slide lock is on the right side of the zero, record a left reading. After taking a direction reading for a target, measure the width of the target in mils using the traversing handwheel to move across the target. Reposition the traversing mechanism before moving to another target.

d. **Obtain Elevation Reading.** Ensure the MK 19 is laid on the center base of the target. Read the elevation from two scales: the first, or major, part from the elevating screw plate scale; the second, or minor, part from the elevating handwheel. Separate the two parts of the elevation reading with a slash (for example, -50/3). An elevation reading may not be valid on any other T&E mechanism than the one from which it was read. If the number of threads is increased or decreased after the data is recorded, accurate fire cannot be placed on the target. For example, if a weapon is freed to engage a secondary sector, and the base of the T&E mechanism is rotated, the data is wrong unless the same number of threads is exposed before and after the move.

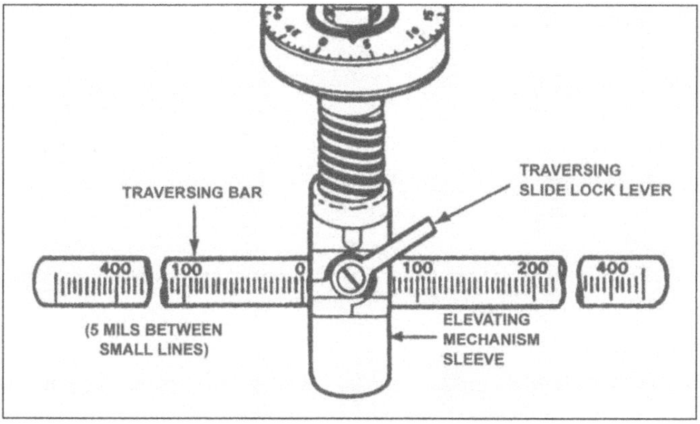

Figure 4-6. Direction reading.

(1) ***Major Reading.*** The elevating scale plate is graduated in 50-mil increments from -250 to +100. Except for the zero, each number has a plus or minus sign above it. All numbers, including zero, have an index line below them. To get the major elevation reading, hold the

T&E mechanism so that the eyes are even with the top of the elevating handwheel. The major reading (for example, -50) is the first number, with an index line, that is visible above the handwheel (Figure 4-7).

(2) ***Minor Reading.*** The elevating handwheel is marked in 50 increments of 1 mil each (Figure 4-7). The minor reading, which has no plus or minus sign, is the number that lines up with the pointer. The minor reading represents the number of mils the gun is laid below the major reading, for example, 3 mils. Verify a direction elevation reading by firing and adjusting on the target. Use the dry-fire method to get data to targets without live firing and adjusting. Set the rear sight on the range to the target, lay the gun on the center base of the target, and take direction and elevation readings. Use the dry-fire method only when firing is not feasible.

NOTE: Correct range setting is critical since any discrepancy causes an error in the elevation when the target is engaged.

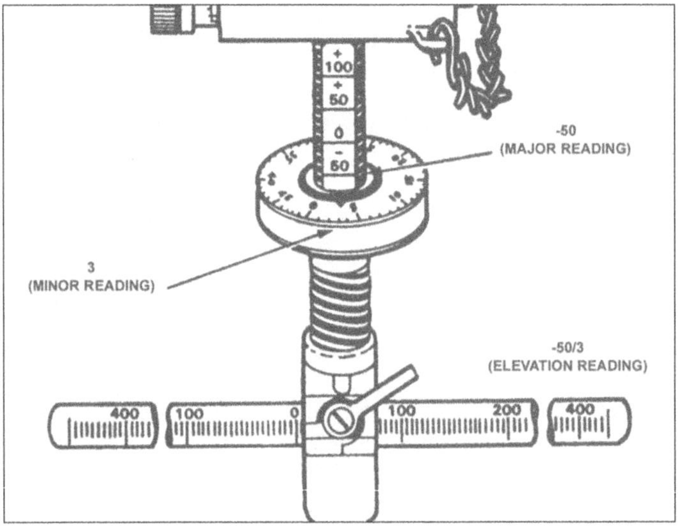

Figure 4-7. Elevation readings.

4-4. EXERCISES

The exercises include sighting and aiming; sight setting and laying; manipulating; determining the range; and preparing a range card.

a. **Sighting and Aiming Exercises.** There are two sighting and aiming exercises. The first calls for the use of the sighting bar.

(1) This exercise teaches how to obtain the correct sight picture, and also teaches alignment methods. Sight alignment means lining up the front and rear sights, with the top center of the front sight blade appearing in the center of the rear sight. Sight picture means lining up the front sight, the rear sight, and the target, with the tip of the front sight blade just touching the bottom center of the target. Gunners must show skill in aligning sights correctly and in getting the correct sight picture, using a sighting bar (Figure 4-8). Gunners must correctly show the point of aim, after sight alignment, 8 out of 10 times.

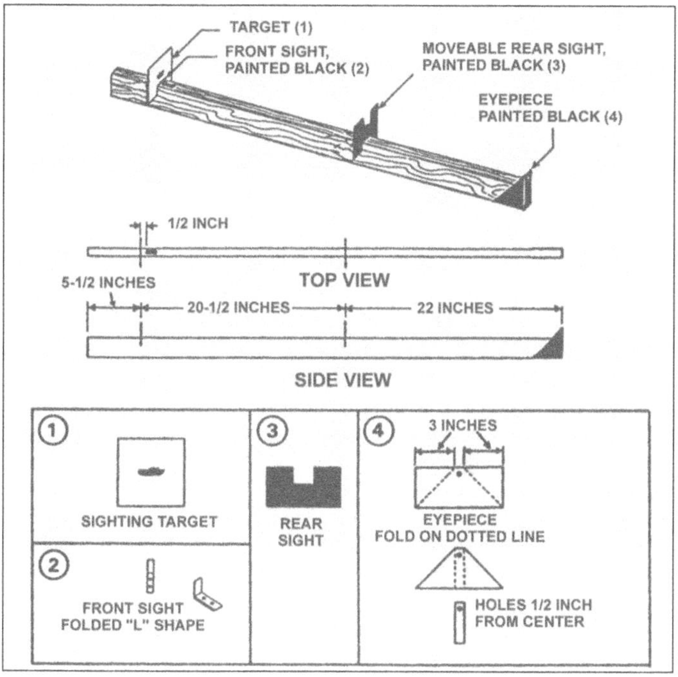

Figure 4-8. Sighting bar.

(2) The second sighting and aiming exercise, used with the MK 19 and target board, allows the gunner to apply lessons learned during the first exercise. Place the target board 10 meters from the muzzle of the MK 19 (Figure 4-9). Use the T&E mechanism to get the correct sight picture on each target that is called out. When the gunner feels he can sight and aim, the coach checks and critiques the gunner's ability to change from the start point on the target board to any new target called by the coach.

NOTE: Continue this exercise until the gunner is skilled in sighting and aiming.

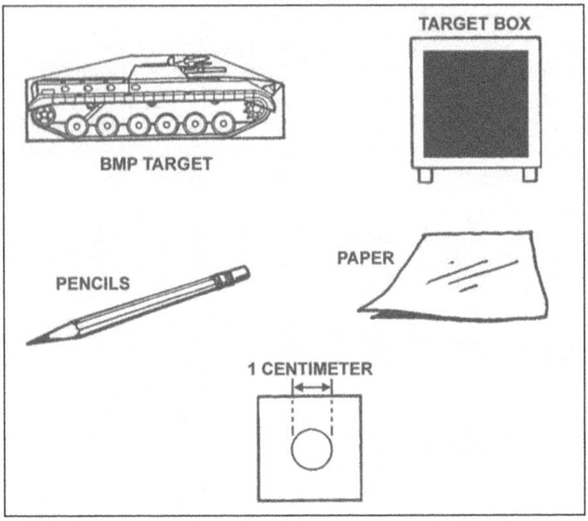

Figure 4-9. MK 19 target board.

b. **Sight Setting and Laying Exercises.** This first exercise teaches the soldier how to set the rear sight quickly and correctly (Figure 4-10). The second builds accuracy and speed in laying the gun on an aiming point, and allows extra practice in sight setting. Explain and show each exercise before beginning instruction.

(1) ***Sight Setting Exercise.*** Explain that, by training the elevating screw knob, the soldier can make minor adjustments in elevation. All major adjustments are made by using the slide release. Have one soldier act as gunner and another act as coach. Announce the range (for example, "EIGHT HUNDRED"). Have the soldier in the gunner's position conduct the following:

- Repeat the range.
- Set the sight at the announced range.
- Assume the correct gunner's position.
- Announce "Up."

The soldier, acting as coach, checks the setting of the slide and points out any errors. The soldiers work in pairs, change roles, and go through the exercise until each one can correctly and rapidly set the sight.

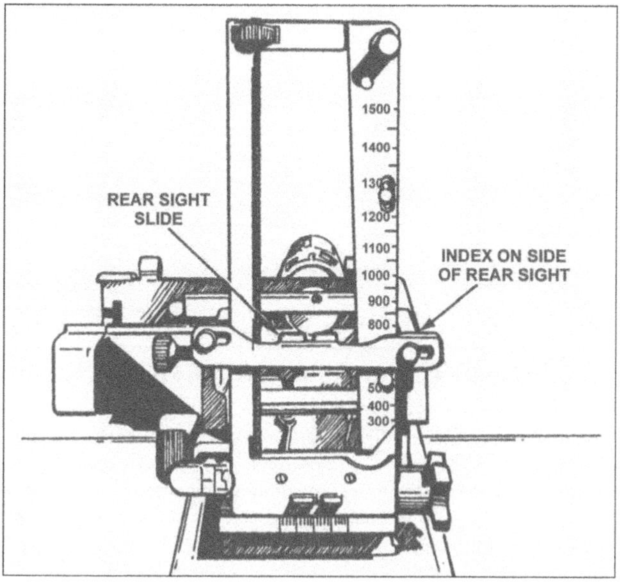

Figure 4-10. Rear sight adjustments.

(2) ***Laying Exercise.*** The MK 19 target board is used in this exercise. Explain that the exercise starts with the sight leaf down and the sight slide at 500 meters. Ranges less than 400 meters or greater than 1,500 meters are not announced. Have one soldier act as the gunner and another act as the coach at the gun. Announce an aiming point and range (for example, "TARGET NUMBER ONE," "ONE TWO HUNDRED"). Pause after each element to allow the gunner time to repeat it. After the gunner repeats each element, have the gunner:

- Set the sights
- Manipulate the gun by turning the T&E handwheels until sights are aligned on the chosen aiming point.
- Assume the correct gunner's position.
- Report "Up."

The soldier, acting as coach, checks the sight setting and lay of the gun. The soldiers work in pairs, changing roles until each is skilled in doing the exercise.

c. **Manipulation Exercise.** Manipulation means shifting the direction of the gun from one point to another. There are two manipulation exercises.

(1) The first exercise gives the gunner practice in manipulation. The coach stands 10 paces in front of the gun and uses hand signals to show the direction in which the gunner is to move the muzzle (Figure 4-11). The gunner works the handwheel to manipulate the gun. The coach observes and makes needed corrections. When the gunner reacts quickly and can manipulate the gun as he has been shown, his instruction may continue.

Figure 4-11. Coach and gunner exercise.

NOTE: Continue this exercise until the gunner is skilled in sighting and aiming.

(2) Once the soldier knows the basics of sighting and aiming, and can assume a good firing position, give him the second manipulation exercise. Show him how to manipulate the weapon to get a correct initial lay and how to skillfully shift the direction of the weapon to successive points. Ensure the soldier understands and knows how to perform the following instructions:

(a) Place the MK 19 target board 10 meters from the muzzle of the gun.

(b) Make large shifts in direction by releasing the traversing slide lock lever and moving the slide to the right or left. Make small changes in direction by turning the traversing handwheel with the left hand. One click on either T&E handwheel moves the strike 1 mil (1 cm on the target).

(c) Manipulate for elevation by rotating the elevating handwheel with the left hand (Figure 4-12).

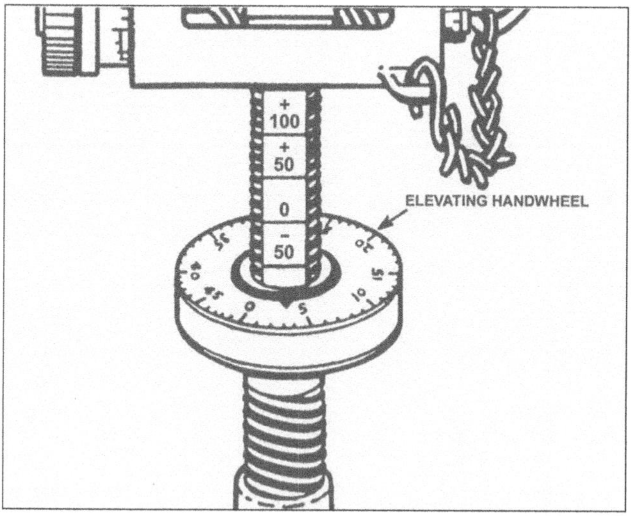

Figure 4-12. Elevating handwheel.

d. **Range Determination Exercise.** Range determination means finding the distance between the gunner's position and his target. The gunner's ability to engage a target depends on his ability to correctly find the range to the target. Under combat conditions, ranges may not be known in advance; the effectiveness of fire depends largely on the accuracy of the range used. There are several methods for finding range. Some of these methods include: estimating visually, firing the weapon, measuring range from a map or aerial photograph, and pacing the distance. Binoculars or laser range finders may be used (see Chapter 5, Techniques of Fire, for more on range determination).

e. **Range Card Exercise.** A range card is a record of the firing data needed to engage predetermined targets within a sector of fire at night or during degraded conditions. The range card may also be used to aid in target engagement during good visibility. It aids the leader in preparing his defense plan. Predetermined targets in the secondary sector are engaged by use of field expedients. A range card has two parts: a sketch section and a data section. Although the sketch is not drawn to scale, the data referring to the targets must be correct. DA Form 5517-R is the standard range card to be used (Figure 4-13).

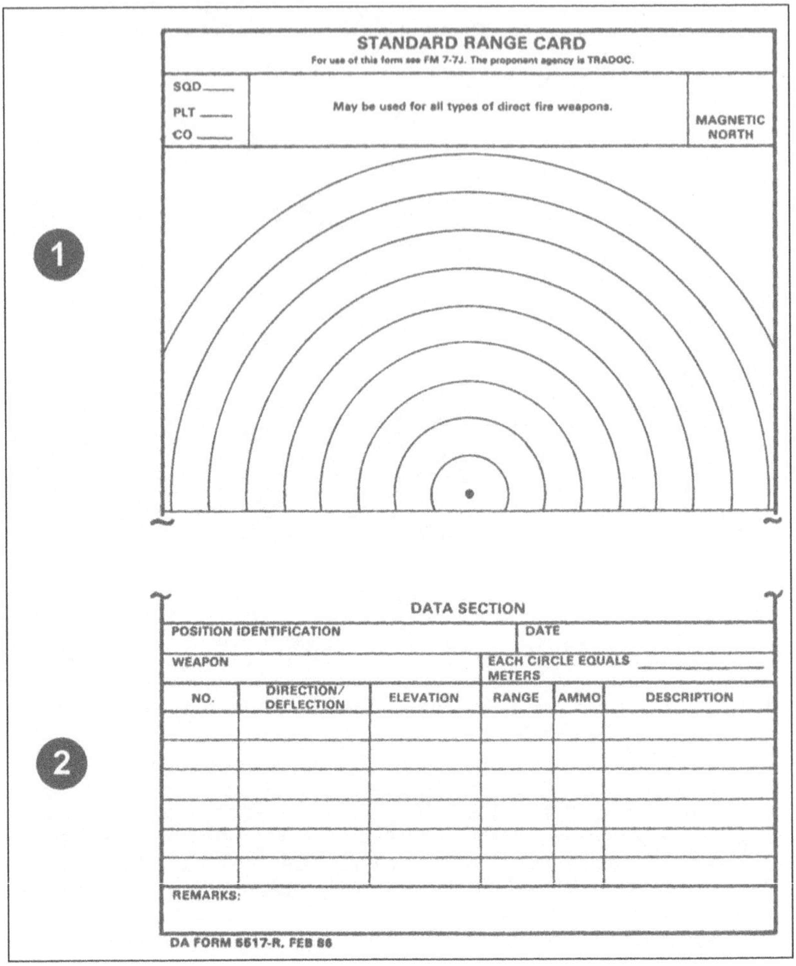

1

STANDARD RANGE CARD

For use of this form see FM 7-7J. The proponent agency is TRADOC.

SQD ____
PLT ____
CO ____

May be used for all types of direct fire weapons.

MAGNETIC NORTH

2

DATA SECTION

POSITION IDENTIFICATION					DATE
WEAPON			EACH CIRCLE EQUALS ______ METERS		
NO.	DIRECTION/ DEFLECTION	ELEVATION	RANGE	AMMO	DESCRIPTION

REMARKS:

DA FORM 5517-R, FEB 86

Figure 4-13. Range card sketch (1) and data sections (2).

(1) Range cards are prepared for primary, alternate, and supplementary positions. They are prepared immediately upon arrival in a position, regardless of the length of stay, and are updated as necessary. Range cards are prepared in duplicate. One copy stays at the gun position, and the other is sent to headquarters. Range cards are prepared on arrival in a position, regardless of the planned length of stay. Range cards are revised often during occupation of a position.

(2) Once the leader has decided where the firing position will be, the gunner prepares the range card.

(3) The range card is prepared as follows:

(a) Draw a sector sketch, covering the entire sector. Make the sketch as large as possible, not to exceed the largest circle. For a large area covered by trees or woods, draw only the outline and label the area "woods" or "orchard" (Figure 4-14).

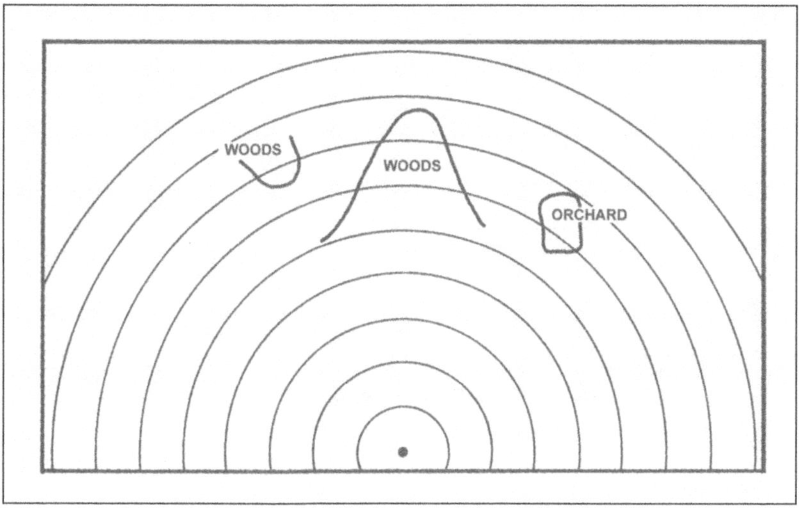

Figure 4-14. Sketch of area on range card.

(b) In the lower center of the sketch section, show the firing position by drawing the symbol for the weapon (Figure 4-15).

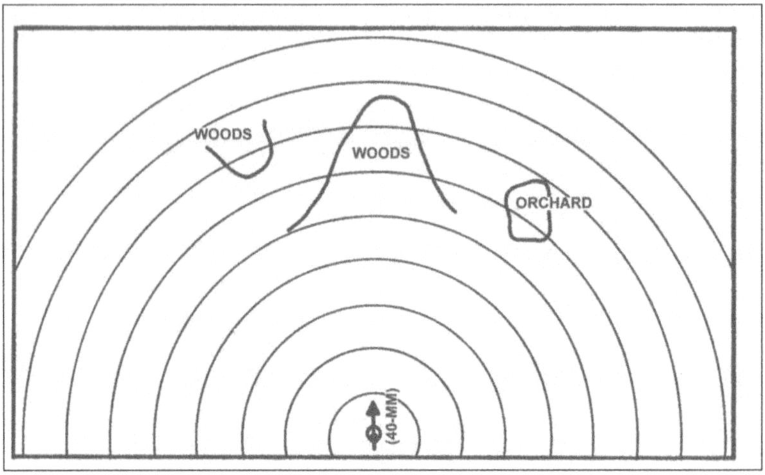

Figure 4-15. MK 19 symbol on range card.

(c) Show the location of the firing position by drawing a sketch of a nearby recognizable terrain feature. Label it and draw an arrow to the weapon symbol. Add the distance and azimuth from the terrain feature to the firing position (Figure 4-16).

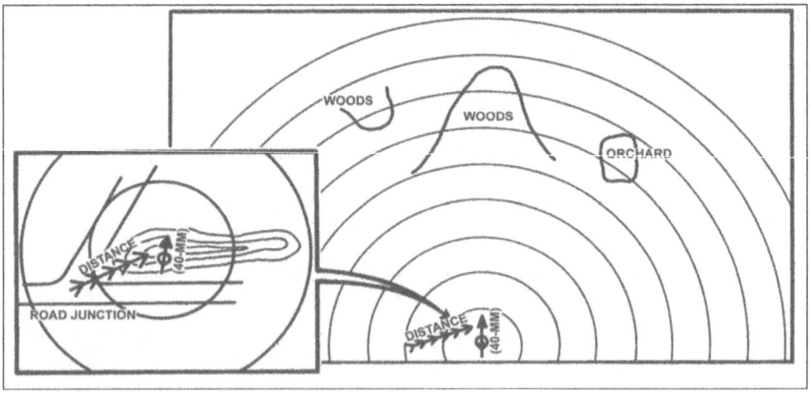

Figure 4-16. Distance and azimuth to terrain feature.

(d) Draw lines from the weapon symbol to reflect the left and right limits (Figure 4-17).

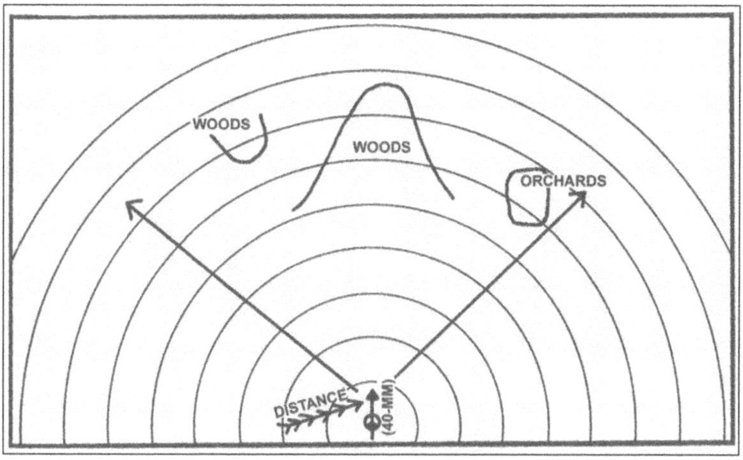

Figure 4-17. Sector of fire.

(e) Number the planned target engagement locations from left to right, and write them on the range card (Figure 4-18).

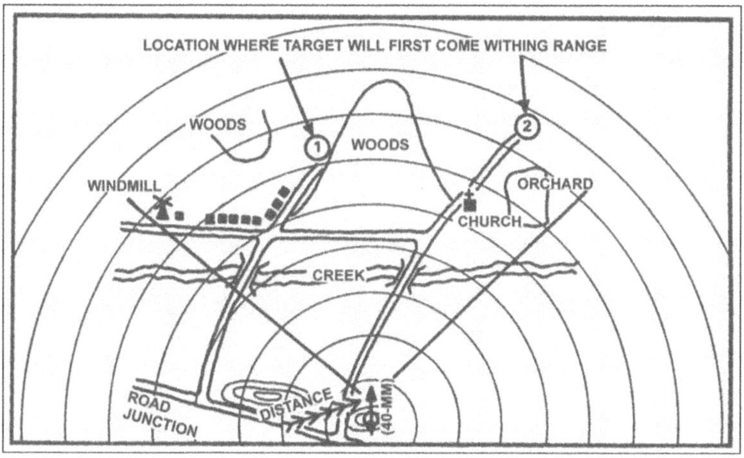

Figure 4-18. Planned target engagement areas.

(f) Number the target reference points (TRPs) and write them on the range card (Figure 4-19).

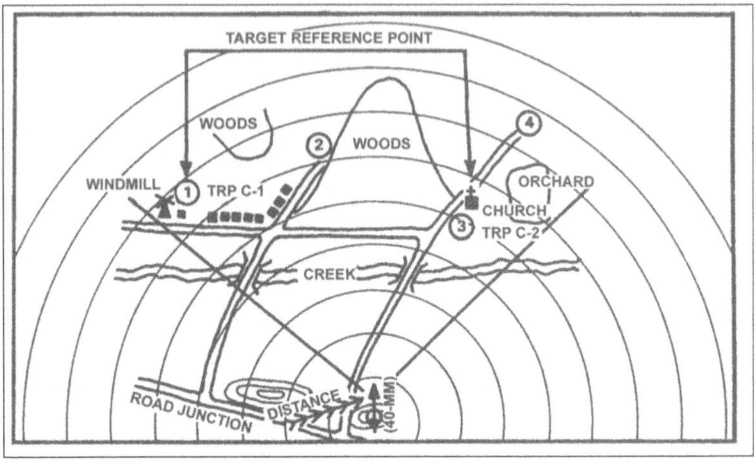

Figure 4-19. Target reference points.

(g) If no limiting factors exist, the maximum engagement line curves and joins the left and right sector-of-fire boundaries at the maximum engagement range (Figure 4-20). If limiting factors exist, draw the maximum engagement line in front of the limiting terrain features (Figure 4-21).

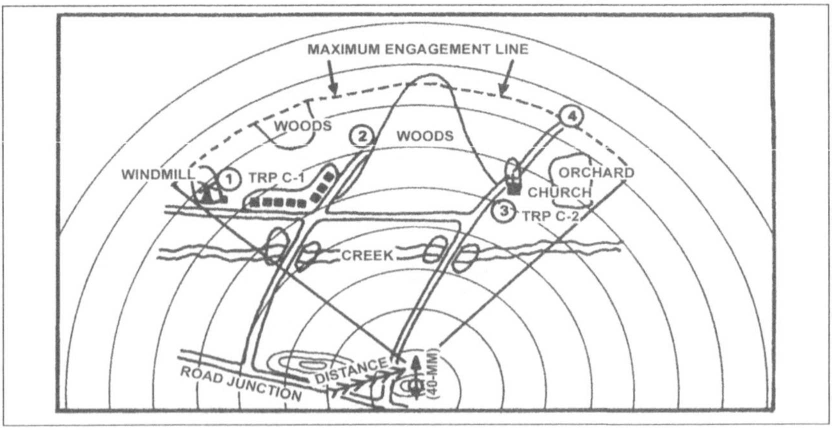

Figure 4-20. Maximum engagement line.

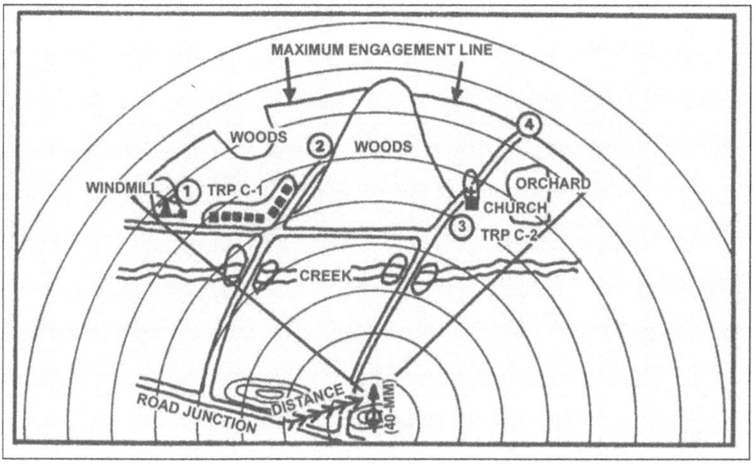

Figure 4-21. Limiting factors on maximum engagement line.

(h) Fill in the marginal information at the top of the card in the following manner (Figure 4-22):

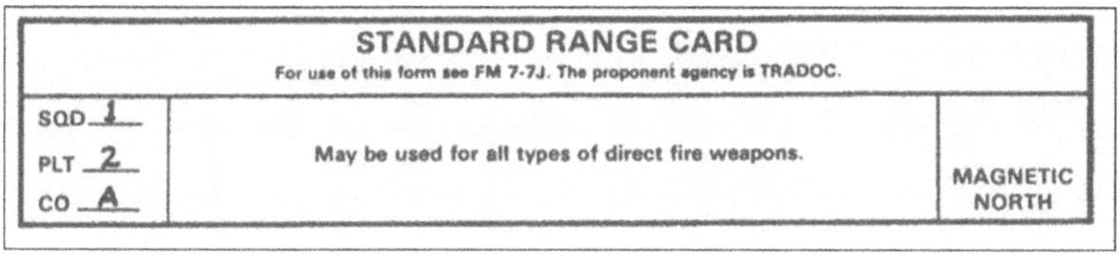
STANDARD RANGE CARD
For use of this form see FM 7-7J. The proponent agency is TRADOC.

SQD 1
PLT 2
CO A

May be used for all types of direct fire weapons.

MAGNETIC NORTH

Figure 4-22. Marginal information.

- **Unit description.** Never show unit higher than company level.
- **Magnetic north.** Orient the range card with the terrain and find the direction of magnetic north with a compass. Draw a magnetic north arrow using the straight edge of the compass.

(i) Fill in the data section in the following manner (Figure 4-23):

- **Position identification.** Write primary, alternate, or secondary.
- **Weapon.** Write MK 19, grenade machine gun.
- **Date.** Write the day and month.

POSITION IDENTIFICATION PRIMARY			DATE 16 DEC		
WEAPON MK19			EACH CIRCLE EQUALS ________ METERS		
NO.	DIRECTION/ DEFLECTION	ELEVATION	RANGE	AMMO	DESCRIPTION
1	L200 MIL	+140/35	2050	M450	WINDMILL/TRP C-1
2	L30 M	+160/35	1750	M430	ROAD
3	R25 M	+140/35	1600	M430	CHURCH/TRP C-2
4	R100 M	+150/10	1400	M430	ROAD
REMARKS:					

Figure 4-23. Data section information.

- **Each circle equals 294 meters.** Write in the distance in meters between the circles. To find the distance, count the intervals from the weapon to the maximum engagement line. Divide the number of intervals into the range. This gives the distance between circles (Figure 4-24). For example:

$$\frac{2{,}212 \text{ meters}}{7.5 \text{ intervals}} = 294 \text{ meters between circles}$$

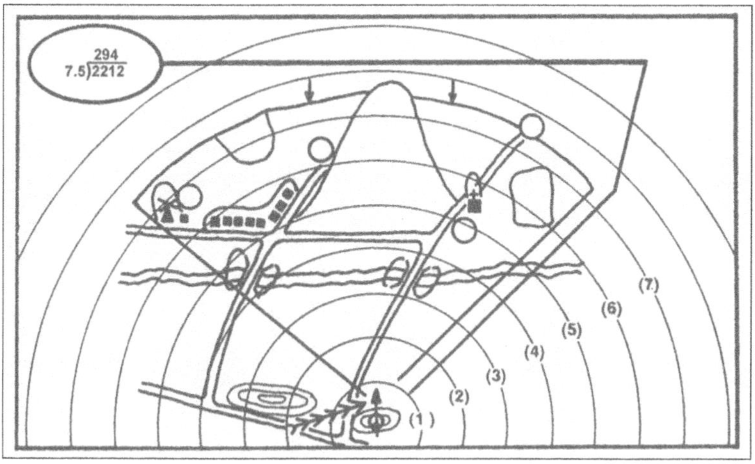

Figure 4-24. Determination of range the circles represent.

- **Number (No.).** Starting with number 1, write the numbers as listed for TRPs and target engagement locations.
- **Direction/deflection.** Write the mils from the traversing bar and handwheel.
- **Elevation.** Write the elevation from the elevating screw plate scale and the elevating handwheel.
- **Range.** Write distance in meters from the weapon to the TRP or target engagement area.
- **Ammunition.** Write the type of round issued for the mission.
- **Description.** List the name of the object; for example: road, windmill, or church. If the item is a TRP, list the TRP number also.

Section II. RANGE FIRE

Range firing gives hands-on MK 19 firing experience to the soldier who has completed the Intermediate Gunnery Test. Use the procedures in this section to set up and conduct range firing. The desired width for each lane is 100 meters at 800 meters from the firing line but can be modified due to local range or terrain limitations. Range firing includes zeroing procedures, observation and adjustment for fire, instructional firing, and qualification firing. Active fighting-force gunners qualify semiannually and participate in exercises and unit-level live fires quarterly. Active supporting-force gunners qualify and participate in instructional fire annually. All reserve component gunners qualify annually. See Appendix E, Annual Gunnery Training Program, for details.

4-5. DESCRIPTION

There are two types of ranges that may be used for MK 19 live-fire practice and qualification. Regardless of which range is used, the MK 19 can be fired from either a tripod or a vehicle.

a. **Multipurpose Gunnery Range for the 40-mm Grenade Machine Gun** (Figures 4-25 and 4-26). This range is designed for the conduct of individual firing exercises and qualifications against hull type targets. Details about this range can be found in TC 25-8.

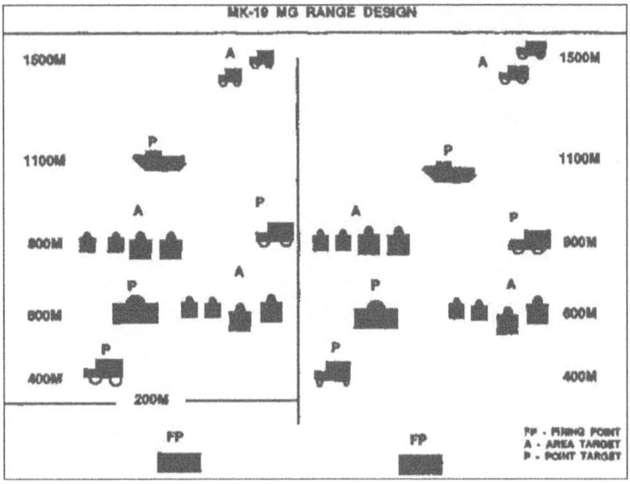

Figure 4-25. Multipurpose gunnery range for the 40-mm grenade machine gun.

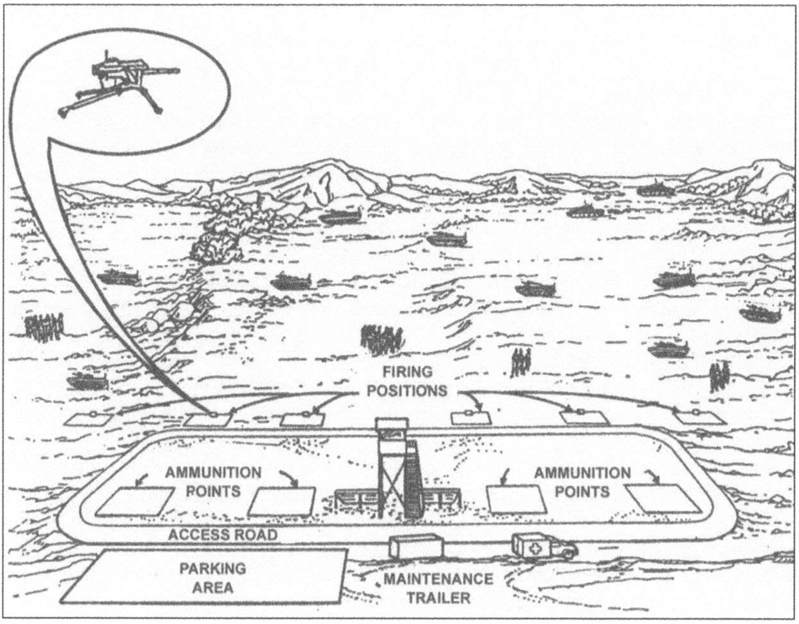

Figure 4-26. Multipurpose gunnery range for the 40-mm grenade machine gun (continued).

b. **Multipurpose Machine Gun Range.** The MPMG is the range suggested for MK 19 live-fire training. It is used to train mechanized, motorized, combat support, and combat service support units. The design and traits of this type of range are described in TC 25-8. Only a non dud-producing practice round is used on this range.

4-6. OPERATION

Use range operation procedures any time LFXs are conducted on the range. These procedures are the same for all range firing exercises.

a. Brief the gun crews on the range setup, fire commands, and safety procedures. Explain exercise needs as defined by the unit commander, or as they are to be done by the firing tables.

b. Divide the crews into groups and assign each group a firing point and an assistant instructor (AI). Ensure they are directed to their firing points and are shown their assigned sectors of fire. Have the AIs ensure that all crews have the equipment needed to perform the task.

c. Conduct the exercise. For example you can command GUNNER, WHEN I SAY BEGIN, PLACE THE MK 19 INTO OPERATION ON THE TRIPOD (ORGANIC CARRIER) WITHIN ____ MINUTES. ARE THERE ANY QUESTIONS? (pause) BEGIN.

At the end of the given time period, direct STOP. AIs, CRITIQUE YOUR CREWS AND GIVE ME YOUR "UP" SIGNAL WHEN YOU HAVE FINISHED.

4-7. COACHES

The instructor may arrange for a coach to be present at each gun during the following preparatory training, exercises, and instructional firing. Successful marksmanship training depends on the coaches. Well-trained and well-rehearsed gunners or crew members may be assigned as coaches. The coaches:

- Require the gunner to inspect his gun and equipment.
- Help the gunner place the gun on target.
- Require the gunner to explain the exercise that he is about to perform.
- Observe the gunner's position, grip, and manipulation during simulated firing.

4-8. FIRE PROCEDURES

Firing the MK 19 is not hard. However, to be effective, the gunner must be able to skillfully manipulate the trigger and T&E mechanism; the weapon must be held down and to one side while aiming and adjusting the T&E mechanism. The following steps are simple, but the gunner must remember to estimate the distance to the target, set the sights for distance, manipulate the T&E mechanism, press the trigger, and fire a single round; since the MK 19 is a fully automatic weapon, this can only be done by "popping" the trigger once. He must spot the impact of the first round and adjust to the target area. He should begin firing six- to nine-round bursts. This whole process can be done in a matter of seconds, the longest part being the flight time of the round. To fire the weapon:

(a) Place the safety on F (FIRE).
(b) Ensure the charger handles are in the forward and upright position.
(c) Place your hands on the control grips, and your thumb(s) on the trigger.
(d) Hold the weapon down and to one side, and check sight picture.
(e) Press the trigger to fire.

WARNING

Because of the gun's recoil, the first burst is the hardest to control. Ensure that the gun's recoil does not cause the barrel of the weapon to drop and rounds to strike short of the target. Keep it elevated.

4-9. ZEROING PROCEDURES

Zeroing procedures are crucial for hitting targets at ranges of 600 meters or more. It is strongly recommended that a target at 400 meters be used to zero. The following is the correct way to zero the MK 19 to the gunner.

a. Loosen the range plate screw. Move the range plate to the midpoint between the two studs. Tighten the range plate screw. Move the rear sight slide to the meter mark that represents the distance to the target. For example, move to the 400-meter mark to zero on a target known to be 400 meters away. Set the windage knob at the zero index line.

b. Align the sights on the base of the target using the T&E mechanism.

c. Fire a single round and spot the impact of the round. If it is on the target, fire another short burst to confirm the zero. If the round is not on target, go on to the next step.

d. To adjust for a round that is not on target, do the following:

(1) If the impact of the round is short or over, adjust the elevation knob. Estimate how short or over the round is. If the round is short, use this estimate to adjust the elevation knob clockwise, which moves the sights up onto the knob counterclockwise to bring the sights down to the target. For example, if the rounds impact 10 mils short, adjust the elevation knob 10 mils up by turning it clockwise. Realign the sights and adjust the gun back on target using the T&E mechanism before the next round is fired.

NOTE: 1. If the adjustment was correct, the second round should be on target. If so, fire the rest of the rounds to confirm the zero. If not, repeat the previous step.
2. If the impacts are not observed, bold adjustments may be necessary.

(2) If the round is to the right or left, adjust the windage knob. Estimate how far to the right or left the sight needs to move to bring the rounds on target. Turn the windage knob clockwise to adjust to the right; turn the windage knob counterclockwise to adjust to the left. For example, if the rounds impact 10 mils to the right, adjust the sight 10 mils to the left by turning the windage knob counterclockwise. Realign and adjust the sights back on target using the T&E mechanism before firing the next round.

(3) Once the zero is completed, align the range plate scale at the exact range of the target used to zero, and tighten it.

e. Point out errors and explain their effect.

f. When the gunner maintains the same sight picture, the type of firing position does not alter the zero.

4-10. OBSERVATION AND ADJUSTMENT OF FIRE

Observation and adjustment gives the soldier a chance to practice adjustment of fire by observing trajectory and impact or by re-laying often on the target using the sights. Live-fire training is held on the firing range after the methods are explained in the classroom.

a. When firing on targets, adjust by moving the burst into the target. Based on the strike of the rounds, adjust the number of clicks for elevation and direction needed to move the center of impact onto the target. This does not call for the use of sights. For example, fire at a target 500 meters away; the rounds impact 20 meters short and 50 meters right. Use the T&E handwheels, moving the muzzle left and up the proper number of clicks, to manipulate the gun onto the target.

b. Another procedure is the adjusted aiming point method. Use the sights and select an aiming point calculated to place the second burst on target. For example, fire at a 500-meter target; the rounds impact 20 meters short and 10 meters right. Rapidly select another aiming point about 20 meters beyond and 10 meters to the left of the target, lay on that aiming point, and fire.

4-11. INDIVIDUAL GUNNER EXERCISES AND QUALIFICATION

The individual gunnery exercises train and qualify MK 19 gunners. There are four scorecards available and they are used based on the type of target (hull or pop-up silhouettes) and whether the practice-qualification is during the day or during limited visibility. Each scorecard has two tables, one for practice and one for qualification. The tables have versions for hull or pop-up silhouette engagements and for the type of NVD used. Sample scorecards are shown in Figures 4-27 through 4-30 (and reproducible scorecards are provided in the back of this manual). MK 19s will be mounted or in the tripod configuration based on the range constraints and the commander's guidance. These tables are *recommended* tables for the infantry MK 19 gunner and crew. The first task in each table is a field zeroing evaluation, which allows the gunner to ensure his weapon is zeroed (even if he boresighted the MK 19). If the gunner fails to zero within four rounds, he is removed from the line and given additional training before attempting the table again. Refer to Section 4-5 for more detailed description of the ranges. The rest of the table consists of firing at individual and multiple targets.

a. **Day and Night Practice and Qualification.** Specific scorecards have been developed for different targets and NVDs. Gunners will only fire one day practice-qualification and one night practice-qualification. Units should select the practice and qualification based on the light conditions, type of targets available, and type of NVDs used. The following table shows which scorecards would be used:

Conditions	Target	Night Vision Devices	Scorecard
Day	Hull	NA	Scorecard I
Day	Pop-Up	NA	Scorecard III
Limited/Night	Hull	**AN/PEQ-2A** mounted on the TWS mounting bracket. **AN/PAS-13** mounted on the TWS mounting bracket. **AN/TVS-5** upgraded with the 3d generation tube mounted on the TWS mounting bracket.	Scorecard II
Limited/Night	Pop-Up or E-Type	All night vision devices.	Scorecard IV
Limited/Night	Any type	No night vision device.	Scorecard IV

Table 4-1. Scorecard matrix.

NOTE: 1. Both the MPMG range and the multipurpose gunnery range (MK 19) can be used for practice and qualification.

2. The MPMG range, modified with stationary armor targets at the required ranges, can be used as pop-up targets if a non dud-producing practice round is used. Scorecards III and IV are used on the MPMG range.

3. The multipurpose gunnery range (MK 19) can be used with any round authorized for the range. Scorecards I and II are used on the multipurpose gunnery range (MK 19).

Every target listed in each task is a point target; however one or two targets between the range of 600 and 900 meters should be changed to an area target. The number of rounds and engagement times should be the same for both point and area targets. The commander's guidance and the local range configuration should determine the location of the area targets.

(1) ***Day Practice and Qualification.*** Due to the types of targets available for practice and qualification, there are two scorecards for day practice and qualification. One (Table I) is used when engaging hull-type targets and the other (Table III) is used when engaging pop-up silhouette targets.

(a) *Hull-Type Targets*. Hull-type targets provide height, width, and depth, and give the MK 19 gunner a realistic target. The engagement ranges for practice and qualifications can therefore be set for the full range of the gun and are set at ranges up to 1,500 meters.

(b) *Pop-Up Silhouette Targets*. Pop-up silhouettes provide a target with width and height but very little depth. Due to the high angle of fall of 40-mm rounds at ranges greater than 800 meters, it is difficult to hit this type of target beyond that range. Therefore, the engagement ranges for practice and qualifications are set at 800 meters or less.

(c) *Engagement Times*. There is a 30 second difference for the completion of each task between the practice and qualification tables. Practice tables allow thirty additional seconds for each engagement.

b. **Night Practice and Qualification.** The MK 19 night practice and qualification tables are shown in Scorecards II and IV. Units with AN/PEQ-2A, AN/PAS-13, and AN/TVS-5 night aiming devices *and* engaging hull-type targets use Scorecard II. Units without a MK 19/sight combination or engaging pop-up silhouettes use Scorecard IV. Gunners do not fire both. Infantry gun crews are required to qualify at night. Other types of units may determine that day qualification is adequate due to their wartime missions.

c. **Scoring.** Scoring is done on a GO/NO GO basis for each task within the practice or qualification table.

(1) Zeroing the gun, the first task in each table, is scored as a GO/NO GO. Giving a score for the zero emphasizes the importance of a proper zero to effectively engage targets at 600 meters and beyond. However, if the gunner fails to zero within four rounds, he is removed from the line and given additional training before attempting the table again. This step reduces the waste of ammunition.

(2) On point target engagements (lightly armored vehicle targets such as BRDMs, threat scout cars, etc.), the gunner receives a GO if he meets or exceeds the engagement standard of one or two rounds hitting the target.

(3) If area targets are included (infantry squads, RPG teams, etc.), the gunner receives a GO when at least the number of rounds stated in the engagement standard for that task impact within ten meters of the area target and thus suppresses it.

(4) At the end of each table, the scorer adds up the number of GOs and NO GOs, places that number in the "Totals block, and checks the box to the left of the appropriate qualification (expert, sharpshooter, marksman, or unqualified).

d. **Range Setup.** Targets should be within the ranges provided on the table scorecard.

(1) Because the gunner has to be able to observe the impact of the round to make adjustments, there should be no dead space within 100 meters of the selected targets.

(2) Every target listed in each task is a point target; however, one or two targets between the range of 600 and 900 meters should be changed to an area target. The number of rounds and engagement time will be the same for both point and area targets. The commander's guidance and the local range configuration should determine the location of the area targets.

(3) For area targets, multiple E type personnel targets may be placed on line or in wedge formations. Multiple personnel targets, indicating area targets, should not be more than 5 meters apart, and not extend more than 30 meters in width or 20 meters in depth.

(4) During night firing using hulls as targets, no modification to the target is needed to assist the gunner in identifying the target. If pop-up silhouette targets are used however, a thermal source is needed on each target to enable the gunner to acquire it with the thermal weapon sight (TWS) and a light source is needed on each target if the AN/TVS-5 is being used. The thermal source can be two chemical lights on targets between 400 meters and 600 meters and three chemical lights on targets over 600 meters.

e. **Grading.** One grader will be required at each firing point.

(1) ***Grading Equipment.*** During the day, the grader will need a set of binoculars. At night equipment will vary according to the type of range being used. With an impact range with hull targets, the grader will need a NVD (examples, the AN/PVS-14,7B with the 3X magnifier, the AN/TAS-4, or the AN/PAS-13 [heavy]) to observe the strike of the round. The same equipment is needed if the pop up targets do not provide feedback. The grader also needs the order in which targets are engaged and a means to provide the gunner with the range to the target for that particular firing point. The grader must be able to identify which target is to be engaged by using, for example, a range card including a diagram of the range with targets numbered and ranges listed.

(2) ***Start and End Time.*** Time will start when the target is exposed and the grader has provided the target range (the graders will provide all information before the target is exposed). If hull targets are used and exposed at all times, then the time will start once the grader has told the gunner which target to engage and provided the range to the target. Time ends when the time indicated for that task expires, the target has been successfully engaged, or the target is no longer exposed.

f. **Ammunition.** Ammunition is broken down by task. The assistant gunner places each belt in its order of use. The number of rounds authorized for each task will be the number of rounds per belt. For example, if there are ten engagements, there should be ten belts of ammunition placed within reach of the assistant gunner in the order they are to be fired.

(1) HE rounds cannot be fired at pop-up silhouette targets because the lift mechanism will be damaged.

(2) Training practice tracer (TPT) rounds can be fired at both types of targets.

(3) The impact of a HE round is much easier to see than that of the TPT round.

g. **Fire Control.** Controlling and observing a target engagement with the MK 19 is not a problem with a range set up with a firing lane for each firing point. Especially with hull targets however, each point will not have an individual firing lane. Some ranges must use the same target for more than one lane, which may be a potential problem for grading. The grader must be able to identify which round impact is from which firing point. This is especially true for the 400-meter target. To prevent this problem, ensure that only one gunner at a time engages each target. The order in which the targets are engaged can be changed to allow more than one gunner to fire at the same time. Engagement start times can also be staggered so that gunners can engage targets at different ranges at the same time. This requires a great deal of coordination and communication between the graders and the personnel controlling the range.

h. **Day Practice and Day Qualification Firing Exercise** (Figures 4-27 and 4-28). Table I contains the tables for day practice and day qualification for hull targets and Table III contains the tables for day practice and day qualification for pop-up targets. Other than a 30 second difference in the engagement times for each task, the practice and the qualification tables are the same. It is held twice a year, or as often as the commander feels is needed to maintain gunner proficiency

(1) The day practice firing exercise allows the gunner to fire on a range engaging hull or pop-up targets to test his skills before qualification firing.

(2) The qualification LFX tests skills practiced during day firing exercise. It is scored on time taken and target hits made based on the firing tables.

(3) During scorecard preparation the grader selects the correct scorecard (Table I for hull targets or Table III for pop-up targets) and enters the gunner's name, rank, and unit in blocks "1" through "3." He also fills in blocks "4" through "7" with the range name, the firing lane, his name, and the date.

(4) The grader positions himself so that he can observe both the gunner and the target. Once live fire commences, he:

- Observes and informs the gunner the strike of each round.
- Observes and records a GO or NO GO for each task.

(5) At the end of the practice, the grader sums the total of GO/NO GOs in the "Totals" block, checks the appropriate qualification in block "9," has the gunner sign the scorecard in block "10," and signs the card in block "11."

(6) During the qualification phase, the grader repeats the steps above by filling in the appropriate blocks, summing the scores, and assigning the correct qualification.

(7) The grader can use the comment section in either table to enter remarks such as the operation of the gun, condition of the targets, and weather conditions to name a few.

MK 19, 40-mm GRENADE MACHINE GUN, MOD 3 FIRING TABLE I
DAY PRACTICE AND QUALIFICATION WITH HULL TARGETS SCORECARD

For use of this form, see FM 3-22.27; the proponent agency is TRADOC.

PRIVACY ACT STATEMENT

AUTHORITY: 10 USC 3012(g)/Executive order 9397
PRINCIPAL PURPOSE: To aid individual training on targets at various ranges.
ROUTINE USES: To evaluate individual proficiency.
DISCLOSURE: Voluntary. However, mass rating and recording require some tracking method.

1a. LAST NAME	1b. FIRST NAME	1c. MI	2. RANK	3. UNIT
PEVOSKI	RICHARD	M	SSG	1/26th IN

TABLE I (A). DISMOUNTED AND MOUNTED DAY PRACTICE

4. RANGE	5. LANE	6. GRADER	7. DATE
DOBOL	5	JOULWAN	4 SEP 03

TASK	RANGE (Meters)	AMMO	TIME (Minutes)	ENGAGEMENT STANDARDS	GO	NO GO
ZERO	400	4	NA	2 ROUNDS HIT	✓	
2	1,100 (+/- 200)	8	2.5	2 ROUNDS HIT		✓
3	1,500 (+/- 200)	10	3.5	2 ROUNDS HIT		✓
4	600 (+/- 100)	6	2	2 ROUNDS HIT	✓	
5	800 (+/- 100)	6	2	2 ROUNDS HIT	✓	
6	400	4	1.5	2 ROUNDS HIT	✓	
MULTIPLE TARGETS						
7	1,100 (+/- 200)	10	4	1 ROUND HIT	✓	
8	600 (+/- 100)			1 ROUND HIT	✓	
9	800 (+/- 100)	14	4.5	1 ROUND HIT	✓	
10	1,500 (+/- 200)			1 ROUND HIT	✓	
				TOTALS	8	

8. COMMENTS: USED HE

9. NUMBER OF ENGAGEMENT MET *(Choose One)*
☐ 10 - EXPERT ☑ 8-7 - MARKSMAN
☐ 9 - SHARPSHOOTER ☐ 6 AND BELOW - UNQUALIFIED

10. GUNNER'S SIGNATURE: Richard M. Pevoski
11. GRADER'S SIGNATURE: George Joulwan

TABLE I (B). DISMOUNTED AND MOUNTED DAY QUALIFICATION

12. RANGE	13. LANE	14. GRADER	15. DATE
DOBOL	8	JOULWAN	4 SEP 03

TASK	RANGE (Meters)	AMMO	TIME (Minutes)	ENGAGEMENT STANDARDS	GO	NO GO
ZERO	400	4	NA	2 ROUNDS HIT	✓	
2	1,100 (+/- 200)	8	2	2 ROUNDS HIT	✓	
3	1,500 (+/- 200)	10	3	2 ROUNDS HIT		✓
4	600 (+/- 100)	6	1.5	2 ROUNDS HIT	✓	
5	800 (+/- 100)	6	1.5	2 ROUNDS HIT	✓	
6	400	4	1	2 ROUNDS HIT	✓	
MULTIPLE TARGETS						
7	1,100 (+/- 200)	10	3.5	1 ROUND HIT	✓	
8	600 (+/- 100)			1 ROUND HIT	✓	
9	800 (+/- 100)	14	4	1 ROUND HIT	✓	
10	1,500 (+/- 200)			1 ROUND HIT	✓	
				TOTALS	9	

16. COMMENTS: USED HG

17. NUMBER OF ENGAGEMENT MET *(Choose One)*
☐ 10 - EXPERT ☐ 8-7 - MARKSMAN
☑ 9 - SHARPSHOOTER ☐ 6 AND BELOW - UNQUALIFIED

18. GUNNER'S SIGNATURE: Richard M. Pevoski
19. GRADER'S SIGNATURE: George Joulwan

DA FORM 7518-R, AUG 2003 APD V1.00

Figure 4-27. Example of completed DA FORM 7518-R, MK 19 day practice and qualification (hull targets).

MK 19, 40-mm GRENADE MACHINE GUN, MOD 3 FIRING TABLE III
DAY PRACTICE AND QUALIFICATION WITH POP-UP TARGETS SCORECARD
For use of this form, see FM 3-22.27; the proponent agency is TRADOC.

PRIVACY ACT STATEMENT

AUTHORITY: 10 USC 3012(g)/Executive order 9397
PRINCIPAL PURPOSE: To aid individual training on targets at various ranges.
ROUTINE USES: To evaluate individual proficiency.
DISCLOSURE: Voluntary. However, mass rating and recording require some tracking method.

1a. LAST NAME	1b. FIRST NAME	1c. MI	2. RANK	3. UNIT
GUNNING	ROBERT	T	SGT	1/26TH IN

TABLE III (A). DISMOUNTED AND MOUNTED DAY PRACTICE

4. RANGE	5. LANE	6. GRADER	7. DATE
SEITZ	2	MULLENIX	9 SEP 03

TASK	RANGE (Meters)	AMMO	TIME (Minutes)	ENGAGEMENT STANDARDS	GO	NO GO
ZERO	400	4	NA	2 ROUNDS HIT	✓	
2	600 (+/- 100)	6	2	1 ROUND HIT	✓	
3	800 (+/- 100)	8	2.5	1 ROUND HIT	✓	
4	400	4	1.5	1 ROUND HIT	✓	
MULTIPLE TARGETS						
5	800 (+/- 100)	12	4	1 ROUND HIT		✓
6	400			1 ROUND HIT	✓	
7	400	10	3	1 ROUND HIT	✓	
8	600 (+/- 100)			1 ROUND HIT	✓	
				TOTALS	7	

8. COMMENTS: CHECK TARGET @ 800M

9. NUMBER OF ENGAGEMENT MET *(Choose One)*
☐ 8 - EXPERT ☐ 6 - MARKSMAN
☑ 7 - SHARPSHOOTER ☐ 5 AND BELOW - UNQUALIFIED

10. GUNNER'S SIGNATURE: Robert T. Gunning
11. GRADER'S SIGNATURE: G. Mullenix

TABLE III (B). DISMOUNTED AND MOUNTED DAY QUALIFICATION

12. RANGE	13. LANE	14. GRADER	15. DATE
SEITZ	5	MULLENIX	9 SEP 03

TASK	RANGE (Meters)	AMMO	TIME (Minutes)	ENGAGEMENT STANDARDS	GO	NO GO
ZERO	400	4	NA	2 ROUNDS HIT	✓	
2	600 (+/- 100)	6	1.5	1 ROUND HIT	✓	
3	800 (+/- 100)	8	2	1 ROUND HIT	✓	
4	400	4	1	1 ROUND HIT	✓	
MULTIPLE TARGETS						
5	800 (+/- 100)	12	3.5	1 ROUND HIT		✓
6	400			1 ROUND HIT		✓
7	400	10	2.5	1 ROUND HIT	✓	
8	600 (+/- 100)			1 ROUND HIT	✓	
				TOTALS	6	

16. COMMENTS: T:6 CAME LOOSE
T 5:6

17. NUMBER OF ENGAGEMENT MET *(Choose One)*
☐ 8 - EXPERT ☑ 6 - MARKSMAN
☐ 7 - SHARPSHOOTER ☐ 5 AND BELOW - UNQUALIFIED

18. GUNNER'S SIGNATURE: Robert T. Gunning
19. GRADER'S SIGNATURE: G. Mullenix

DA FORM 7520-R, AUG 2003 APD V1.00

Figure 4-28. Example of completed DA FORM 7520-R, MK 19 day practice and qualification (pop-up targets).

i. **Night Practice and Night Qualification Firing Exercises** (Figures 4-29 and 4-30). Table II contains the tables for night practice and night qualification for hull targets using the AN/PEQ-2A, AN/PAS-13, or the AN/TVS-5 NVDs. Table IV contains the tables for night practice and night qualification with pop-up targets or hull targets without using NVDs. Other than a 30 second difference in the engagement times for each task, the practice and the qualification tables are the same. It is held twice a year, or as often as the commander feels is needed to maintain gunner proficiency.

(1) The night practice firing exercise allows the gunner to fire on a range engaging hull or pop-up targets to test his skills before qualification firing.

(2) The qualification exercise tests skills practiced during night firing exercise. It is scored on time taken and target hits made based on the firing tables.

(3) The grader selects the correct scorecard (Table II for hull targets using the NVDs listed on the scorecard, or Table IV for pop-up targets or hull targets without a NVD) and enters the gunner's name, rank, and unit in blocks "1" through "3." He also fills in blocks "4" through "7" with the range name, the firing lane, his name, and the date. He also checks the appropriate NVD in block "8" (Table II only).

(4) The grader positions himself so that he can observe both the gunner and the target. Once live fire commences, he:

- Observes and informs the gunner the strike of each round.
- Observes and records a GO or NO GO for each task.

(5) At the end of the practice, the grader sums the total of GO/NO GOs in the "Totals" block , checks the appropriate qualification in block "9," has the gunner sign the scorecard in block "10," and signs the card in block "11."

(6) During the qualification phase, the grader repeats the steps above by filling in the appropriate blocks, summing the scores, and assigning the correct qualification.

(7) The grader can use the back of the form or the comment section to enter remarks such as the operation of the gun, condition of the targets, and weather conditions to name a few.

MK 19, 40-mm GRENADE MACHINE GUN, MOD 3 FIRING TABLE II
NIGHT PRACTICE AND QUALIFICATION WITH HULL TARGETS SCORECARD
For use of this form, see FM 3-22.27; the proponent agency is TRADOC.

PRIVACY ACT STATEMENT

AUTHORITY: 10 USC 3012(g)/Executive order 9397
PRINCIPAL PURPOSE: To aid individual training on targets at various ranges.
ROUTINE USES: To evaluate individual proficiency.
DISCLOSURE: Voluntary. However, mass rating and recording require some tracking method.

NOTE: Use these tables if the targets being used are hulls, and any of the following applies:
AN/PEQ-2A mounted on the TWS mounting bracket.
AN/PAS-13 mounted on the TWS mounting bracket.
AN/TVS-5 with the 3d generation tube mounted on the TWS mounting bracket.

1a. LAST NAME	1b. FIRST NAME	1c. MI	2. RANK	3. UNIT
PEVOSKI	RICHARD	M	SSG	1/26 IN

TABLE II (A). DISMOUNTED AND MOUNTED NIGHT PRACTICE

4. RANGE	5. LANE	6. GRADER	7. DATE
DOBOL	2	CROSS	2 SEP 03

TASK	RANGE (Meters)	AMMO	TIME (Minutes)	ENGAGEMENT STANDARDS	GO	NO GO
ZERO	400	4	NA	2 ROUNDS HIT	✓	
2	1,100 (+/- 200)	8	2.5	2 ROUNDS HIT	✓	
3	1,500 (+/- 200)	10	3.5	2 ROUNDS HIT		✓
4	600 (+/- 100)	6	2	2 ROUNDS HIT	✓	
5	800 (+/- 100)	6	2	2 ROUNDS HIT		✓
6	400	4	1.5	2 ROUNDS HIT	✓	
MULTIPLE TARGETS						
7	1,100 (+/- 200)	10	4	1 ROUND HIT		✓
8	600 (+/- 100)			1 ROUND HIT	✓	
9	800 (+/- 100)	14	4.5	1 ROUND HIT	✓	
10	1,500 (+/- 200)			1 ROUND HIT		✓
				TOTALS	6	

8. TYPE DEVICE *(Choose One)*
- [ ] AN/PEQ-2A
- [ ] AN/PAS-13
- [✓] AN/TVS-5

9. NUMBER OF ENGAGEMENT MET *(Choose One)*
- [ ] 10 - EXPERT
- [ ] 9 - SHARPSHOOTER
- [ ] 8-7 - MARKSMAN
- [✓] 6 AND BELOW - UNQUALIFIED

10. GUNNER'S SIGNATURE: Richard M. Pevoski
11. GRADER'S SIGNATURE: Arnie Cross

TABLE II (B). DISMOUNTED AND MOUNTED NIGHT QUALIFICATION

12. RANGE	13. LANE	14. GRADER	15. DATE
DOBOL	3	CROSS	2 SEP 03

TASK	RANGE (Meters)	AMMO	TIME (Minutes)	ENGAGEMENT STANDARDS	GO	NO GO
ZERO	400	4	NA	2 ROUNDS HIT	✓	
2	1,100 (+/- 200)	8	2	2 ROUNDS HIT	✓	
3	1,500 (+/- 200)	10	3	2 ROUNDS HIT		✓
4	600 (+/- 100)	6	1.5	2 ROUNDS HIT	✓	
5	800 (+/- 100)	6	1.5	2 ROUNDS HIT		✓
6	400	4	1	2 ROUNDS HIT	✓	
MULTIPLE TARGETS						
7	1,100 (+/- 200)	10	3.5	1 ROUND HIT		✓
8	600 (+/- 100)			1 ROUND HIT	✓	
9	800 (+/- 100)	14	4	1 ROUND HIT	✓	
10	1,500 (+/- 200)			1 ROUND HIT		✓
				TOTALS	6	

16. TYPE DEVICE *(Choose One)*
- [ ] AN/PEQ-2A
- [ ] AN/PAS-13
- [✓] AN/TVS-5

17. NUMBER OF ENGAGEMENT MET *(Choose One)*
- [ ] 10 - EXPERT
- [ ] 9 - SHARPSHOOTER
- [ ] 8-7 - MARKSMAN
- [✓] 6 AND BELOW - UNQUALIFIED

18. GUNNER'S SIGNATURE: Richard M. Pevoski
19. GRADER'S SIGNATURE: Arnie Cross

DA FORM 7519-R, AUG 2003 APD V1.00

Figure 4-29. Example of completed DA FORM 7519-R, MK 19 night practice and qualification with hull targets and NVDs.

MK 19, 40-mm GRENADE MACHINE GUN, MOD 3 FIRING TABLE IV
NIGHT PRACTICE AND QUALIFICATION WITH POP-UP TARGETS SCORECARD
For use of this form, see FM 3-22.27; the proponent agency is TRADOC.

PRIVACY ACT STATEMENT

AUTHORITY:	10 USC 3012(g)/Executive order 9397
PRINCIPAL PURPOSE:	To aid individual training on targets at various ranges.
ROUTINE USES:	To evaluate individual proficiency.
DISCLOSURE:	Voluntary. However, mass rating and recording require some tracking method.

NOTE: Use this table if you do not have a MK 19/sight combination that applies to Table II, and you are using hull targets OR if targets are pop-up/E type silhouettes.

1a. LAST NAME	1b. FIRST NAME	1c. MI	2. RANK	3. UNIT
GUNNING	ROBERT	T	SGT	1/26th IN

TABLE IV (A). DISMOUNTED AND MOUNTED NIGHT PRACTICE

4. RANGE	5. LANE	6. GRADER	7. DATE
SEITZ	2	MULLENIX	9 SEP 03

TASK	RANGE (Meters)	AMMO	TIME (Minutes)	ENGAGEMENT STANDARDS	GO	NO GO
ZERO	400	4	NA	2 ROUNDS HIT	✓	
2	600 (+/- 100)	6	2	2 ROUNDS HIT	✓	
3	800 (+/- 100)	6	2	2 ROUNDS HIT	✓	
4	400	4	1.5	2 ROUNDS HIT	✓	
MULTIPLE TARGETS						
5	800 (+/- 100)	10	3.5	1 ROUND HIT		✓
6	400			1 ROUND HIT		✓
7	400	14	2.5	1 ROUND HIT	✓	
8	600 (+/- 100)			1 ROUND HIT	✓	
				TOTALS	6	

8. COMMENTS: OVERCAST BUT TARGETS VISIBLE.

9. NUMBER OF ENGAGEMENT MET *(Choose One)*
☐ 8 - EXPERT ☑ 6 - MARKSMAN
☐ 7 - SHARPSHOOTER ☐ 5 AND BELOW - UNQUALIFIED

10. GUNNER'S SIGNATURE	11. GRADER'S SIGNATURE
Robert T. Gunning	J. Mullenix

TABLE IV (B). DISMOUNTED AND MOUNTED NIGHT QUALIFICATION

12. RANGE	13. LANE	14. GRADER	15. DATE
SEITZ	4	CORLEY	9 SEP 03

TASK	RANGE (Meters)	AMMO	TIME (Minutes)	ENGAGEMENT STANDARDS	GO	NO GO
ZERO	400	4	NA	2 ROUNDS HIT	✓	
2	600 (+/- 100)	6	1.5	2 ROUNDS HIT	✓	
3	800 (+/- 100)	6	1.5	2 ROUNDS HIT	✓	
4	400	4	1	2 ROUNDS HIT		✓
MULTIPLE TARGETS						
5	800 (+/- 100)	10	3	1 ROUND HIT	✓	
6	400			1 ROUND HIT		✓
7	400	14	2	1 ROUND HIT	✓	
8	600 (+/- 100)			1 ROUND HIT	✓	
				TOTALS	6	

16. COMMENTS:

17. NUMBER OF ENGAGEMENT MET *(Choose One)*
☐ 8 - EXPERT ☑ 6 - MARKSMAN
☐ 7 - SHARPSHOOTER ☐ 5 AND BELOW - UNQUALIFIED

18. GUNNER'S SIGNATURE	19. GRADER'S SIGNATURE
Robert T. Gunning	Thomas Corley

DA FORM 7521-R, AUG 2003 APD V1.00

Figure 4-30. Example of completed DA FORM 7521-R, MK 19 night practice and qualification with pop-up targets or without NVDs.

CHAPTER 5
TECHNIQUES OF FIRE

Firing techniques allow the gunner to deliver controlled, effective fire on target. This chapter covers the characteristics and classes of fire; range determination; application of fire; firing under degraded conditions; and predetermined fires.

Section I. CHARACTERISTICS AND CLASSES OF FIRE

To effectively employ the MK 19, gunners must understand the definitions of the terms used to describe the actions and effects of firing the MK 19. This section discusses the characteristics and classes of fire.

5-1. CHARACTERISTICS OF FIRE

The following are definitions of the characteristics of fire:

a. **Trajectory.** The path of the projectile in flight. The path is curved due to gravity and elevation of the barrel. The trajectory increases as the sights are elevated for far targets.

b. **Line of Sight.** An imaginary line from the gun to the target, as seen through the sights.

c. **Ordinate.** The vertical distance that occurs anywhere between the line of sight and the trajectory.

d. **Maximum Ordinate.** The highest point of trajectory, at which the vertical distance between the trajectory and line of sight is greatest.

e. **Danger Space.** The area in which the impact of the round or the dispersal of fragmentation causes injuries to personnel or destruction of the target.

f. **Dead Space.** The area(s) in which personnel or targets are safe from the gun's fire. Dead space can be, ditches, depressions, or ravines.

g. **Cone of Fire.** The dispersion of the rounds as they leave the gun.

h. **Beaten Zone.** The area in which the cone of fire strikes the ground or target. Terrain slope and range to the target affect the beaten zone. If the cone of fire falls on upward-sloping ground, the beaten zone is shortened, and vice versa. As range increases, the beaten zone is shortened and widened.

i. **Center of Impact.** The area where the heaviest concentration of fire impacts.

5-2. CLASSES OF FIRE

Fire is classified with respect to the ground, target, and the MK 19.

a. **Respect to the Ground.** Normally, this means either plunging or grazing fire. However, since grazing fire is not practical for use with the MK 19, only plunging fire will be considered. Plunging fire strikes the target from a high angle and confines the danger space to the beaten zone. For example, when fired from the top of a hill, projectiles follow an arcing trajectory and land in the valley (Figure 5-1).

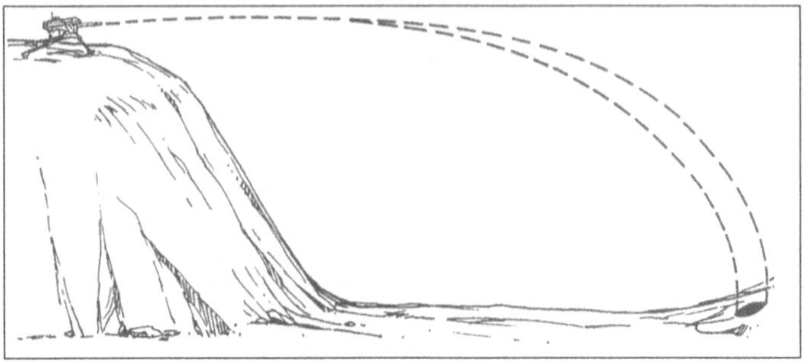

Figure 5-1. Plunging fire.

b. **Respect to the Target.** This class of fire is divided into four types of fire (Figure 5-2a and 5-2b):

(1) ***Frontal.*** The long axis of the beaten zone is at a right angle to the long axis of the target.

(2) ***Flanking.*** This type of fire is delivered against the flank of the target.

(3) ***Oblique.*** This type of fire is directed at a target moving at any angle other than directly toward or perpendicular to the gun.

(4) ***Enfilade.*** This type of fire occurs when the long axis of the beaten zone coincides with the long axis of the target. Enfilade fire may be frontal or flanking, depending on which way the target is facing. For example, frontal enfilade occurs if the MK 19 is in the middle of a road and the target is approaching on the same road. Flanking enfilade occurs if the target is moving either way, left or right, yet is still in a long axis configuration away from the MK 19.

Figure 5-2a. Classes of fire with respect to the target: frontal, flanking, and oblique fires.

Figure 5-2b. Classes of fire with respect to the target: flanking and frontal enfilade fires.

c. **Respect to the MK 19.** Fire from the MK 19 may be conducted in six ways (Figure 5-3).

(1) ***Fixed.*** Fire is delivered against a target with one aim point, which concentrates the beaten zone.

(2) ***Traverse*** Fire is moved from left to right or right to left, with no range change. It may be used against frontal or flanking targets.

(3) ***Search.*** Fire is directed against a deep target. Elevation changes are made, but direction changes are not. Searching fire is used against enfilade targets.

(4) ***Traverse and Search.*** This is a combination of traversing and searching fires used against a target with depth and width (most likely an oblique target).

(5) ***Swinging Traverse.*** This fire is slightly different from traversing and searching. Although it is delivered against a wide target, with major changes in direction, no elevation changes are made. To deliver fire using a swinging traverse, the gunner releases the traversing slide lock, allowing the gun to travel freely across the traversing bar.

(6) ***Free Gun.*** Fire is delivered with the T&E mechanism removed and is used when quick changes in direction and depth are needed to engage moving targets.

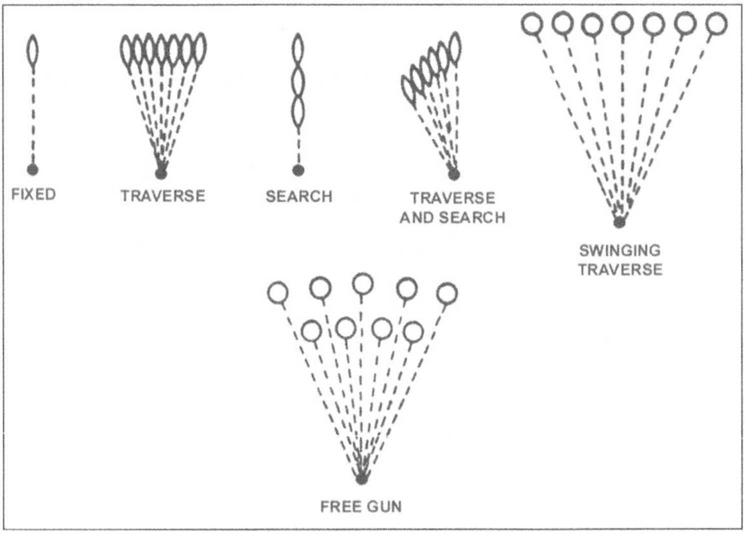

Figure 5-3. Classes of fire with respect to the MK 19.

Section II. RANGE DETERMINATION

Range determination is the process of finding the distance between the gunner's position and his target. The gunner's ability to engage a target effectively depends on determining the correct range to the target.

5-3. MEASUREMENT BY VISUAL ESTIMATION

Range is often determined by this method. There are two ways to estimate range visually.

a. Using the 100-meter unit-of-measure method, visualize 100 meters on the ground (this takes practice) and determine how many units of 100 meters there are to the target (Figure 5-4). For targets more than 500 meters (5 units) away, pick a point about halfway, estimate the range to that point, and double it (Figure 5-5).

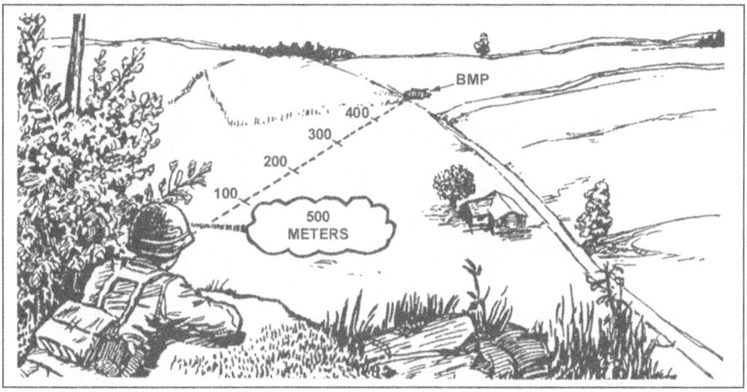

Figure 5-4. Application of the 100-meter unit-of-measure method for ranges up to 500 meters.

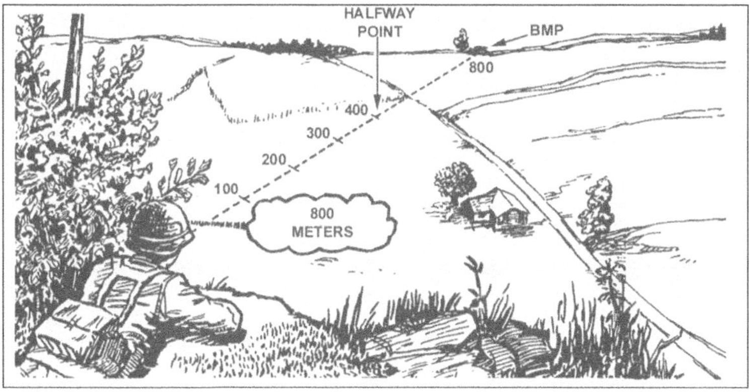

Figure 5-5. Application of the 100-meter unit-of-measure method for ranges greater than 500 meters.

b. Use the appearance-of-objects method, which is fairly reliable, to determine range (this also takes practice). Memorize the sizes and shapes of familiar objects at different ranges. Consider the factors in Table 5-1, which affect the appearance of objects.

FACTORS AFFECTING RANGE ESTIMATION	FACTORS CAUSING UNDERESTIMATION OF RANGE	FACTORS CAUSING OVERESTIMATION OF RANGE
The clarity of outline and details of the target. Nature of terrain or position of the observer. Light and atmosphere.	When most of the target is visible and a clear outline can be seen. When looking across a depression that is mostly hidden from view. When looking downward from high ground. When looking down a straight, open road or along a railroad. When looking over uniform surfaces like water, snow, desert, or grain fields. In bright light or when the sun is shining from behind the observer. When the target is in sharp contrast with the background or is silhouetted because of its size, shape, or color. When seen in the clear air or high altitudes.	When only a small part of the target can be seen or the target is small in relation to its surroundings. When looking across a depression that is totally visible. When looking from low ground toward high ground. When vision is narrowly confined as in streets, draws, or forest trails. In poor light such as dawn and dusk; in rain, snow, fog; or when the sun is in the observer's eyes. When the target blends into the background or terrain.

Table 5-1. Factors affecting visual range estimation.

5-4. MEASUREMENT FROM A MAP

Another way range may be determined is by using a military map. The gunner locates his position and the target's position on the map. He measures the distance and uses the legend scale at the bottom of the map to find the range.

5-5. MEASUREMENT BY PACING

When using this method, the leader ensures there is no immediate danger to the gun's crew. The crew sets up its position, and the gunner selects the target. The assistant gunner walks to the target in a straight line, counting the number of paces it takes to reach the target. As the distance is paced off, the gunner can determine dead space when the assistant gunner disappears from view.

5-6. MEASUREMENT USING BINOCULARS

Binoculars can be used to determine range. The recently adopted M19 binocular, unlike other models, is a lightweight, compact instrument intended for use in general field observation and fire direction. The binocular's left lens includes horizontal and vertical reticles graduated in 10-mil increments (Figure 5-6). When using binoculars to determine range, the soldier must understand the mil relationship (Figure 5-7).

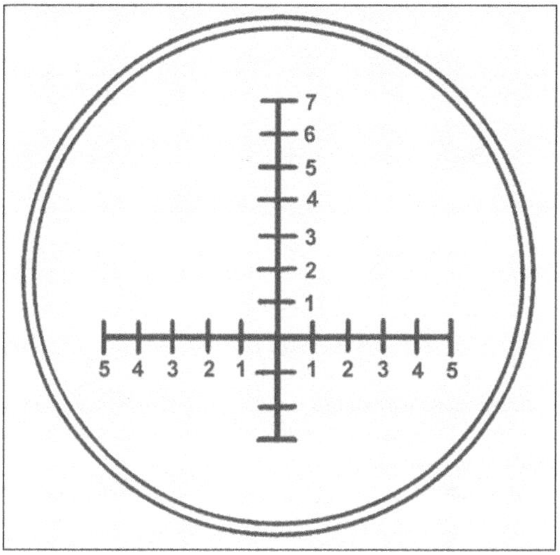

Figure 5-6. M19 binocular reticle.

a. The mil (₥) is the unit of angular measurement used in adjusting fire. A circle has 6,400 mils. At a distance of 1,000 meters, an object 1 meter wide measures 1 mil. Change mils to meters by multiplying the number of mils times the range (distance) in thousands of meters. Obtain the unknown width or range to an object using a similar method (Figure 5-7).

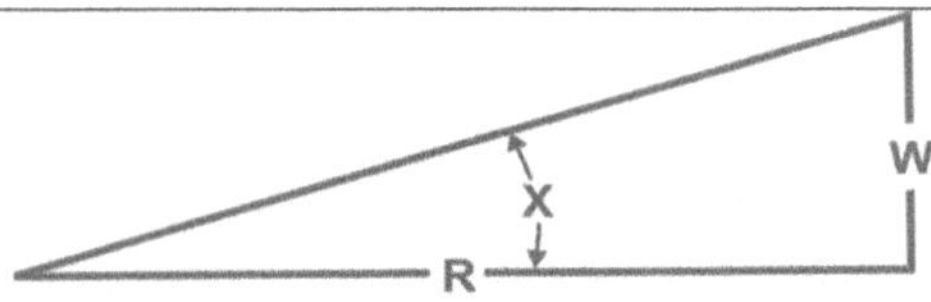

X - IS THE ANGULAR WIDTH OF THE OBJECT IN MILS (₥).

W - IS THE WIDTH OF THE OBJECT IN METERS.

R - IS THE RANGE OR DISTANCE IN THOUSANDS OF METERS.

FORMULA: $X = \frac{W}{R}$

Figure 5-7. Mil relationship.

b. The mil relationship, as shown in Figure 5-7, is X = W ÷ R; where X is the angular width of the object in mils (₥), W is the width of the object in meters, and R is the range or distance in thousands of meters.

(1) To find <u>W</u>idth when the known <u>R</u>ange is 4,000 meters and the object is 15 ₥ wide, multiply R times X ₥ or 4 (range in thousands) times 15 (mils). The answer is 60 (meters), or 4 x 15 = 60 meters.

W = <u>R times</u> <u>X₥</u>

W = <u>4 times 15</u>

W = 60

(2) To find <u>R</u>ange when the known <u>W</u>idth in meters (between two bursts or two objects) is 60 and the angular measurement for the same width, when measured with binoculars, is known to be 15 mils, divide (W: ₥) 60 (meters) by 15 (mils).

The answer is 4,000 meters:

$$R = \frac{W}{X \not{m}}$$

$$R = \frac{60}{15}$$

R = 4 (in thousands of meters) = 4,000 meters

(3) To find the angle A in $\not{m}$ when the known Width in meters between a reference point and the target is 60 meters and the known Range to the target is 4,000 meters, divide W by: R or 60 (meters) by 4 (range in thousands) and the answer is 15 mils (mils):

$$A \not{m} = \frac{W}{R}$$

$$A \not{m} = \frac{60}{4}$$

$$A \not{m} = 15$$

c. The observer uses the mil relation in computing direction shifts as indicated in Figure 5-8.

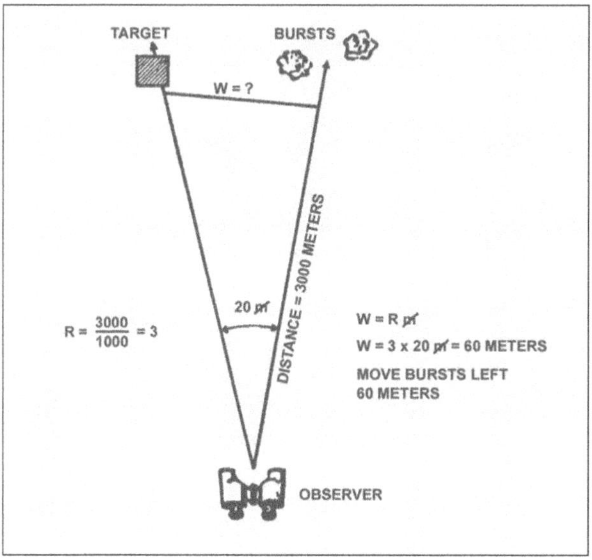

Figure 5-8. Computation of direction shift.

5-7. MEASUREMENT USING AN/GVS-5 LASER RANGE FINDER

The handheld AN/GVS-5 laser range finder provides observers and small-unit leaders the ability to more accurately determine range to targets and other known points. This range finder allows first-round fire for effect, which results in increased enemy casualties and saves ammunition. Friendly units can use the range finder to precisely locate themselves using intersection and modified resection techniques. The range finder aids in the determination of sector depth, range to suspect enemy locations, distance to targets along avenues of approach, and information for overlapping fires. When an AN/GVS-5 is available, it should be used to construct range cards. Information about placing the AN/GVS-5 into action can be found in TM 11-5860-201-10.

5-8. MEASUREMENT BY FIRING

The gunner can determine range by firing a zeroed MK 19. The gunner fires a burst and uses the T & E mechanism to "walk" the rounds on target. The gunner sets and reads the sights on the target, and notes the range. This measurement method is used frequently in combat situations.

5-9. MEASUREMENT OF LATERAL DISTANCE

The gunner can use the finger method to find distance. He extends his arm toward the target with his palm out, elbow locked, one eye closed, and index finger raised (Figure 5-9). He sights along the edge of the finger, adding extra fingers to fill in the space (the average finger is 30 mils wide). One finger equals 30 mils, two fingers equals 70 mils, three fingers equals 100 mils (Figure 5-10).

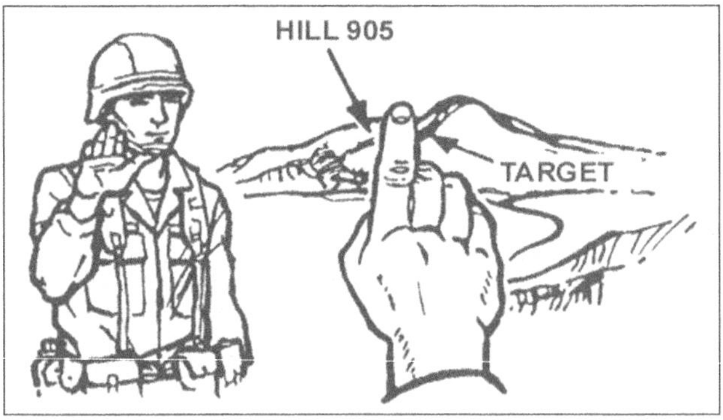

Figure 5-9. Use of fingers to measure deviation.

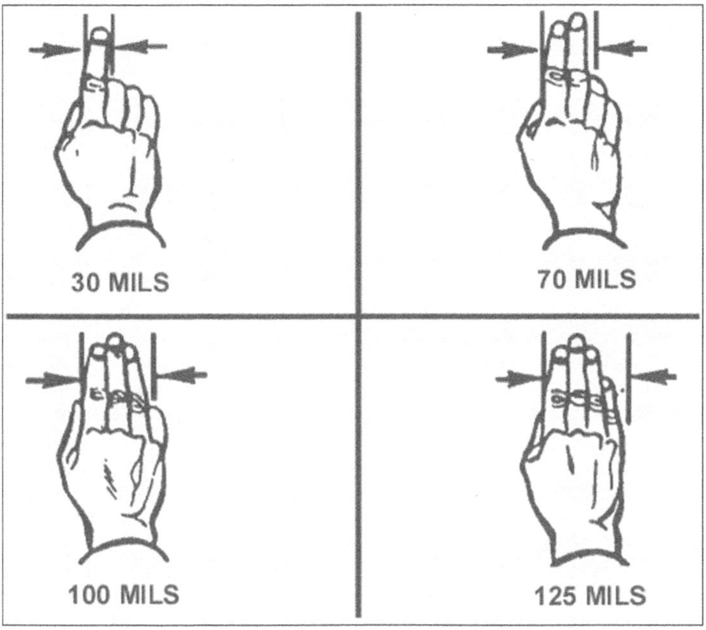

Figure 5-10. Hand-and-finger value estimates.

Section III. APPLICATION OF FIRE

This section includes information about MK 19 fire control and target engagement. Fire control includes all operations connected with the preparation and application of fire to a target.

5-10. FIRE CONTROL

To correctly exercise fire control, the gunner must be able to:

- Open fire at the instant desired.
- Adjust fire on the target.
- Regulate the rate of fire.
- Shift fire from one target to another.
- Cease fire.

Failure to exercise correct fire control results in danger to friendly troops, loss of surprise, premature disclosure of position, misuse of fire on unimportant targets, loss of time in securing adjustments, and waste of ammunition. The main factors to consider for effective fire control are as follows:

a. **Sector of Fire.** This is an area of terrain that an assigned unit covers. It has specific, designated boundaries. Sectors of fire vary in size but are generally limited areas that can be engaged without movement of the tripod.

b. **Rate of Fire.** Surprise and shock effect may be obtained by firing all MK 19s at the same time, using the rapid rate for at least the first few bursts. Engage fleeing targets as soon as possible and with all available fire. Use the rapid rate for the initial delivery of fire to make adjustment of fire easier. Also, whenever the rate of fire is not stated, use the rapid rate. In all cases, unless otherwise ordered, fire the first few bursts at the rapid rate; thereafter, use the prescribed rate.

c. **Adjustment of Fire.** Adjustments may be made before or after firing.

(1) ***Initial Adjustment.*** Set the sights on the range to the target, lay on the target, fire an aimed burst, and observe the impact. When the initial burst is correct, continue to fire (manipulating if necessary) until the target is covered. When the initial burst is not correct, determine from its impact the amount of traverse and search required to place the next burst on the target. Manipulate the gun with the T&E handwheels, making large range corrections by resetting the sights and re-laying on the target.

(2) ***Subsequent Correction and Adjustment.*** Observation and adjustment of fire is the most important element of fire control. It is continuous throughout the action. The gunner is trained to observe and adjust fire without command and to check the lay of the gun frequently. The gunner is also trained to foresee the action of the enemy after he opens fire, and to shift fire to cover any changes in the formation or location of the target. If the gunner fails to do these things, the leader promptly corrects him by announcing or signaling subsequent fire commands. This responsibility to adjust fire continues up through the chain of command. When subsequent fire commands are issued, the gunner makes the required adjustments. When the gun is fired from the tripod mount, subsequent commands are given as adjustments to the elevation or deflection at which the last burst was fired. The adjustments are given in mils and are announced or signaled as ADD, DROP, RIGHT, or LEFT so many mils. Changes in deflection and elevation such as RIGHT 2 or ADD 5 are made with the T&E handwheels.

5-11. FIRE COMMANDS

Fire commands are the means by which leaders control fires. A fire command is a set of instructions, given in a prescribed format that enables the gun crew to properly engage the desired targets. There are two types of fire commands: initial and subsequent. Initial fire commands are used for engagement of targets and the shifting of fire to new targets. Subsequent fire commands are used to adjust fire, change the rate of fire, and cease-fire. The explanation below is based on two gun crews. A good fire command is brief and clear, and in the proper sequence. It is given clearly at a rate that is easily understood by the gunner. Gunners repeat each element to ensure understanding. It is unlikely for a complete initial fire command to be issued during a fire fight. The leader determines which elements of a fire command are obvious to the gunners and which elements must be given to them. Some targets may be engaged with only the alert, range, and command of fire. For example: FIRE MISSION, SEVEN HUNDRED, FIRE. The procedures outlined below are used to accustom gunners to issuing and executing instructions in a logical sequence. The use of complete fire commands in training prepares the gunners to receive fragmentary fire commands in combat.

a. **Initial Fire Command.** This contains the following elements:

- ALERT.
- DIRECTION (as needed).
- RANGE.

- DESCRIPTION (as needed).
- ASSIGNMENT/METHOD (division, manipulation, rate; all only as needed).
- CONTROL.

(1) ***Alert.*** The alert is the first element of the initial fire command. Its purpose is to choose the gun crews and ready them to receive and execute the fire command. FIRE MISSION is announced for all targets. When two or more guns are to fire, the leader announces, FIRE MISSION. If only one gun is to fire, then NUMBER 1 (2) FIRE MISSION is given. When the leader wants to alert all guns, but only wants one gun to fire, the announcement is FIRE MISSION, NUMBER 1 (2).

(2) ***Direction.*** There are several ways to designate target direction:

(a) *Oral.* When the target is not obvious, the leader must tell the gunners where to look. The leader gives direction as: FRONT, RIGHT FRONT, LEFT FLANK, and so on. The leader may identify an indistinct target by the use of a reference point. The selected reference point, announced as REFERENCE, must be a distinctive terrain feature or object in or near the target area. The word TARGET precedes the target description when a reference point is used.

- When the reference point is within the target area, the leader may describe the target as extending so many mils, meters, or fingers from that point. He announces range to the reference point. For example:

  REFERENCE: BUNKER, FIVE ZERO ZERO.

  TARGET: TROOPS EXTENDING RIGHT TWO FIVE, LEFT TWO FIVE.
- If the selected reference point is outside the target area, the leader announces the range to the target. For example:

  REFERENCE: LONE TREE.

  TARGET: MACHINE GUN IN EDGE OF WOODS; FIVE ZERO ZERO.
- The leader may identify the direction to a target that is not obvious by selecting an obvious feature, and (by naming successive reference points) leading the gunner to the target step by step. For example:

  REFERENCE: RED-ROOFED HOUSE, RIGHT OF HOUSE, HEDGE, CENTER OF HEDGE, GATE, ABOVE GATE.

  TARGET: MACHINE GUN.
- For a tripod-mounted MK 19, the interval between the reference and the target is measured by laying the gun on the reference point and manipulating the gun a given number of mils to the target.

(b) *Firing a MK 19.* Pointing out an obscure target by firing a MK 19 is simple, fast, and accurate. However, it may cause loss of surprise or premature disclosure of the gun's position. The leader announces the general direction of fire, if it is not obvious. He then lays one gun on the target, commands: WATCH MY BURSTS, and fires one or more on the target. The leader completes the designation orally using the target's midpoint or extremes: RIGHT FLANK, MIDPOINT, or NEAR END. Firing rifle tracers at the target also makes it easier for the gunners to find.

(c) *Laying a MK 19.* Laying the gun on a target is a simple and accurate method that does not sacrifice surprise. The leader goes to each gun, lays it on the target, and has the gunners check the lay. The gunners open fire at the same time on command.

(3) ***Range.*** This element follows the target description. The leader announces it in even digits, hundreds or thousands. For example: FOUR FIVE ZERO; THREE HUNDRED; ONE ONE HUNDRED; ONE THOUSAND.

(4) ***Description.*** The leader gives a brief target description to inform the gunners of the nature of the target. Examples of target descriptions include TROOPS, which refers to any dismounted enemy personnel; MACHINE GUN, which refers to any automatic gun; and TANK, which refers to any armored vehicle. If several targets are in view, the leader may describe the target or part of a target to be engaged as LEADING TRUCK, RIGHT BUILDING, FAR END, HALTED COLUMN, and so on. If the target is obvious, no description is necessary.

(5) ***Assignment Method.*** This element is used only when specific assignments are required to divide or subdivide the target, assign class of fire, or announce a rate of fire.

(a) *Division (or subdivision).* The leader gives this element only when it is required (paragraph 5-12). Division may be announced as follows:

NUMBER 1 RIGHT HALF, NUMBER 2 LEFT HALF.
NUMBER 1 RIGHT TWO-THIRDS, NUMBER 2 LEFT TWO-THIRDS.
NUMBER 1 RIGHT TWO-THIRDS, NUMBER 2 LEFT ONE-THIRD.

(b) *Manipulation.* The leader gives this element to prescribe the class of fire with respect to the gun, and gives it only if the required manipulation is not obvious. For example:

FIXED.
TRAVERSE.
SEARCH.
TRAVERSE AND SEARCH.
SWINGING TRAVERSE.

(c) *Rate.* The greatest effect results from having two guns open fire at the rapid rate, at the same time. Regardless of the rate announced, gunners open and adjust fire at the rapid rate, and use the announced rate (sustained, rapid, or cyclic) thereafter. Size of target, type of target, and ammunition supply are the factors that influence selection of a rate of fire. Rates of fire are distinguished as follows:

- **Sustained.** The sustained rate of fire is measured in rounds for each minute and is the rate at which a MK 19 may be fired indefinitely without damage from overheating (40 rounds per minute). The leader announces SUSTAINED.
- **Rapid.** The rapid rate is measured in rounds for each minute and is the rate at which a MK 19 may be fired for a limited time without danger of overheating (60 rounds per minute). This rate permits a high volume of fire to be delivered for a short, set time. Gunners use the rapid rate unless another rate is announced.
- **Cyclic.** The cyclic rate is the rate at which a MK 19 can be fired for a limited time only, at the fastest rate the gun will fire (325 to 375 rounds per minute). This rate may be used when the target or area needs the greatest possible suppression. The leader announces CYCLIC.

(6) ***Control.*** The leader uses this element to give the command to open fire. For immediate engagement of the target, the leader gives the command, FIRE, or the arm-and-hand signal to fire without pause. Fire is often withheld for surprise and effect, and both guns open fire at the same time. To ensure this, the leader may preface the command or signal to fire with the words AT MY COMMAND or ON MY SIGNAL. When the gunners are ready

to engage the target, they report UP or announce: NUMBER 1 (2) UP, READY, and the leader gives the command or signal to fire.

b. **Subsequent Fire Commands.** The leader uses these to repeat or correct a fire command, to adjust fire, to cease or commence fire, or to terminate the alert. The deflection correction must always be given first.

(1) ***Repeat Fire Commands.*** If the gunner fails to understand any element of the fire command, a repetition of the element may be requested by announcing the misunderstood element with rising inflection to denote a question. When repeating any portion of the fire command, the leader will preface it with the words THE COMMAND WAS.

(2) ***Correct Fire Commands.*** The leader changes or corrects a portion of the fire command by announcing CORRECTION and giving a command. For example, to change an incorrect range command of 500 meters to 600 meters, the command is CORRECTION, SIX HUNDRED.

(3) ***Adjust Fire.*** The leader adjusts fire if the initial burst was not on target. The leader does this by announcing the direction and elevation needed to get on target ADD 100, RIGHT 20; or DROP 200, LEFT 50.

(4) ***Cease and Commence Fire.*** CEASE FIRE is announced if the leader wishes to interrupt for any reason. This type of subsequent fire command informs the MK 19 crew that it will remain on the alert and that more instructions will follow. Firing is renewed by announcing a subsequent fire command or by announcing a new fire command. Firing is resumed with the same data by using the command FIRE.

(5) ***Terminate the Alert.*** To allow the MK 19 crew to relax between fire missions, termination of the alert is announced as CEASE FIRING, END OF MISSION.

5-12. TARGET ENGAGEMENT

The method chosen depends on terrain, target presentation, type of target, and tactical situation.

a. **Distribution of Fire.** To be effective, fire must be distributed over an entire target. Improper distribution results in gaps between beaten zones and allows some of the enemy to escape or to use guns without effective opposition.

(1) ***Factors Affecting Distribution of Fire.*** No fixed rule can be given as to the widest target that a single MK 19 may effectively engage. Ideally, the target should be no more than 100 mils wide. The traversing screw on the T&E mechanism limits the gun's traverse unless the traversing slide lock lever is unlocked. Wider targets require more traversing time, which prevents the gunner from placing a continuous volume of fire on the whole target. The engagement of a wide target by a single MK 19 requires excessive ammunition.

(2) ***Fire Unit.*** This consists of a pair of MK 19s. If possible, at least two guns should be assigned to the same mission, although sometimes a single MK 19 may be effective. The assignment of a pair to a single mission ensures continuous fire in case either gun is put out of action. Two guns can provide a great volume of fire on the target, and can reduce the time required to cover it.

(3) ***Manipulation of the Tripod-mounted MK 19.*** Traversing fire is moved in 5-mil increments. The MK 19s are fired after each manipulation to ensure the beaten zones overlap. Searching fire is often used on level or evenly sloping ground. When the ground is irregular, however, the amount of search to apply between bursts in order to ensure overlap of the beaten zones is determined by observation.

b. **Engagement of Point Target.** Any target no larger than the beaten zone is a point target, and is engaged by fixed fire. The command is FIXED. MK 19 crews are trained to follow any movement or change in formation made by the enemy after the initial burst of fire. An example of a fire command for a point target is as follows:

FIRE MISSION.
FRONT.
MACHINE GUN.
SIX HUNDRED.
FIXED.
RAPID.
FIRE.

c. **Engagement of Linear Target.** There are several ways to engage a linear target. The method used is chosen based on the number of MK 19s available and whether or not the entire target is visible to the gunner.

(1) ***Linear Engagement with One MK 19.*** A single MK 19 engages the target the same as either one of a pair. The MK 19 is laid just outside either flank (or on a reference point within the target area) and the gunner adjusts fire on the flank (or point). The gunner traverses back and forth across the entire area until told to cease firing. The leader may designate where he wants initial fire. For example, if he directs a single gunner to engage a target with width he may announce:

FIRE MISSION.
FRONT.
TROOPS EXTENDING FROM DEAD TREE RIGHT 20 MILS.
FIVE HUNDRED.
TRAVERSE.
RAPID.
AT MY COMMAND.
FIRE (given after the gunners announce "Up").

(2) ***Linear Engagement with Two MK 19s.*** A linear target may be engaged two ways when a pair of MK 19s is used, depending on the width of the target in mils.

(a) *Target 100 mils or Less in Width.* The normal traversing method is used in this case. Each MK 19 delivers the initial burst of fire on its corresponding flank of the target. Fire is adjusted on that point. Each MK 19 is traversed across the target to the other flank, covering the entire target, until the gunners are told to cease firing (Figure 5-11). The command for this type of fire is TRAVERSE. An example of a fire command used for this situation follows:

FIRE MISSION.
RIGHT FRONT.
TROOPS, EXTENDING FROM DEAD TREE RIGHT TO CLEARING.
SEVEN FIVE ZERO.
TRAVERSE.
RAPID.
AT MY SIGNAL.
FIRE (given after the gunner announces "Up").

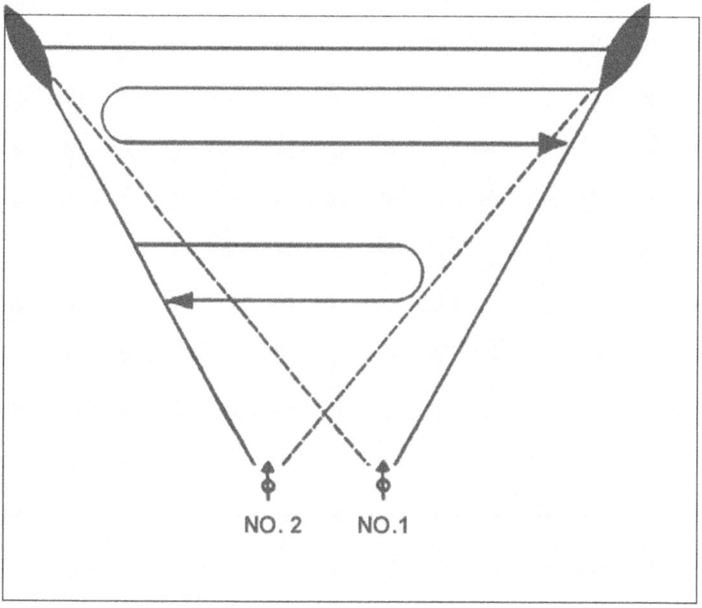

Figure 5-11. Traversing method, pair of MK 19s, target 100 mils or less in width, both flanks visible.

(b) *Target More than 100 mils in Width.* In this case, the leader assigns part of the target to each MK 19 (Figure 5-12). One part may be smaller than the other to ensure it receives a heavier concentration of fire. Each gun fires an initial burst on its respective flank, and covers its assigned part of the target as described in subparagraph (a). An example of a fire command follows:

FIRE MISSION.
RIGHT FRONT.
TROOPS, EXTENDING FROM DEAD TREE RIGHT TO CLEARING, ROW.
EIGHT HUNDRED.
NUMBER 1, RIGHT ONE-THIRD; NUMBER 2, LEFT TWO-THIRDS.
TRAVERSE.
RAPID.
AT MY COMMAND.
FIRE (given after the gunners announce "Up").

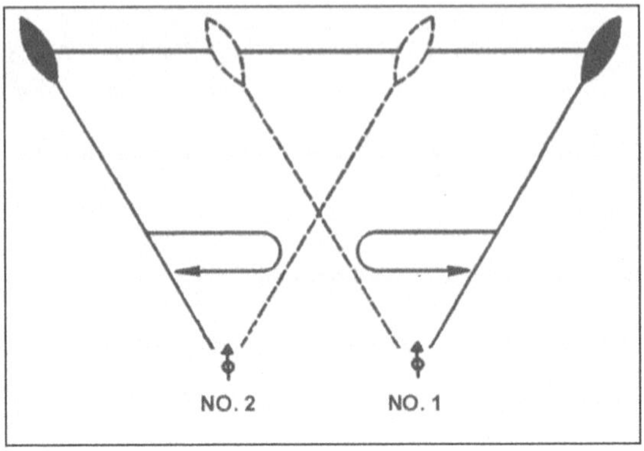

Figure 5-12. Traversing method, pair of MK 19s, target more than 100 mils in width.

(3) ***Linear Engagement with Four MK 19s.*** Four guns may be needed to concentrate a heavy volume of fire into a wide target area. Methods used are similar to the ones described for a pair of MK 19s. The width of the target in mils determines which type of linear engagement is used.

(a) *Target 100 to 200 mils in width*:

- Each pair of MK 19s engages the whole target (Figure 5-13).

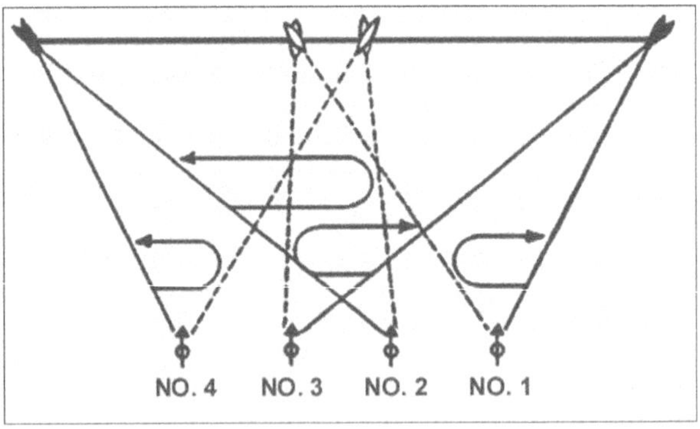

Figure 5-13. Four MK 19s, target 100 to 200 mils in width.

- Each pair of MK 19s engages half the target (Figure 5-14).

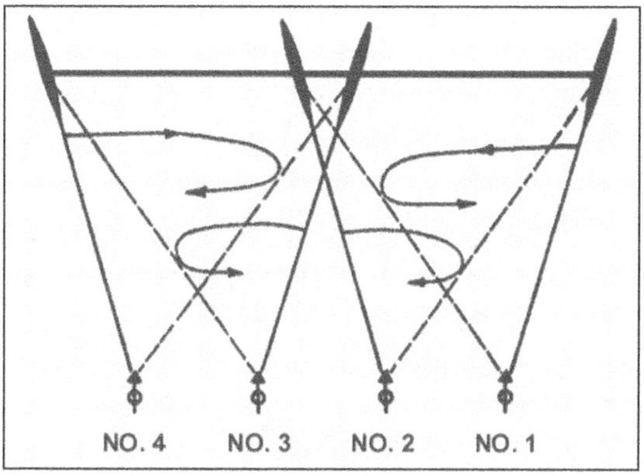

Figure 5-14. Two pairs of MK 19s engaging each half of a target 100 to 200 mils in width.

(b) *Target Greater than 200 mils in Width.* The leader may assign part of the target to each gun in this case. The target may be divided in half for each fire unit or in four parts, one for each MK 19 (Figure 5-15).

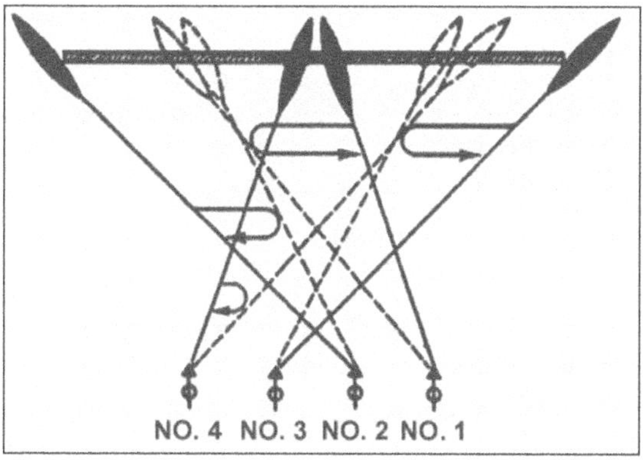

Figure 5-15. Four MK 19s, each pair engaging part of target with extreme width.

(4) ***Linear Engagement with the Flanks of Target Not Identifiable.*** The target may be located by firing tracers from an observer's gun, laying MK 19s, or using a reference point. The reference point may be in or near the target area.

(a) When a reference point is outside the target area, the leader may direct the gunner to it by announcing the interval (right or left, long or short) between the reference point and the target. With the MK 19 on a tripod mount, the gunner measures the interval right or left between the reference point and the flank of the target by laying on the reference point, with the sights set at the range to the target, and then shifting the MK 19 the designated number of mils. The interval long or short may be measured in mils using the computed search method, or it may be estimated in meters. When the gunners lay each MK 19 on its respective flank, they adjust fire and engage the target.

(b) When a reference point is in the target area, the leader may identify the flanks of an obscure target as extending so many mils from the reference point. In this case, gunners use the swinging traverse. With the swinging traverse each MK 19 is laid on the announced reference point (initial aiming point), adjusted for fire, and traversed the given distance to its corresponding flank and back, firing after each manipulation (Figure 5-16). Each gunner continues traversing back and forth across the entire target until told to cease-fire. Each gunner stops firing while traversing past the reference point. Examples of fire commands that may be used are as follows:

- The leader designates the target by firing one gun:

  FIRE MISSION.
  LEFT FLANK.
  WATCH MY BURSTS (OR TRACERS).
  (Lays and fires MK 19 at left flank) LEFT FLANK.
  (Lays and fires MK 19 at right flank) RIGHT FLANK.
  TROOPS.
  NINE HUNDRED.
  TRAVERSE.
  RAPID.
  AT MY COMMAND.
  FIRE (given after gunners announce "Up").

- The leader designates the target by using a reference point outside the target area (MK 19s on tripod mounts):

  FIRE MISSION.
  RIGHT FRONT.
  REFERENCE: DEAD TREE.
  RIGHT SIX ZERO MILS; DROP THREE FIVE ZERO METERS
  (or drop so many mils, if using computed search).
  TARGET: TROOPS EXTENDING RIGHT FIVE ZERO MILS.
  SIX HUNDRED.
  TRAVERSE.
  RAPID.
  AT MY COMMAND.
  FIRE (given after the gunners announce "Up").

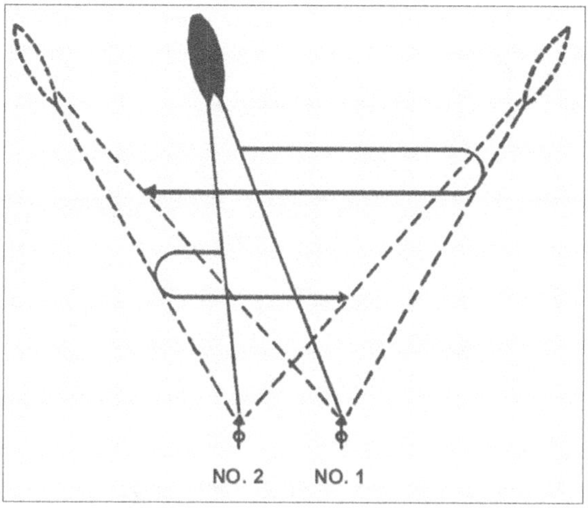

Figure 5-16. Two MK 19s, deep enfilade target, ends not visible.

- The leader designates the target by using a reference point within the target area (MK 19s on tripod mounts):

  FIRE MISSION.
  FRONT.
  REFERENCE: LONE TREE.
  TARGET: TROOPS EXTENDING RIGHT TWO ZERO MILS, LEFT, THREE ZERO MILS.
  SEVEN HUNDRED.
  TRAVERSE.
  RAPID.
  AT MY COMMAND.
  FIRE (given after the gunners announce "Up).

(5) ***Linear Engagement with Swinging Traverse.*** As previously described, a single gun uses this method against massed or rapidly moving targets at short ranges. The leader's fire command for a swinging traverse can be:

FIRE MISSION.
LEFT FRONT.
TRUCKS.
SWINGING TRAVERSE.
FIRE.

d. **Engagement of Deep Targets Using Searching Fire.** There are several ways for leaders to use two guns to engage a stationary deep target, depending on whether the ends of the target are visible to the gunner. If the target location is unknown, the computed search method may be used to engage the target. A deep target that is stationary or that has limited mobility may be engaged with searching fire.

(1) ***Target Ends Visible to Gunner.*** The Number 1 gun is adjusted first on the near end and the Number 2 gun on the far end. Each gunner searches to the opposite end and back again repeatedly, until told to cease firing (Figure 5-17). Target depth is considered in determining sight settings. The command is SEARCH.

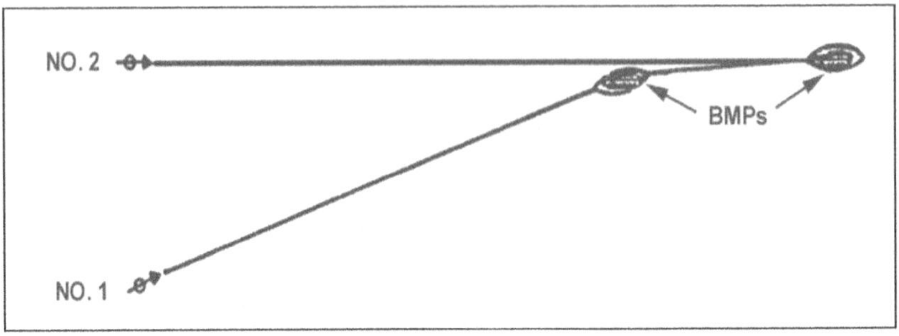

Figure 5-17. Two MK 19s, deep enfilade target, ends visible.

(a) *Target Less than 200 meters Deep.* The leader announces the range to the midpoint of the target for both MK 19s, using the length of the beaten zones to ensure the initial bursts impact on the target. An example of a fire command for this situation follows:

FIRE MISSION.
FRONT.
HALTED COLUMN.
SEVEN HUNDRED (midrange).
SEARCH.
RAPID.
AT MY COMMAND.
FIRE (given after gunners announce "Up").

(b) *Target Depth Greater than 200 meters.* The leader announces the range to the near end for the Number 1 gun, and to the far end for the Number 2 gun. An example of a fire command for this situation follows:

FIRE MISSION.
LEFT FRONT.
TROOPS ALONG RIGHT EDGE OF WOODS.
NUMBER 1, SIX HUNDRED; NUMBER 2. NINE HUNDRED.
SEARCH.
RAPID.
AT MY COMMAND.
FIRE (given after gunners announce "Up").

(2) ***Target Ends Not Visible to Gunner.*** The target may be located by having an observer fire tracers at the target, by laying the MK 19s, or by using a reference point. The reference point must be visible to the gunner and in the vicinity of the target. When a selected reference point is outside the target area, the leader may direct the gunner to the target and announce the interval to the right or left (long or short) between the reference

point and the target. If the MK 19 is on a tripod, the gunner lays on the reference point with the sights set at the range to the midpoint of the target; he measures the interval and shifts the appropriate number of mils. Using the computed search method, the gunner may measure the long or short interval in mils, or he may estimate the interval in meters. When the gunners lay their MK 19s on the right or left flank (near or far end), respectively, they adjust fire and engage the target. When the gunners lay their MK 19s on the announced release point (initial aiming point), they set their sights at the range to the reference point, and adjust fire. The gunner on the Number 1 gun searches down the designated number of meters (or mils), firing after each manipulation until the gun is set on the near end of the target. The gunner on the Number 2 gun searches up the designated number of meters (or mils), firing after each manipulation, until the gun is set on the far end of the target. When each gun has reached the near or far end, respectively, the gunners reverse the direction and both search up and down between the two limits, covering the entire target, until told to cease firing.

e. **Engagement of Deep Targets Using Computed Search Method.** When the depth of a target must be described in mils (search), the leader may compute the depth from the reference point using the computed search method. The leader determines the ranges to the near and far ends of the target and finds the required AE (angles of elevation) in mils for both ranges (Appendix F, Firing Table). The leader then computes the difference between the two. This is the amount of search required when the MK 19 and target are at the same elevation on level ground. If the fire is plunging, the amount of search should be increased. For example, when using M430 ammunition, a deep target has been sighted (the ends of which are not visible to the gunners). A reference point is also visible on the target at a range of 1,100 meters, the depth of the target is estimated at 200 meters, and the reference point appears to be midway between the ends of the target:

$\frac{200 \text{ meters}}{2}$ = 100 meters from the reference point to each end of target

1,100 + 100 = 1,200 meters to far end of target

1,100 - 100 = 1,000 meters to near end of target

AE 1,200 meters = 175 mils

<u>-AE 1,000 meters</u> = <u>132 mils</u>

depth of target 43 mils

$\frac{43 \text{ mils}}{2}$ = 21.5 mils from release point to each end of target

To search in 5-mil increments, the gunners cover the target by searching 25 mils long and 25 mils short of the reference point, or a total of 50 mils. The leader describes the target to the gunner as TARGET; TROOPS EXTENDING LONG 25 MILS; SHORT 25 MILS. The following are examples of fire commands:

- Reference point in target area, ends of target not visible to the gunner.

  FIRE MISSION.
  RIGHT FRONT.
  REFERENCE: LONE BUSH.
  TARGET: TROOPS EXTENDING SHORT ONE HUNDRED YARDS, LONG TWO HUNDRED YARDS.
  SIX HUNDRED (range to reference point).
  SEARCH.
  RAPID.

AT MY COMMAND.
FIRE (given after gunners announce "Up").

- Reference point outside target area, gun on tripod.

FIRE MISSION.
LEFT FRONT.
REFERENCE: DEAD TREE.
RIGHT THREE ZERO MILS; ADD TWO HUNDRED METERS
(or add so many mils, using computed search).
TARGET: TROOPS EXTENDING OVER TWO HUNDRED METERS
(or long so many mils, using computed search).
NINE HUNDRED.
SEARCH.
RAPID.
AT MY COMMAND.
FIRE (given after gunners announce "Up").

f. **Engagement of Moving Targets.** Due to the slow flight of the MK 19 round, it is not practical to engage moving targets using the track-and-lead methods of engaging moving targets. In order to effectively engage moving targets with the MK 19, use the trapping method of engaging targets. The gunner chooses an aiming point forward of the target and on the target path, and presses the trigger before the target comes into the sights. After starting to fire, the gunner moves the MK 19 slowly toward the target. The target moves into the impact area, and is "trapped." The gunner must begin firing before the target is in the sights or in the impact area (Figure 5-18).

(1) ***Engagement of Deep Target Moving Away from the MK 19 Position.*** If the target is moving rapidly away, both gunners lay their MK 19s on the far end of the target with the range set to that point, and search up.

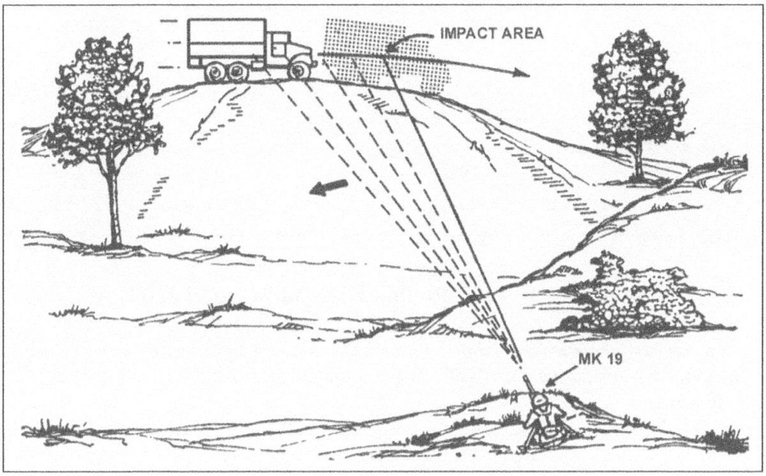

Figure 5-18. Target-trapping method.

(2) ***Engagement of Deep Target Moving toward the MK 19 Position.*** If the target is moving rapidly toward the MK 19 position:

(a) *One MK 19.* The gunner lays the MK 19, adjusts on the near end of the target with a range to the target's midpoint, and searches up and down the entire target. When a reference point within the target area is announced, the gunner lays on it with the range to that point. The gunner adjusts fire and searches up and down the entire target, beginning with the near end, until told to cease firing.

(b) *Two MK 19s.* The gunners lay both guns on the near end of the target, set the sights on the range to that point, and search down. The manipulation element of the fire command for a rapidly approaching or receding target is BOTH WEAPONS; FAR (NEAR) END; SEARCH.

(c) *Four MK 19s.* Each pair of gunners engages the target as if they were acting alone. If one pair must be switched to a different target, the other pair continues to cover the original target. A deep target should never be subdivided since the elevating mechanism on the M3 tripod allows enough search to cover any deep target within the maximum effective range for direct lay (1,500 meters).

g. **Engagement of Linear Target with Depth Using Traversing and Searching Fire.** The method used depends on the number of MK 19s available and whether the flanks of the target are visible to the gunner.

(1) ***One MK 19, Flanks Visible.*** The gunner lays initially on the near flank of the target with range set to the midpoint, adjusts fire on the near flank, and traverses and searches back and forth, covering the entire target, until told to cease-fire.

(2) ***Two MK 19s, Flanks Visible.*** Gunners use the same method as for engagement of a deep linear target: they must search to keep the center of impact on the target. The leader determines the range for the initial fire command the same as for a deep target (Figure 5-19).

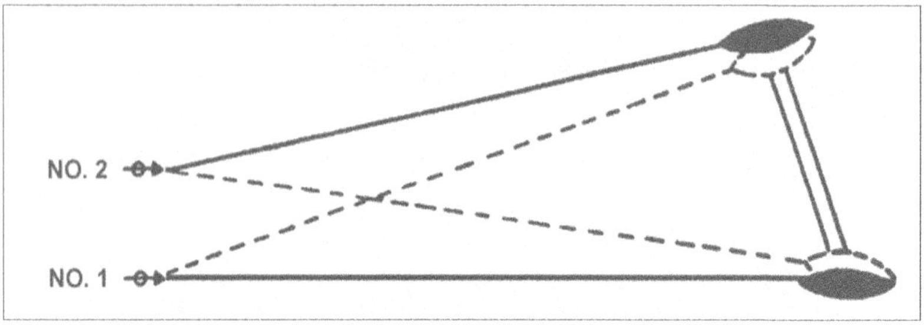

Figure 5-19. Two MK 19s, linear target with depth.

(3) ***Four MK 19s, Flanks Visible.*** Gunners use the same method for two pairs as is used for two guns. If one pair must be switched to another target, the other continues to cover the original target.

(4) ***Flanks Invisible.*** The leader may designate the target using a reference point. However, a reference point in the target area is not used as an initial aiming pointbecause of the difficulty in describing an oblique target.

(5) ***Fire Commands.*** The leader may use the following fire commands for linear targets with depth.

- When the target is not wide enough to require subdivision, is 200 meters or less in depth, and is to be engaged by either one or two MK 19s:
  FIRE MISSION.
  RIGHT FRONT.
  TROOPS EXTENDING FROM BLACK STUMP RIGHT TO LONE PINE.
  SEVEN HUNDRED (midrange).
  TRAVERSE AND SEARCH.
  RAPID.
  AT MY SIGNAL.
  FIRE (given after gunners announce "Up").
- When the target is 100 mils wide or less, is 200 meters deep or less, and is to be engaged by four MK 19s:
  FIRE UNIT, FIRE MISSION.
  FRONT.
  TROOPS TO THE RIGHT OF RED BANK.
  SEVEN HUNDRED (midrange).
  TRAVERSE AND SEARCH.
  RAPID.
  AT MY SIGNAL.
  FIRE (given after gunners announce, "Up").

- When the target is wide enough to require subdividing, is more than 200 meters in depth, and is to be engaged by either two or four MK 19s:
  FIRE MISSION.
  FRONT.

TROOPS EXTENDING FROM TRUCK, RIGHT TO BRIDGE.
NUMBER 1 (and 3 if four MK 19s are engaging the target),
SEVEN HUNDRED.
NUMBER 2 (and 4 if four MK 19s are engaging the target),
ONE THOUSAND.
NUMBER 1 (and 3), RIGHT HALF.
NUMBER 2 (and 4), LEFT HALF.
TRAVERSE AND SEARCH.
RAPID.
AT MY SIGNAL.
FIRE (given after gunners announce "Up").

h. **Engagement of Target Using Combined Sights.** Leaders may choose this expedient way to engage a linear target with depth. They may lengthen the beaten zone of a two- to four-gun fire unit by having gunners set the sights differently on each gun. The leader gives one gunner a setting 150 meters less than the estimated range, and the other a setting 50 meters more than the estimated range. When four guns are used, one pair of gunners sets sights 150 meters less than the range, the other 150 meters more than the range. This technique sacrifices precision for speed, but may be necessary for fast-moving or fleeing targets. A sample fire command for the use of combined sights follows:

FIRE MISSION.
RIGHT FRONT.
TROOPS EXTENDING FROM CROSSROADS TO HOUSE.
1,300 METERS (midrange).
COMBINED SIGHTS; NUMBER 1, 1,350 METERS; NUMBER 2,1,250 METERS.
TRAVERSE.
RAPID.
FIRE (given after gunners announce "Up").

i. **Engagement of Area Target.** This applies to a target that cannot be covered by either traversing fire, searching fire, or traversing and searching fire. Area target engagement requires that the leader mass fires from four to six guns. Examples of area targets include:

- Large troop or mechanized formations.
- Targets that must be suppressed (if the exact positions are unknown or are not visible).
- Large kill zones or engagement areas for which predetermined fires have been planned.

j. **Engagement of Target Using Massed Fires.** Leaders can mass fires in both offensive and defensive operations, depending on the time available to plan and conduct the engagement.

(1) ***Defensive Operations.*** There are two types of defensive operations.

- **Defense, Time Not a Factor.** Massing fires into an engagement area has advantages over other types of area fire engagements. Once TRPs are designated, gunners may fire upon them to register or rehearse the MK 19s. After the target has been subdivided and TRPs registered, the gunners traverse, search up, traverse back, and search down through their assigned parts of the target to assure full coverage of the target (Figure 5-20). The same method applies regardless of how many MK 19s are used. If four or six guns are used, the leader must subdivide the target and assign more TRPs than for a pair of MK 19s. An example of a fire command for this situation follows:

  FIRE MISSION.
  FRONT TRPs 1, 2, 3, 4.
  (MK 19s are laid on respective TRPs)
  AREA: ENEMY COMPANY ENTERING EA GOLD.
  (Range omitted MK 19s on TRPs.)
  TRAVERSE.
  SEARCH UP 50 MILS.
  RAPID.
  AT MY COMMAND.
  FIRE (given after the enemy is completely within the engagement area and the gunners announce "Up").

- **Defense, Time a Factor.** Leaders may not have the luxury of registering their MK 19s on TRPs before chance contacts due to time factors or the elements of surprise. However, the lethal effect of massed MK 19 fire on an area target should not be overlooked. Units should develop and practice SOPs that enable them to engage large-area targets on quick notice. For example, the leader may give each MK 19 a section or quadrant for all quick area engagements. He may assign the upper left side of the target to the Number 1 gun, the upper right to Number 2, and so on.

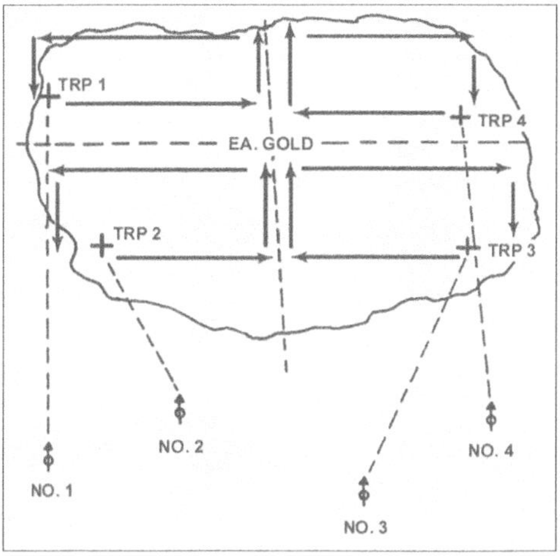

Figure 5-20. Use of massed fires, time not a factor.

(2) ***Offensive Operations.*** Massing fires into an area target can also be useful for offensive operations. During a movement to contact or an attack where time is a factor, massed fires may be used to suppress a suspected or actual enemy position, allowing the unit to flank or bypass the enemy. Although fast target acquisition and volume of fire are primary goals, the leader should subdivide the target for control, and adjust fires for greatest effect. Massed fires are especially useful in support of an attack where time is not a factor, which may require the use of overhead fire.

WARNING

Overhead fire should not be used unless the vertical interval between friendly soldiers and the target line is great enough to ensure safety.

5-13. DEFILADE POSITION

The MK 19 is most effective when laid directly on a target. Depending on the tactical situation, however, defilade could be the most desirable position.

a. **Definition.** A MK 19 is in defilade position when the gun and its crew are hidden by a landmass such as the crest of a hill from enemy ground (Figure 5-21). The position can be on the reverse slope of a hill, on the forward slope of the next high ground to the rear of a hill, or in a small fold in the ground. A defilade firing position does not necessarily reduce the effectiveness of fire against a stationary enemy target or preclude fire over the heads of friendly troops. Defilade may also be useful in providing predetermined fire into an EA.

b. **Advantages.** Some advantages of using the defilade position are that the MK 19 and a crew have cover and concealment from direct fire guns; the crew has some freedom of movement near the MK 19 position; control and supply are facilitated; and the characteristic smoke and flash of the MK 19 are partly concealed from observation.

c. **Disadvantages.** Disadvantages of using the defilade position are that targets close to the mask usually cannot be engaged, and (because adjustment of fire must be made through an observer) rapidly moving ground targets are not easily engaged.

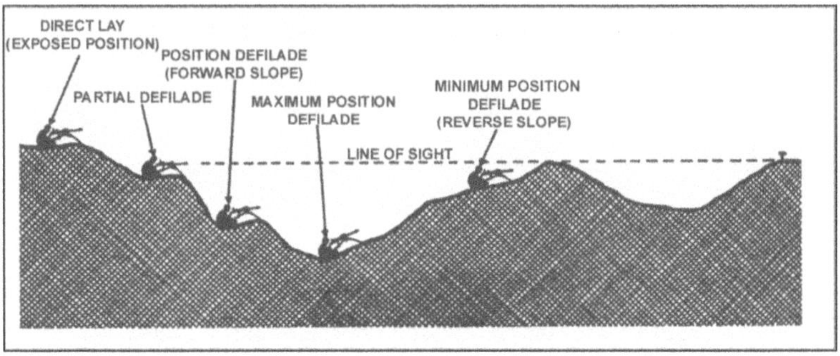

Figure 5-21. Defilade positions (and direct lay).

d. **Options.** A defilade position can either be on a forward or reverse slope.

(1) ***Reverse Slope Positions.*** These positions afford some protection from high-angle enemy fire and direct-fire cannon.

(a) *Partial Defilade Position.* The MK 19 is in partial defilade position when it is sited at the highest point on a slope on which it is defiladed. Partial defilade position provides great flexibility in engagement of new targets, although it does not provide the best cover. The MK 19 crew positioned in partial defilade can be grazed by enemy machine gun fire, but is concealed from enemy observation near the target area.

(b) *Maximum Defilade Position.* The MK 19 is in maximum defilade position when it is sited at the lowest point on a slope from which it can engage a target. The MK 19 crew positioned in maximum defilade has relatively good cover but lacks the flexibility to engage new targets.

(2) ***Forward Slope Defilade Position.*** A forward slope defilade position is not vulnerable to grazing fire from the target area, but it is open to attack or damage from mortar and artillery fire. The MK 19 is in partial defilade on a forward slope if the gun and gunner have some protection from direct fire, and if the gunner is able to engage the target using direct lay. Partial defilade is used only when a fire mission cannot be completed from defilade. The MK 19 is held in defilade and moved into partial defilade to fire.

e. **Considerations.** Direction, elevation, mask clearance, and adjustment of fire are four essential elements that must be considered when using the MK 19 in defilade position.

(1) ***Direction.*** There are four ways to lay the gun for direction.

(a) *Direct-alignment Method.* The observer selects a position on the gun-target line from which the target can be seen. The gunner loosens the traversing slide and, directed by the observer, moves the MK 19 right or left until it is aligned on the target.

(b) *Aiming-point Method.* For an aiming point, the gunner selects a prominent landmark visible through the sights (Figure 5-22). An aiming point on the gun-target line and at an equal or greater range than the target is desirable; however, the gunner may use an aiming point on the mask.

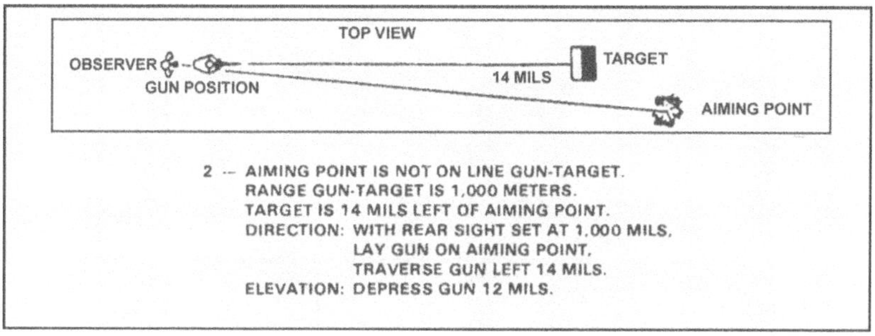

Figure 5-22. Aiming-point method.

- When the aiming point is on the gun-target line, the gunner lays the MK 19 on the aiming point, which aligns it on the target.
- When the aiming point is not on the gun-target line, the gunner measures the deflection with binoculars or compass, and adjusts the lay of the MK 19.

(c) *Aiming-stake Method.* When no natural aiming point is available, the assistant gunner sets out an aiming stake, and the gunner aligns the gun on the target.

(d) *Map-and-compass Method.* The leader locates the MK 19 position and target on a map; draws a line between the two points; orients the map to the terrain; and places the line of sight on the compass along the gun-target line drawn on the map. The leader then announces the magnetic azimuth at the compass index to the gunner as the direction of lay (Figure 5-23). Using this method with terrain-profiling techniques permits the MK 19 to be used in various defilade positions.

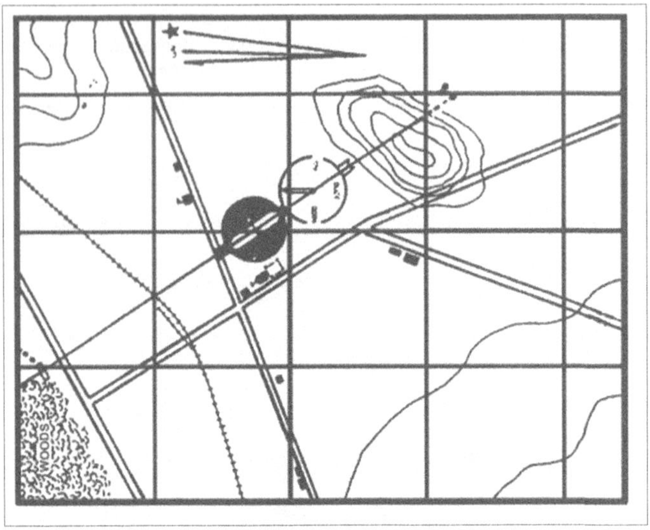

Figure 5-23. Map-and-compass method.

(2) ***Elevation.*** To lay the gun for elevation when engaging targets from defilade, knowledge of the trajectories of 40-mm rounds and of indirect lay is necessary. The MK 19 mounted on the M3 tripod with the T&E mechanism is not equipped to deliver indirect fire in the traditional sense like a mortar. Because the MK 19 has a high-arc trajectory when in the direct-fire mode, it can be fired effectively from defilade if the gunner positions it properly with the help of a well-trained observer. Discussion of laying the MK 19 for elevation includes those techniques and procedures that do not require a lot of data and calculations.

(a) The AE is the vertical angle between the bore line and the line of sight, when the gun and target are at the same elevation (Figure 5-24). The AE is always a positive (plus) and increases as the range increases. The AE for 40-mm ammunition, for each 100 meters of range up to 2,700 meters, is contained in Appendix F, Firing Table. For example, to hit a target at a range of 1,000 meters with M430 HEDP ammunition, the MK 19 must have an AE of +131.9 mils to the line of sight. Is a discussion on leveling the tripod necessary or can the gunner "eye-ball" it?

(b) When the MK 19 and target are not at the same elevation, an additional angle must be taken into consideration. The angle of sight (AS) is the vertical angle formed by the line of sight and a horizontal line from the base of the gun. When the target is at a higher elevation than the gun, the AS is positive (plus) (Figure 5-24). When the target is lower than the gun, the AS is negative (minus).

(c) The angle of quadrant elevation (QE) is formed between the bore line and the horizontal line through the base of the gun (Figure 5-24). The QE is positive (plus) whenever the gun is aimed above the horizontal, and negative (minus) whenever the gun is aimed below the horizontal. It is the algebraic sum of the AE and the AS; that is, if the angle of sight is positive, it is added to the AE; if the AS is negative, it is subtracted from the AE.

QE = AE + AS (target above horizontal line)
QE = AE - AS (target below horizontal line)

Other methods of elevation are as follows:

- **Computed quadrant elevation method.** The leader must determine the correct range to the target. Using the range, the leader finds the corresponding AE from Appendix F, Firing Table. The leader must find the AS using binoculars, by measuring in mils the vertical interval between the target and the estimated horizontal. The leader may assume the distant horizon to be at a zero AS, or at the same elevation as the MK 19 position. QE may be determined by algebraically adding this data as previously described.
- **Measured quadrant elevation method.** The gunner should locate the MK 19 in partial defilade and lay it on the target using direct-lay methods. The leader measures the QE with the M2 compass. The gunner moves the MK 19 into defilade position and places the measured QE on the gun. For each meter difference in elevation between the position in partial defilade and the firing position, the gunner adds 1 mil to the QE when firing at a range of 1,000 meters; 1/2 mil when firing at 2,000 meters, and so on.
- **Aiming-point method.** The gunner selects an aiming point visible from the MK 19 position; preferably a point at a greater range than the target and at a higher elevation than the target, and the leader finds the range to the target. Using binoculars, the leader measures the vertical angle in mils from the aiming point to the base of the target. The leader has the gunner lay the MK 19 on the aiming point, with the sight set to hit the aiming point, and then directs the gunner to manipulate the gun through the number of mils measured from the aiming point to the target. For example, the range to the target is 1,000 meters (Figure 5-25). The angle read with the binoculars from the aiming point down to the base of the target is 12 mils. The sight should be set at 1,000 meters, with the MK 19 laid on the aiming point, and the muzzle then depressed 12 mils.

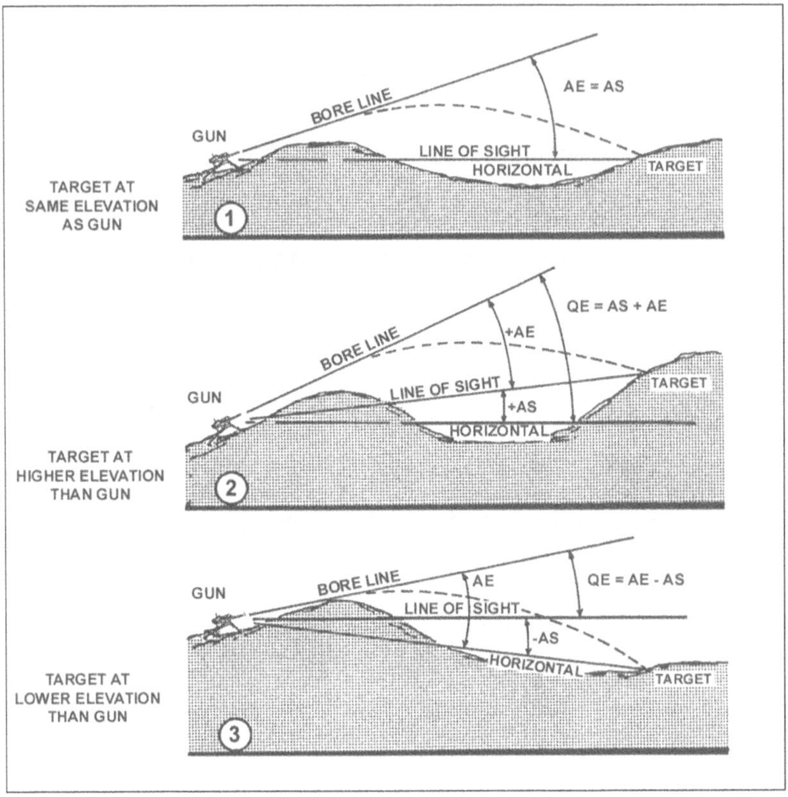

Figure 5-24. Angle of elevation, angle of sight, and quadrant elevation.

(3) ***Mask Clearance.*** After the gun(s) has been laid, the leader must determine whether or not the entire cone of fire will clear the mask, if mask clearance is not obvious. In order to ensure mask clearance, the leader has the gunner sight on the crest of the mask and re-lays the MK 19 for elevation on the target. The gunner ensures that the difference in elevation from the MK 19 to the target and from the MK 19 to the mask is at least +10 mils. The gunner checks this by sighting along the bottom of the receiver and the barrel.

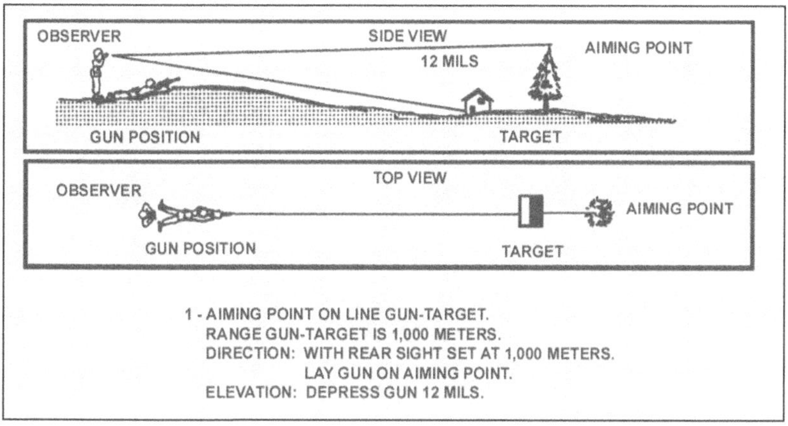

Figure 5-25. Aiming-point method for elevation.

NOTE: The procedures already discussed aid in the initial lay of the gun. The best tactical application of fire from the defilade position is in defense when time is not a factor, especially in a reverse slope defense. Leaders should try to register their MK 19s on selected TRPs, as previously discussed. Use trial and error to adjust for accurate fire from the defilade position. While the MK 19s are being registered, every precaution must be taken to guarantee troop safety.

(4) ***Adjustment of Fire.*** Under field conditions, the MK 19 that is quickly laid on the target seldom results in an initial impact directly on the target. Rapid adjustment of fire is essential. This is assured by thoroughly training the observer to estimate range and lateral distance. The observer should be in a position to see the initial impact. To ensure this, the burst (three to five rounds) should be long. Except for long ranges, or when for other reasons visual observation is hard, the observer does not ordinarily use binoculars for the initial burst. If the initial lay is incorrect, the binocular's limited field of view can cause the observer to miss the impact of the round. Adjustment of fire should be bold and aggressive. The observer should command large corrections and avoid creeping fire toward the target. When a burst is incorrect for deflection and elevation, the observer's next command should correct both.

f. **Construction of a Terrain Profile.** The leader may use a side view or cross section of the ground along a selected line or direction to determine where friendly and enemy forces can see each other. The leader can also use it to plan fires; that is, to locate dead space, to plan a defilade position, and so on. To construct a terrain profile, locate the gun position on the map and determine the direction of fire, or profile line (Figure 5-26).

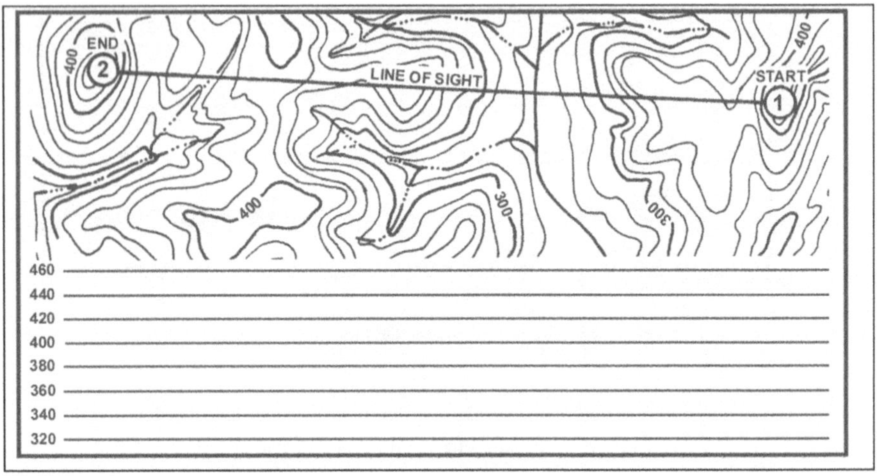

Figure 5-26. Profile line (map contour lines are 20 meters).

(1) Place the edge of a lined piece of paper along the profile line. Tick-mark all points of known elevation on the edge of the paper. These points are the contour lines. Any paper with evenly spaced horizontal lines, such as graph or notebook paper, may be used. The wider the spacing of the lines, the greater the vertical exaggeration in the profile; however this does not affect the information.

(2) Draw perpendicular lines, down across the horizontal lines for each marked point, and identify the high and low points along the profile (Figure 5-27).

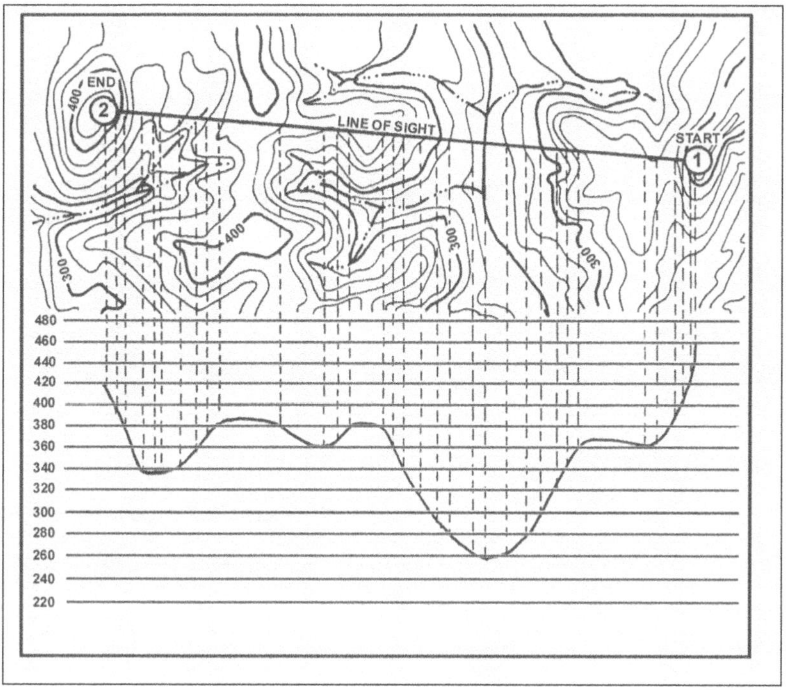

Figure 5-27. Marking of elevation.

(3) Connect all the points with a smooth curve (Figure 5-28).

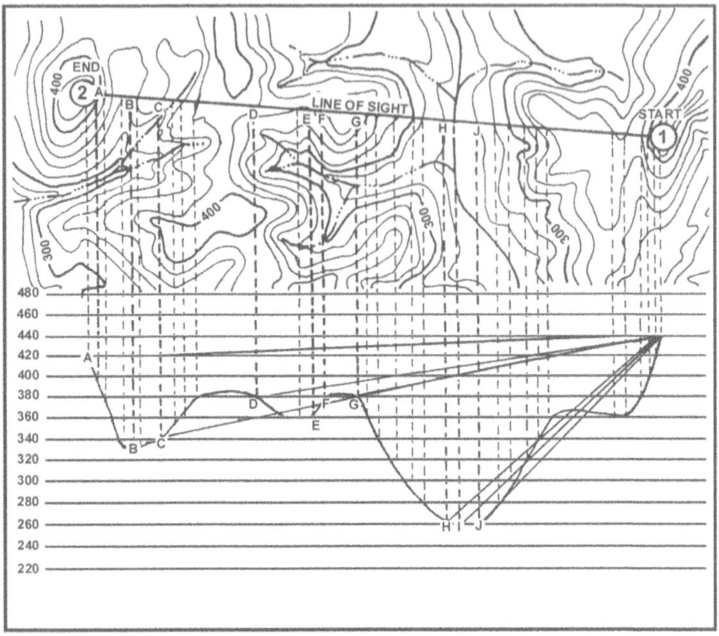

Figure 5-28. Terrain profile.

Section IV. DEGRADED CONDITIONS

The MK 19 is fired the same way under degraded conditions as it is fired under good conditions, except that degraded conditions limit the gunner's ability to observe the impact of rounds in relation to the target and to make adjustments. Refer to Appendix G, Aiming Devices, for additional information.

5-14. NIGHT FIRING

The use of range cards and predetermined fires are important at night or during degraded conditions because the targets or engagement areas are chosen and marked in advance, which enables the gunner to adjust onto targets quickly and easily. The gunner must know how to read his traversing slide bar and T&E mechanism. This task can be difficult at night or during degraded conditions. The gunner can mark targets on the slide bar with luminous tape, use mini-chemical lights, or use flashlights with red lens covers.

a. Firing with the use of illumination has proven to be much the same as firing during normal daylight conditions. Estimation of distances and adjustment of rounds are difficult at night or during degraded conditions. Gunners should be able, using illumination, to effectively engage targets out to 800 meters as well as they can in the daylight. Due to long shadows and dispersion of light past this distance however, visibility is poor even with illumination.

b. The AN/TVS-5 can be mounted on the MK 19 and used effectively to spot and adjust rounds to targets out to ranges of 800 meters. Past this range, the elevation of the scope on the gun is too high to see the impact of the rounds, so it can be handheld by an observer who calls out adjustments to fire.

5-15. FIRING IN A NUCLEAR, BIOLOGICAL, OR CHEMICAL ENVIRONMENT

Firing the MK 19 during NBC conditions is no different than firing in normal conditions. Wearing the protective mask during firing does not change the way the gunner sights and aims the MK 19 to engage targets. The assistant gunner should take care to avoid catching his clothes or gloves while feeding ammunition. The gunner should also avoid contact with the hot barrel.

5-16. FIRING IN FOG OR SMOKE

Neither the degraded conditions of fog or smoke change the way the gunner fires the MK 19. The only limitations are observation of rounds and adjustment onto targets. The use of predetermined fires, trip flares, PEWS, or forward observers can cut down on these limitations.

Section V. PREDETERMINED FIRES

Predetermined fires are planned to cover target areas such as enemy avenues of approach, likely sites for enemy guns, and probable enemy assault routes. Laying the MK 19 on predetermined targets by either of the following methods may be verified by firing the MK 19 and adjusting it on target.

5-17. T&E MECHANISM METHOD

When the MK 19 is laid on target in the primary sector, the direction and elevation is taken from the traversing bar and the T&E mechanism. Both direction and elevation, as well as the range to each target, are recorded on the range card. Each target in the sector of fire must be numbered and laid on in sequence.

5-18. FIELD-EXPEDIENT METHOD OF LAYING THE MK 19

Another method for laying the MK 19 on predetermined targets is to use field expedients (Figure 5-29). Field expedients must be used in the secondary sector, and consist of aiming and base stakes. They can be used in the primary sector to aid the gunner.

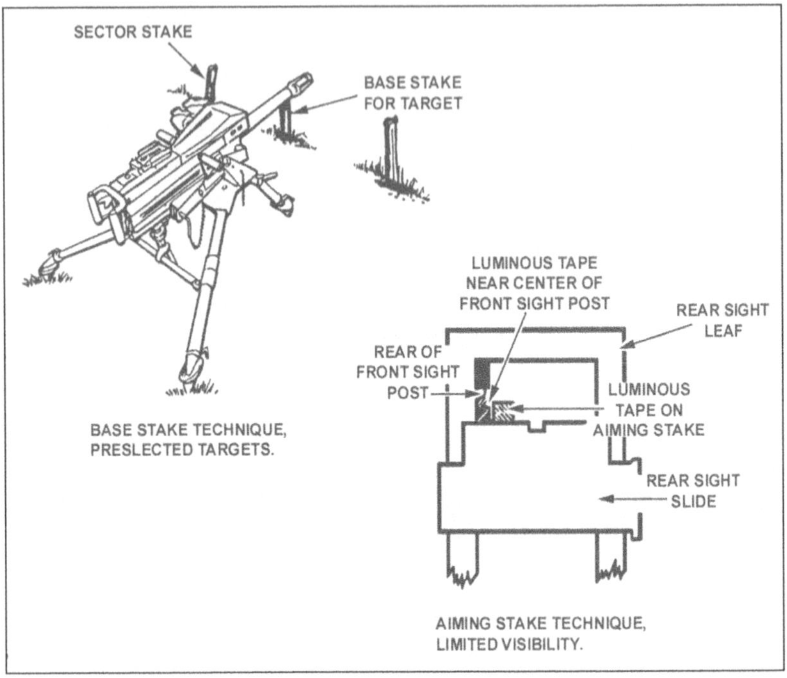

Figure 5-29. Techniques for laying the MK 19 using field-expedient method.

CHAPTER 6
INFORMATION FOR INSTRUCTORS

This chapter provides information to aid instructors in training MK 19 crews to become skilled teams. This chapter is organized to lead the trainer in a progressive manner through the material necessary to aid him in training units. Information is presented beginning with preliminary subjects; these areas include mechanical training, knowledge of the gun's capabilities, and the principles and fundamentals of MK 19 gunnery.

6-1. PRESENTATION

To present instruction in the time allotted, the instructor must be prepared and rehearsed. Assistant instructors and demonstrators must also be trained and rehearsed so they can conduct demonstrations using training aids that clearly illustrate the information. Use public address systems if the group is larger than a platoon.

a. **Organization.** Divide the soldiers into groups of appropriate size, while also considering the phase of instruction and the facilities available. If available, assign an assistant instructor to each crew; otherwise, position the assistant instructors so that they can closely supervise the crew.

b. **Presentation.** Explain and demonstrate each portion of the instruction, following with closely supervised practical work. Emphasize practice throughout training in order to develop skill and teamwork. Secure and critique all MK 19 firing during training.

c. **Examination.** Ensure the soldiers make good progress throughout each period of instruction. Those who do not progress well must receive more instruction under close supervision. Give a formal or informal participation-type examination following each phase of instruction to measure effectiveness of instruction.

6-2. MECHANICAL TRAINING

During mechanical training, the gun is mounted on the M3 mount, or installed on a table. Nomenclature charts (or mats) help the crewmembers learn the names of parts. If possible, assign one assistant instructor for each crew. Present a brief history of the gun. Stress the gun's combat role and assigned mission. Emphasize the purpose, scope, and importance of the instruction to be presented. Be sure to clearly define the term stoppage during the introduction, and emphasize the need for correct care and cleaning procedures.

a. **Organization.** Organize the class into groups of appropriate size. If enough assistant instructors are available, assign one to each crew; otherwise, position the assistant instructors so they can closely supervise the crew.

b. **Presentation.** Table 6-1 can be used as a guide in presenting instruction in stoppages and immediate action. Stress precision in detecting the cause and reducing stoppages. As the soldiers progress, concentrate on speed in applying immediate action and other methods of reducing a stoppage. Train the crew to function as a team.

STOPPAGES	INSTRUCTION PREPARATION
Misfire due to defective primer.	Place a dummy cartridge in the belt.
Short round.	Place a short round in the belt.
Bulged round.	Insert a bulged round in the belt.
Belt loaded improperly.	Pull a cartridge partly out of the belt.
Battered or thick-rimmed cartridge.	Place a battered or thick-rimmed cartridge in the belt.
Failure to remove a round from the chamber.	Place a dummy cartridge, with the rim filed off, in the chamber.
Separated case that is removed from the chamber by the new round when the bolt is pulled.	Drive the front portion of a cartridge securely on a dummy cartridge. Pull the bolt to the rear and place the cartridge properly on the face of the bolt. Ease the bolt forward.

Table 6-1. Preparation for instructing stoppages and immediate action.

6-3. CREW DRILLS

The demonstration crew must be well-trained and equipment must be complete. Use the following equipment for each crew: basic table of organization and equipment (TOE) and distinctive helmet markings that identify each crewmember according to his duties, and normal individual equipment. For more information on crew drills see Appendix H.

a. **Organization.** Present initial instruction and demonstration to one crew. For practical work, divide the class into crews of four men for each MK 19. To aid in control and supervision, mark off equipment with engineer tape or wire, and line up the crewmembers behind their equipment.

b. **Presentation.** Stress the fact that a crew drill is the first step in developing teamwork, and that precision is required of every member for the crew to function smoothly. During initial practice periods, work on precision. Talk each soldier through his duties as a crewmember. Require the soldiers to do each operation when directed with speed and precision.

c. **Advanced Training.** In advanced training, use a speed drill to create interest and stimulate competition. This tactic also furthers teamwork and tests individual crew performance. A well-trained crew should be able, without error, to place the gun into action, or take it out of action in 25 seconds.

6-4. MARKSMANSHIP

With gun and mount, the instructor makes arrangements for a demonstration crew to be available. He ensures that he has a sighting bar, an illustration of a correct sight picture, a sighting target (panel with holder), a stopwatch, control flags, cleaning rods, and patches.

a. **Organization.** Organize the crew and the training area. Organize each crew with four members, if possible. Use a centralized area, preferably with bleachers, for conferences and demonstrations. The area should be fairly level, and large enough for the guns to be placed 10 meters from the targets.

b. **Presentation.** Ensure marksmanship basics are learned and practiced. Make sure the soldiers know that marksmanship basics must be learned and practiced during the preparatory phase of the marksmanship training, in order for them to satisfactorily complete the marksmanship course. With the help of coaches and assistant instructors, ensure that steps are done correctly during the preparatory phase of marksmanship. Give a brief description of the operation, general data, and exterior nomenclature of the MK 19. Ensure that assistant instructors disassemble and assemble the MK 19 while the procedure is being explained.

NOTE: When the crews can skillfully disassemble and assemble the gun, demonstrate this phase of instruction to them blindfolded, using an assistant. Doing this encourages crews to practice during their free time, and develops individual skill and initiative.

Ensure the crew does not assemble or disassemble the MK 19 under stress against time because there is a danger of damaging the gun's parts.

c. **Operation.** Each crew requires a MK 19 mounted on an M3 mount and a belt of five dummy rounds. Emphasize the safety factors involved. Stress how the feed throat must be attached before loading the MK 19. Demonstrate the method of letting the bolt go forward.

d. **Functioning.** Training aids are useful if the class is platoon-sized; otherwise, training aids may be used for study and discussion during breaks. Use one gun for each crew, as in previous mechanical training instruction. Divide functioning into six phases (see Chapter 2, Operation and Function). Explain and demonstrate each phase of functioning. Assistant instructors should duplicate each demonstration on the guns within each crew. With these setups at the guns, each crewmember can see how the parts work. Setups may be used for each phase of functioning.

APPENDIX A
TRAINING STRATEGY

This appendix discusses the two primary components of training strategies: initial and sustainment training.

A-1. COMPONENTS

Training strategies have two primary components: *initial* and *sustainment* training. Both may include individual and collective skills, but *initial training is critical*. With initial training a soldier retains a correctly taught task that he has learned well. With such training, he can easily sustain or quickly regain his performance. Thus, leaders must conduct initial training correctly and well.

a. Trainers use standard, common, and current procedures. Regardless of how well skills are taught or learned, individual and unit skills decay over time. How fast and how much skills decay depends on many factors. Personnel turnover is a primary factor in the decay of collective skills. Other factors include the difficulty and complexity of the tasks.

b. When the unit loses critical crewmembers, it must quickly train new gunners to restore unit proficiency. If too much time passes between initial and sustainment training periods or if the unit alters the established training doctrine, skills erode and the unit might have to completely retrain its MK 19 gunners.

c. A critical step in the Army's overall training strategy is to train leaders and trainers first. Leader courses and unit publications help develop officer and noncommissioned officer (NCO) proficiencies needed to plan and conduct marksmanship training. These resources also help leaders learn how to evaluate the effectiveness of unit marksmanship programs. Proponent schools provide training support materials. These materials include field manuals, training aids, training devices, simulators, and programs of instruction that serve as the doctrinal foundation and guidance for training the force.

A-2. INITIAL TRAINING

The MK 19 training strategy starts with leader training and goes on to unit training. Because "MK 19 gunner" is a Skill Level 2 position, the MK 19 gunner receives most of his MK 19 training within the unit. The training institution provides the unit with a soldier who has received training in the basics of the MK 19. The unit plans minimal MK 19 training for advanced individual training (AIT) soldiers.

A-3. SUSTAINMENT TRAINING

Basic MK 19 gunnery skills and gunner qualification continues in units. Once soldiers learn individual and crew skills, the unit must help them sustain those skills. To do this, leaders follow the annual gunnery-training program provided in Appendix E, Annual Gunnery Training Program. In addition to sustaining skills, the unit incorporates the MK 19 into collective training exercises such as platoon and squad live-fire situational training exercises (STXs).

a. The strategy for sustaining basic MK 19 gunnery skills includes annual or semiannual instructional and qualification range firing. However, the unit must set up a year-round sustainment program. Key elements of the sustainment program include training the trainers, refreshing preliminary skills, and using devices to remedy the training of soldiers who have trouble qualifying on the MK 19.

b. The unit also trains the gunner in other skills such as night firing; mission-oriented protective posture (MOPP) firing; engaging moving targets, using camouflage, cover and concealment; firing and maneuvering; and preparing or selecting a fighting position. The unit also integrates these skills into tactical training. The references at the back of this manual contain additional guidance on the training for these skills.

c. The unit integrates individual and leader MK 19 tasks into collective training such as squad, section, and platoon drills and STXs as well as company LFXs and field training exercises (FTXs). Organizational mission training plans (MTP) and drill books include the tasks, conditions, and standards for the collective tasks in these exercises. These books also explain how to plan and conduct such collective tasks. Based on the type of organization, the unit evaluates collective exercises in leader and trainer after-action reviews (AARs). Objectively evaluating both individual and unit proficiency indicates the readiness and future training requirements.

d. Trainers schedule live fire after the soldiers demonstrate preliminary skills. For initial fire, the trainer conducts an instructional exercise. This improves the soldiers' ability to acquire targets with the MK 19. Once soldiers understand the gun and demonstrate the skills needed to acquire a target, the trainer conducts additional live-fire training and target-detection exercises at various ranges. To develop proficiency, soldiers have to master targets and scenarios of increasing difficulty. After initial individual training, the trainer conducts qualification fire on a range to evaluate each soldier's skills. When conducted in accordance with Chapter 4, Marksmanship, this evaluation also gives leaders some insight into the unit's skill and training effectiveness.

APPENDIX B

OPERATIONS CHECKLIST

This appendix consists of an operations checklist of procedures to be followed before, during, and after use of the MK 19.

WEAPON PART	BEFORE	DURING	AFTER
1. Bore	Ensure it is clear and clean.	________	Clean and oil lightly.
2. Moving parts	Oil lightly and test for worn or broken parts. The moving parts should function without excessive friction.	Lubricate working parts. Observe the functioning of the gun to anticipate failures.	Inspect, clean, and oil lightly.
3. Ammunition	a. Ensure the correct type of ammunition is used. b. Ensure the ammunition is clean and dent free. c. Ensure all ogives are tight. d. Have an adequate supply of ammunition on hand.	Keep correctly aligned in the feedtray. Check resupply. Protect from sun, moisture, and dirt. Watch for link stoppage.	Clean, store carefully, and replenish supply.
4. Top cover	Inspect the top cover for dents or damage.	Keep closed and locked down.	Lube properly after cleaning.
5. Line of fire	Ensure the line of fire is clear of all obstructions.	Cease fire if any obstruction appears in the line of fire.	________

Table B-1. Operations checklist.

APPENDIX C

STOPPAGES, MALFUNCTIONS, IMMEDIATE ACTIONS, AND REMEDIAL ACTIONS

A stoppage is any interruption in the cycle of operation caused by faulty action of the weapon or ammunition. A malfunction is a failure of the weapon to function properly. Neither defective ammunition nor improper operation of the gun by a crewmember is considered a malfunction of the MK 19. This appendix lists the actions to correct common stoppages and malfunctions. For more complete information on troubleshooting common malfunctions and stoppages, refer to TM 9-1010-230-10 Troubleshooting Index.

Section I. STOPPAGES, IMMEDIATE ACTIONS, AND REMEDIAL ACTIONS

A stoppage is cleared through immediate or remedial action. Immediate action is the action taken to reduce the stoppage without investigating the cause. The gunner takes remedial action if immediate action does not end the stoppage. Remedial action has three parts: the gunner first unloads and clears the MK 19, identifies the problem, and then applies corrective action to fix the problem. If the MK 19 still does not fire, or cannot be repaired at the user level, contact the unit armorer.

C-1. IMMEDIATE ACTION

Immediate action is different during peacetime and combat.

a. **Immediate Action During Peacetime and Training.** The gunner does the following immediate action to reduce a stoppage during peacetime.

> **CAUTION**
> Both charger handles must be forward and up for firing. If either handle is down, the gun will not fire.

(1) The gunner ensures the safety of other personnel and clears the weapon of ammunition by performing the following steps in sequence:

(a) Clear the area of personnel.

(b) Pull bolt to the rear. If the bolt does not go to the rear, go to BOLT JAMMED.

(c) Catch the live round as it is ejected.

(d) Push the charging handles forward and up.

(e) Put the gun on S (SAFE) and check for any bore obstructions.

(f) If bore is clear, move safety switch from the S (SAFE) to the F (FIRE) position, and attempt to fire.

(g) If nothing happens, put gun on S (SAFE) and wait 10 seconds.

(h) Pull the bolt to the rear.

(i) Catch the live round as it is ejected.

(j) Open the top cover and clear the ammunition.

(k) Check bore for any obstructions.

(2) Charge the MK 19 in the normal manner and check to see the primary pawls have clicked up behind the cartridge in front of the bolt face and that the secondary pawls have clicked up behind the next round before firing. The feed pawls should click up within the 1-inch of charging handle travel.

(3) If the primary and secondary pawls do not click up within the last 1-inch of charging travel, follow these procedures:

- Turn the MK 19 into unit level maintenance.
- Record ammo lot number, type of ammo, number of rounds fired, serial number of the MK 19 and indicate whether ammo is linked with one-piece or two-piece links.

WARNING

1. Do not use combat misfire procedures during peacetime or training. Serious injury can result if precautions are not observed.
2. Do not relink or fire ammunition that has been cycled through the weapon.

b. **Immediate Action During Combat.** To correct a misfire during combat, perform the following immediate actions:

(1) Press charger handle locks and rotate charger handles down.

(2) Pull charger handles to the rear until the bolt sears.

(3) Push charger handles forward and rotate charger handles up and lock.

(4) Relay and fire.

(5) Turn in live rounds that cycled through the weapon as instructed by current directives.

C-2. REMEDIAL ACTIONS TO CLEAR STOPPAGES

If the weapon still does not fire, the gunner clears the weapon, identifies the problem and then takes corrective action to fix or report the problem.

a. **Bad Ammunition.** Bad ammunition can cause a stoppage and is checked first.

(1) ***Primer on the Ejected Round is Indented.*** Clear the weapon and check to see if the primer is indented. If the primer is indented, it is a bad round and dispose of it as authorized. Reload the weapon and continue with the exercise.

(2) ***The Primer is Not Indented.*** Clear the weapon and check to see if the primer is indented. If the primer is not indented, then the firing pin can be bad. Turn the MK19 into unit level maintenance.

(3) ***Round on Bolt Face.*** Clear the weapon. If the round remains on the bolt face, remove the round and place the weapon on S (SAFE) with the bolt to the rear. Check the primer and see if it is indented. See the aforementioned steps for the corrective action.

b. **Ammunition Jammed in Feeder.** A misfire might occur due to ammunition jammed in feeder. Clear the weapon and put gun on S (SAFE).

(1) ***Feed Throat Improperly Attached.*** Check for proper attachment of the feed throat. If the feed throat is attached incorrectly, then re-attach it and reload. Try to recharge and fire the weapon. If charging is not possible, go to BOLT JAMMED.

(2) ***Rounds Crooked or Not Seated Firmly.*** Clear the weapon. If the rounds are crooked or not seated firmly, remove the linked ammunition from feeder. Ensure the link band is even and adjacent to copper band all around ammo. Reload and continue with the exercise.

(3) ***Broken Link.*** A broken link can cause the weapon to stop firing. Clear the weapon. If the link is inside the weapon, remove the link. If it is still on the round, then remove the round and dispose it as authorized. Reload and continue with the exercise.

(4) ***Link Off Rotating Band.*** Clear the weapon. Remove round from belt, dispose of round as authorized. Reload and continue with the exercise.

(5) ***Female Link Not First.*** A round loaded with the female link not first will cause a stoppage. Clear the weapon. Reload the weapon and ensure that the female link is first.

c. **Bad Firing Pin.** A stoppage might occur due to a bad firing pin or firing pin spring. A possible bad firing pin can be determined while checking for a bad round.

(1) ***Primer on Ejected Round Not Indented.*** Clear the weapon. If the primer on the ejected round is not indented, then the firing pin may be bad. Report the defect to the unit armorer.

(2) ***Firing Pin Tip Fails to Spring Forward.*** Clear the weapon. If the firing pin tip fails to spring forward then the firing pin spring can be bad. Report the defect to the unit armorer.

d. **Defective Feeder or Feed Slide Assembly.** A misfire might occur due to a defective feeder or feed slide assembly.

CAUTION

Clear feeder of ammunition. Make sure gun is on S (SAFE).

(1) ***Broken or Worn Feed Pawls.*** Clear the weapon and check to see if the primary and secondary pawls are broken, worn, without spring action, or the pin is missing or dislodged. If they are, turn the MK 19 into unit level maintenance.

(2) ***Badly Worn or Gouged Link Guides.*** Clear the weapon and check to see if the guides are worn or gouged. If they are, turn the MK 19 into unit level maintenance.

(3) ***Binding Feed Slide Assembly.*** Clear the weapon. Remove the feed slide assembly and tray. Clean, inspect and lubricate feed assembly and tray. If the feed slide assembly continues to bind, report the defects to armorer or support maintenance.

e. **Bolt Won't Pick Up the Round.** A misfire might occur because a bolt won't pick up the round.

(1) ***Charging Handles are Down.*** Clear the weapon. Raise the charging handles before firing. If the weapon still does not fire, report defects to the armorer or support maintenance.

(2) ***Dirt-Clogged, Weak, or Damaged Extractors.*** Clear the weapon and remove obstruction. Clean, lubricate, and reload weapon. Continue with the exercise.

(3) ***Feed Slide Out of Adjustment.*** Clear the weapon. The feed slide is out of adjustment if:

- Round fails to feed.
- Round drops.
- Extractors won't pick up round.
- Round stubs on face of chamber.

Report defects to armorer or support maintenance.

f. **Bolt Drops a Round Before Firing.** A stoppage may occur because a bolt drops a round before firing.

(1) ***Weak or Damaged Extractors or Bolt Fingers.*** Clear the weapon and check for weak or damaged extractors or bolt fingers. Report any defects to the armorer or support maintenance.

(2) ***Binding Receiver Rails.*** A misfire might occur because of binding receiver rails. Clear the weapon and pull the bolt to the rear without stopping or pausing. Ease it forward (holding onto one charging handle while you press the trigger) and check for binding. Place weapon on S (SAFE) and remove backplate pin. Lift up slightly on the backplate assembly and pull the bolt and backplate assembly to the rear. Remove the chargers. Check the charger rails and receiver rails for burrs. Report defects to armorer or support maintenance.

WARNING

Be sure bolt is forward before removing backplate pin assembly. If not, serious injury could result.

CAUTION

When installing bolt and backplate, ensure cocking lever is in the forward position. Damage to equipment could result if cocking lever is to the rear.

g. **Bad Cocking Lever.** Clear the weapon and remove the bolt and backplate assembly. Examine the cocking lever on the left side of the bolt for wear or damage. Report defects to armorer or support maintenance.

h. **Unknown Cause.** If the MK 19 stops firing and the procedures above do not identify and correct the problem, then turn it into the armorer or support maintenance.

Section II. MALFUNCTIONS, EMERGENCY ACTIONS AND REMEDIAL ACTIONS

A malfunction is a failure of the weapon to function properly. Neither defective ammunition nor improper operation of the gun by a crewmember is considered a malfunction of the MK 19. The two most common MK 19 malfunctions are sluggish action and runaway gun.

CAUTION

Do not try to break the ammunition belt; injury could result. Lower one charging handle to stop the gun.

WARNING

1. Before performing any non-firing procedure, ensure the weapon is clear of any ammunition.
2. Ensure all ammunition and non-essential personnel are at least 65 meters to the rear of the weapon.
3. If the bolt jams during firing, do not let the bolt slam forward as the top cover is being opened because a round could fire.
4. Be sure to put bolt in forward position before removing the backplate pin assembly. Serious injury can result if the pin assembly is removed with the bolt to the rear.
5. Be prepared to catch dropped/ejected live round from weapon.

C-3. SLUGGISH OR ERRATIC FIRING

Excessive friction from dirt, carbon buildup, lack of lubrication, or burred parts usually causes sluggish or erratic action. Once the gunner realizes that the gun is operating sluggishly or erratically, he should cease-fire and clear the weapon. The two primary corrective actions are cleaning or turning the weapon into the armorer. There are several reasons why a weapon may exhibit sluggish or erratic firing.

a. **Dirty Bore or Chamber.** A dirty bore or firing chamber can cause sluggish or erratic firing. Clear the weapon and clean its bore and chamber. Reload and continue with the exercise.

b. **Recoil Springs or Guide Rods.** Clear the weapon and remove the bolt and backplate assembly from gun. Push against springs to test for weakness. Note bent rods. Report defect to armorer or support maintenance.

c. **Bolt Sear Timing Adjustment.** Clear the weapon and turn it into direct support maintenance.

C-4. RUNAWAY GUN

A runaway weapon continues to fire after the trigger has been released. Worn parts or short recoil of the bolt assembly may cause a runaway gun. Consider the amount of ammunition left and the type of MK 19 mount used when determining the best way to stop the weapon.

a. **Emergency Action.** This paragraph addresses emergency actions that must be taken to stabilize uncontrolled automatic fire. The following procedures should be taken to control a runaway gun:

(1) If ammunition is not low and the MK 19 is used in the free gun mode, keep rounds on target until the all the rounds on the belt have been fired.

(2) If the MK 19 is mounted on either the M3 tripod or on a vehicle with the T&E mechanism attached, hold the grip with one hand. At the same time, press the charger handle lock and lower one charger handle. This action interrupts the cycle of operation, causing the MK 19 to cease firing. The gunner therefore:

(a) Keeps gun pointed downrange and slightly elevated.

(b) Presses charger handle locks.

(c) Lowers the charging handle(s) so the gun will stop firing.

(d) Places the gun on S (SAFE).

WARNING

Never try to break the ammo belt with your hands. Injury could result. Lower one charger handle to stop gun from firing.

b. Once the gun has stopped firing, clear the weapon and report the condition to the armorer or support maintenance.

C-5. BORE OBSTRUCTION

A bore obstruction is indicated by a muffled sound of round firing, excess smoke out of the chamber, and/or excess debris/gases below the gun.

a. **Emergency Action.** The emergency actions for an obstructed bore are the following:

(1) Place weapon on S (SAFE).

(2) Notify range safety officer (during training).

(3) Depress feed pawls, release ammunition belt, and clear feed.

(4) Move the ammunition belt and can to a safe area.

(5) Remove the empty case catch bag.

(6) Charge gun and hold bolt to rear.

WARNING

Do not relink or fire any ammunition that has been cycled through the weapon.

(7) Holding the bolt to the rear, insert a cleaning rod through the receiver rail to the top of the shell casing and as close to the face of the bolt as possible.

(8) Place left hand underneath as close to the round as possible. Raise cleaning rod upward forcing the round off the bolt face into the hand. Remove round to designated area for explosive ordinance disposal (EOD).

b. **Correcting an Obstructed Bore.** To correct an obstructed bore follow these procedures:

(1) Place selector lever on F (FIRE) and ease the bolt forward.

(2) Remove the backplate pin bolt, backplate assembly, vertical cam assembly, and primary drive lever.

(3) Check for any type of obstruction.

(4) Check and remove any case or round from bolt face.

(5) Insert bore obstruction detector into bore to check for a live round.

(6) Remove obstruction per round removal procedures (for exact bore obstruction instructions using the round removal tool, refer to WP 0014 00).

C-6. GUN FIRES TOO SOON

If the weapon fires too soon, or when the trigger is not engaged, the gunner applies emergency action and the weapon is taken to support maintenance. The gunner must not continue to fire the weapon. The following emergency actions must be taken if a weapon fires too soon:

> **WARNING**
>
> **Do not attempt to clear the weapon if the weapon fires too soon. Do not attempt to clear or fire the weapon until it is fixed.**

(1) Cease fire.

(2) Place weapon on S (SAFE).

(3) Clear area of personnel and ammunition.

(4) Notify the range safety officer.

(5) Check barrel for lodged round using the bore obstruction detector (BOD).

(6) After proper personnel have cleared weapon, evacuate weapon to support maintenance.

C-7. BOLT JAMMED

The following actions are taken if the gunner cannot pull the bolt to the rear.

> **WARNING**
>
> **The following procedures must be performed in sequence to open the top cover. The bolt could spring forward suddenly and fire a round causing severe injury. Be prepared to catch ejected round.**

a. **Emergency Action and the Bolt Can be Pulled to the Rear.** The following emergency actions must be taken if the weapon initially jams but the bolt can then be pulled to the rear.

(1) Put gun on S (SAFE).

(2) Press charger handle locks and rotate charger handles down.

(3) Pull charger handles to the rear as far as possible without stopping or pausing, until bolt locks. Ensure bolt will stay to the rear before releasing charger handles.

(4) Maintain rearward pressure on charging handles while assistant lifts top cover.

(5) Insert cleaning rod section through slot in side of receiver. Prepare to the catch ejected, live round.

(6) Raise cleaning rod to force live round down. Catch live round as it is ejected.

(7) Remove ammo belt from feeder.

(8) Reposition ammo belt in feeder.

(9) Put gun on F (FIRE).

(10) Ride the bolt forward by grasping one charging handle and depressing the trigger.

(11) Ensure feed slide assembly is to the left.

(12) Ensure secondary drive lever is engaged with the feed slide pin. If not, engage forked end with feed slide pin.

(13) Close top cover gently.

(14) Charge weapon and attempt to fire.

(15) If bolt still jams, repeat first seven steps. Put weapon on S (SAFE), and evacuate to support maintenance.

b. **Gunner's Actions.** The gunner performs the following action once emergency actions are completed:

(1) Clear the jam and make sure there is no bore obstruction.

(2) Charge the MK 19 in the normal manner, and check to see if the primary pawls have clicked up behind the cartridge in front of bolt face and the secondary pawls have clicked up behind the next round before firing. The feed pawls should click up within 1-inch of the charging handle travel.

(3) Check the ammo link to ensure it is even and touches the copper band all around the ammo. If the primary and secondary pawls do not click up within the last 1 inch of the charging handle then:

- Turn MK 19 into unit level maintenance.
- Record ammo lot number, type of ammo, number of rounds fired, serial number of the MK 19 and indicate whether ammo is linked with one-piece or two-piece links.

(4) If the MK 19 does not require feed slide adjustment and there appears to be no other deficiencies that would prevent the weapon from firing, it should be turned into the armorer or support maintenance.

C-8. SHORT RECOIL

A short recoil occurs when the bolt does not fully return to the rear.

> **WARNING**
>
> **When firing HE or TP ammunition, the gunner should observe downrange and attempt to determine if the round left the barrel and also be alert to these three danger signals:**
>
> **1. A muffled report from the gun.**
> **2. Smoke and debris from the bottom of the receiver.**
> **3. Failure of the projectile to leave the muzzle.**
>
> **Any of these three symptoms mean a bore obstruction. Do not attempt to clear a bore obstruction.**

a. **Emergency Actions after a Short Recoil.** The gunner does the following actions after a short recoil:

(1) Place weapon on S (SAFE).

(2) Clear area of personnel and ammunition.

(3) Notify range safety officer.

(4) Pull the charger handles to the rear, without stopping or pausing, until the bolt locks. Ensure that the bolt stays to the rear before releasing the charger handles.

(5) Insert cleaning rod section through slot in side of receiver. Prepare to catch the ejected live round.

(6) Raise cleaning rod to force live round down. Catch the live round as it is ejected.

(7) Raise top cover.

(8) Remove ammo belt from feeder.

(9) Check for bore obstruction using bore obstruction detector.

(10) If bore is obstructed, refer to the section above on clearing bore obstructions or refer to WP 0014 00 for round removal procedures.

(11) If there is no obstruction, reposition belt in feeder.

(12) Ensure feed slide assembly is to the left.

(13) Charge weapon and attempt to fire.

b. If a short recoil occurs again, repeat the first eight steps. Put weapon on S (SAFE) and turn it into support maintenance.

C-9. TOP COVER WILL NOT CLOSE

The gunner checks the following so that the top cover can be closed.

a. **Improper Position of the Feed Slide Assembly.** Move feed slide assembly all the way left. The spring should touch the cover.

b. **Bolt is Locked to the Rear.** Ride the bolt forward.

c. **Misaligned Ammunition.** Ensure rounds are straight and firmly seated in the feeder. Ensure links are evenly aligned in the link guide and on rounds. Clean dirt from the feeder.

APPENDIX D

GUNNERY TESTS

There are three gunnery tests: a diagnostic gunnery test, an intermediate gunnery test, and a gunnery skills test. The instructor is responsible for the conduct of all tests and must ensure that sufficient scorecards or answer sheets are available.

D-1. DIAGNOSTIC GUNNERY TEST

This written test, given before and during preparatory training, evaluates trainees' knowledge of the MK 19. It also serves as an evaluation tool for the instruction. Training units prescribe procedures for the test, and announce scores of those who satisfactorily complete the test. If this test is used, it should be given before range firing. To satisfactorily complete the test, the trainee must earn 80 out of 100 points.

		POINTS
1.	Name the five major assemblies of the MK 19.	5
2.	Explain how to clear and disassemble the MK 19.	15
3.	Explain how to assemble the MK 19.	15
4.	Explain how to perform a function check.	5
5.	Explain how to apply immediate action to a runaway gun.	5
6.	Explain how to apply immediate action when a malfunction occurs.	10
7.	Explain one phase of functioning (directed by the examining officer).	10
8.	Explain why the weapon should not be fired with the top cover assembly open.	10
9.	Explain how to inspect ammunition before loading (belt to have at least one each of the following: short round, bent round, loose bullet, broken link or round not pushed fully into its link).	10
10.	Explain how to load to half-load position.	5
11.	Explain and demonstrate sight alignment.	10
	TOTAL	100

D-2. INTERMEDIATE GUNNERY TEST

This written test, given at the end of preparatory marksmanship training, is used to evaluate the instruction as well as the soldier's gunnery skills. This test has 21 questions: 20 multiple-choice and 1 true or false. Each soldier must correctly answer 17 of the 21 in order to move on to the next phase of training. The instructor is responsible for the conduct of the test. The following is a sample of the standard Intermediate Gunnery Test.

INTERMEDIATE GUNNERY TEST

1. What are the MK 19's five major groups (field strip groups)?
   a. Receiver assembly, top cover assembly, sear assembly, bolt assembly, and feed slide assembly.
   b. Top cover assembly, receiver assembly, barrel, sear assembly and feed tray, and bolt and backplate assembly.
   c. Receiver assembly, feed slide assembly and feed tray, top cover assembly, sear assembly, and bolt and backplate assembly.
   d. Sear assembly with chargers, receiver assembly, feed tray assembly, top cover assembly, bolt and backplate assembly.

2. Where on the weapon is the safety located?
   a. On the right side of the receiver assembly.
   b. On the left side of the receiver assembly.
   c. On top of the receiver assembly.
   d. On the sear assembly at the bottom rear of the receiver assembly.

3. In a nonfiring situation to clear the weapon, in which order are the steps below taken?
   a. 2, 4, 3, 1.
   b. 1, 3, 4, 2.
   c. 3, 1, 2, 4.
   d. 1, 2, 4, 3.

   1. Ensure that the safety switch is in the S (SAFE) position, and the bolt is forward. Open the top cover by rotating the top cover latch counterclockwise. Lift the cover.
   2. Return the bolt to the full forward position. Rotate the charger handles upward.
   3. Inspect the feed tray and extractors on the top of the bolt to make sure no round is in the "pickup" position.
   4. Unlock and pull the charging handles to the rear two to three inches. Inspect the face of the bolt and the chamber.

4. In which substance(s) should you never immerse the bolt and backplate assembly?
   a. Cleaning solvent.
   b. Lubricant.
   c. Bore cleaner.
   d. All of the above.

5. Under which conditions should you apply a generous second coat of lubricant when cleaning the weapon?
   a. Hot, wet, salt air, sea spray.
   b. Hot, dry, sandy, dusty.
   c. Cold, icy, snow.
   d. All of the above.

6. Under which conditions should you use a second coat of lubricant?
   a. Hot, wet, salt air, sea spray.
   b. Hot, dry, sandy, dusty.
   c. Cold, icy, snow.
   d. All of the above.

7. During reassembly, what is the first major group you should place back onto the weapon?
   a. Sear assembly.
   b. Bolt and backplate assembly.
   c. Top cover assembly.
   d. Receiver assembly.

8. To ensure the primary and secondary drive levers align properly when you close the top cover of the weapon, in what positions should you place the feed slide assembly and bolt?
   a. Left, to the rear.
   b. Right, to the rear.
   c. Left, forward.
   d. Right, forward.

9. What is the maximum range of the weapon, in meters?
   a. 1,500.
   b. 2,212.
   c. 2,000.
   d. 1,800.

10. What is the maximum effective range of the weapon, in meters, for point targets?

a. 1,500.
b. 2,212.
c. 2,000.
d. 1,800.

11. What is the cyclic rate of fire for the weapon, in rounds per minute?
    a. 225 to 275.
    b. 275 to 325.
    c. 325 to 375.
    d. 375 to 425.

12. What is the main characteristic(s) of the MK 19?
    a. Fully automatic.
    b. Blowback operated.
    c. Air-cooled.
    d. All of the above.

13. Which mount(s) should you avoid attaching the weapon to?
    a. M3 tripod without the T&E mechanism.
    b. M3 tripod with the T&E mechanism.
    c. M4 pedestal.
    d. All of the above.

14. How much armor plating, in inches, can a M430 HEDP round penetrate when fired at zero degrees obliquity?
    a. 1.
    b. 2.
    c. 3.
    d. 4.

15. Within how many meters must the 40-mm HE round impact to cause casualties to exposed enemy personnel?
    a. 5.
    b. 10.
    c. 15.
    d. 20.

16. What steps must you take to stop a runaway weapon?
    a. Release the trigger and step away from the weapon.
    b. Grab and twist the ammunition belt to break it off.

c. Unlock and lower one charging handle while keeping control of the weapon.
d. None of the above.

17. When loading the weapon, place the female link end of the ammunition belt through the feed throat and draw the ammunition into the weapon feed area.
    a. True.
    b. False.

18. What term describes an interruption in the functioning cycle caused either by faulty action of the gun or by faulty ammunition?
    a. Blockage.
    b. Breakage.
    c. Malfunction.
    d. Stoppage.

19. What steps must you take to perform corrective action?
    a. Unload and clear the weapon.
    b. Inspect the weapon and ammunition to find the cause of the problem.
    c. Correct the problem, reload, and try to fire the weapon.
    d. All of the above.

20. When you engage a target, how many rounds should you fire in each burst?
    a. 3 to 5.
    b. 5 to 8.
    c. 7 to 10.
    d. 9 to 12.

21. In which position(s) must you place both charger handles before the weapon will fire?
    a. Forward and down.
    b. Forward and up.
    c. To the rear and down.
    d. None of the above.

D-3. GUNNERY SKILLS TEST

This is a guide for unit commanders to standardize and maintain MK 19 gunnery skills. It is a nonfiring, hands-on test given at the end of MK 19 training. The commander may authorize indoor testing if the facilities are available. Seven stations are used for the skill tests. The test is graded on a GO (pass) or NO-GO (fail) basis.

a. **Instructions for the Gunner's Examination.** The following is a guide for giving gunners the skill test.

(1) To organize for testing, assemble the entire unit in one area and orient them as follows:

> "During the next _____ hours, you will take part in a performance-type examination to test your knowledge of the MK 19 machine gun. Stay with your assigned group during the entire examination. Initially, we will assign each group to a specific station. At each station, an assistant instructor will explain exactly what task or tasks you will perform. After beginning the task, if you do not know a step or steps, ask the assistant instructor for help. He will tell you how to perform the step. He will deduct the necessary points from your grade, and then you may continue to work. You must achieve a score of seventy percent in order to pass the test."

(2) Organize the unit into seven equal groups that rotate from station to station until all individuals have been tested. Gunners must wear their helmet and load-bearing equipment (LBE) to be tested. Assign enough assistant instructors to each station to grade and critique each gunner's performance. For ease in marking, give score sheets to each grader and collect the sheets immediately after the test.

(3) At the end of testing for each group, have each grader assemble the soldiers and give them a thorough five-minute critique.

b. **Testing Stations.** There are seven stations for gunnery skills testing. All stations except one, six, and seven have intermediate training objectives.

c. **Scoring.** For tasks at stations one, six, and seven, score the soldier GO if all steps are passed (P). Score the soldier NO-GO if any steps are failed (F). If the soldier fails any steps, show him what was done wrong and how to do it correctly. For stations two through five, score the soldier GO on a task if he receives no more than one intermediate training objective NO-GO.

STATION ONE

TASK: Identify the characteristics of the MK 19.

CONDITIONS: In any environment, the gunner is given a complete tripod-mounted MK 19.

STANDARDS: The gunner will know characteristics of the MK 19.

PERFORMANCE MEASURES	GO	NO-GO
1. General description (nomenclature).		
2. Weapons sight.		
3. Weight of gun.		
4. Maximum effective range for a point target.		
5. Maximum effective range for an area target.		
6. Cyclic rate of fire.		
7. Muzzle velocity.		
8. Types of ammunition.		
9. Weight of the MK 64, MOD 7, gun mount.		
10. Types of mounts.		

STATION TWO

Station Two has seven intermediate training objectives.

TASK: Perform operator's maintenance on the MK 19.

CONDITIONS: In any environment, the gunner is given a complete MK 19, cleaning equipment, tools, workspace, and TM 9-1010-230-10.

STANDARDS: Disassemble the MK 19 within four minutes. Use the cleaning materials as specified in the operator's TM to clean the MK 19. Inspect, lubricate, and assemble the MK 19 within four minutes, and perform a function check.

Intermediate Training Objective 1: The gunner must disassemble the MK 19.

CONDITIONS: The gunner is given a location in which to work, an assembled MK 19, weapon-cleaning equipment, lubricant, and TM 9-1010-230-10.

STANDARDS: The gunner must disassemble the MK 19 within four minutes.

PERFORMANCE MEASURES	GO	NO-GO
1. Clear the gun.		
2. Remove the bolt and backplate assembly.		
3. Remove the secondary drive lever.		
4. Remove the feed slide assembly.		
5. Remove the top cover assembly.		
6. Remove the feed tray.		
7. Remove the primary drive lever and vertical cam.		
8. Remove the alignment guide assembly.		
9. Remove the ogive plunger.		
10. Remove the round-positioning block.		
11. Remove the charger assemblies (left hand [LH], right hand [RH]).		
12. Remove the sear assembly.		
13. Complete within four minutes.		

Intermediate Training Objective 2: Clean the MK 19.

CONDITIONS: The gunner is given a disassembled MK 19, with the proper cleaning equipment.

STANDARDS: The gunner must clean the MK 19.

PERFORMANCE MEASURES	GO	NO-GO
1. Receiver assembly: a. Wipe or brush away dirt from all parts, especially the interior of the receiver housing, receiver rails, and feeder area. b. Use a rag to apply the cleaning solvent. c. Swab out the bore and chamber using a bore brush and RBC. d. Wipe dry.		
2. Sear assembly: a. Wipe or brush away dirt. b. Use a rag to apply the cleaning solvent. c. Wipe dry.		
3. Alignment guide assembly: a. Wipe or brush away dirt. b. Soak assembly in cleaning solvent. c. Wipe dry.		
4. Ogive plunger assembly: a. Wipe or brush away dirt. b. Use a rag to apply the cleaning solvent. c. Wipe dry.		
5. Round-positioning block: a. Wipe or brush away dirt. b. Soak in cleaning solvent.		
6. Charger assemblies: a. Wipe or brush away dirt. b. Use a rag to apply cleaning solvent. c. Wipe dry.		
7. Vertical cam assembly: a. Wipe or brush away dirt.		

PERFORMANCE MEASURES	GO	NO-GO
b. Soak in cleaning solvent.		
c. Wipe dry.		
8. Primary drive lever:		
a. Wipe or brush away dirt.		
b. Soak in cleaning solvent.		
c. Wipe dry.		
9. Secondary drive lever:		
a. Wipe or brush away dirt.		
b. Soak in cleaning solvent.		
c. Wipe dry.		
10. Feed slide assembly:		
a. Wipe or brush away dirt.		
b. Use a rag to apply the cleaning solvent.		
c. Wipe dry.		
11. Feed tray:		
a. Wipe or brush away dirt.		
b. Soak in cleaning solvent.		
c. Wipe dry.		
12. Top cover assembly:		
a. Wipe or brush away dirt.		
b. Use a rag to apply the cleaning solvent.		
c. Wipe dry.		
13. Bolt and backplate assembly:		
a. Wipe or brush away dirt.		
b. Use a rag to apply the cleaning solvent.		
c. Wipe dry.		
14. Use only authorized cleaning materials.		

Intermediate Training Objective 3: Inspect the MK 19.

CONDITIONS: The gunner is given a disassembled and cleaned MK 19.

STANDARDS: The gunner must inspect the MK 19.

PERFORMANCE MEASURES	GO	NO-GO
1. Receiver assembly:		
a. Inspect the receiver housing for cracks (even hairline), rust, burrs.		
b. Inspect the receiver rails for burrs, deformity, or unusual wear.		
c. Inspect the feeder pawls for spring action, burrs, or broken pawls.		
d. Inspect the barrel for carbon buildup or pitting in the bore or chamber.		
e. Inspect the flash suppressor for dents, cracks, or loose pin.		
f. Inspect the rear sight for rust, and the sight scale for readability.		
2. Sear assembly: Inspect for burrs, rust, or wear on sear face, or binding in the safety switch movement.		
3. Alignment guide assembly:		
a. Inspect for deformed or cracked spring.		
b. Inspect for cracks around the pin.		
c. Inspect for loose pin.		
d. Inspect for burrs, dents, and cracks.		
4. Ogive plunger assembly: Inspect for cracks in the housing.		
5. Round-positioning block:		
a. Inspect for burrs, dents, or cracks.		
b. Inspect for broken, deformed, or missing springs.		
c. Inspect for bent, cracked, or broken parts.		
6. Charger assemblies (LH, RH):		
a. Inspect for burrs on grooved edges.		

PERFORMANCE MEASURES	GO	NO-GO
b. Inspect for chipped, cracked, broken, or missing handle assemblies.		
c. Inspect for spring tension in the charger handle lock.		
d. Inspect for spring tension in the lock plunger.		
e. Inspect for bent charger housing.		
7. Vertical cam assembly:		
a. Inspect for burrs or buckling.		
b. Inspect for scratches, chips, or aluminum buildup on the chromed edge.		
c. Inspect for lock binding.		
8. Primary drive lever: Inspect for burrs around the pivot post.		
9. Secondary drive lever:		
a. Inspect for burrs, cracks, chips, or buckling.		
b. Inspect for missing retaining pin.		
10. Feed slide assembly:		
a. Inspect for burrs or cracks.		
b. Inspect for broken feed pawls, binding, or no spring action.		
c. Inspect for deformed shuttle-spring tip.		
d. Inspect for loose mounting screws.		
11. Feed tray:		
a. Inspect for burrs or cracks.		
b. Inspect for lack of spring action.		
12. Top cover assembly:		
a. Inspect for burrs, cracks, or rust.		
b. Inspect for loose or missing latch.		
c. Inspect for bent pins, and for missing or broken cross pins.		
13. Bolt and backplate assembly:		
a. Inspect overall for burrs, rust, or cracks.		

PERFORMANCE MEASURES	GO	NO-GO
b. Inspect handles for cracks. c. Inspect for broken or bent recoil springs. d. Inspect for bent or broken guide rods. e. Inspect for missing safety wire. f. Inspect for broken or worn cocking lever. g. Inspect for cracked or chipped bolt sear. h. Inspect for cracked or binding cam followers. i. Inspect for cracked or chipped bolt fingers. j. Inspect for chipped or broken firing pin tip. k. Inspect for chipped, cracked, or missing extractors. l. Inspect for bent or cracked backplate or missing retaining ring.		

Intermediate Training Objective 4: Lubricate the MK 19.

CONDITIONS: The gunner is given a disassembled, cleaned, and inspected MK 19, and lubricant and brush.

STANDARDS: The gunner must lubricate the weapon.

PERFORMANCE MEASURES	GO	NO-GO
1. Receiver assembly: Apply a light coat of lubricant on the receiver rails, feeder, and feeder pawls. 2. Sear assembly: Apply a light coat to all parts. 3. Alignment guide assembly: Apply a light coat to all parts. 4. Round-positioning block: Apply a light coat to the springs. 5. Charger assemblies (LH, RH): Apply a light coat to all parts, especially the grooved edges of the rails. 6. Vertical cam: Apply a light coat to all parts, especially the chromed edge. 7. Primary drive lever: Apply a light coat to all parts, especially the pivot posts. 8. Secondary drive lever: Apply a light coat to all parts, especially the pivot post and forked end. 9. Feed slide assembly: Apply a light coat to all parts, especially the feed pawls. 10. Feed tray: Apply a light coat to all parts, especially the pawl and guide rails. 11. Top cover assembly: Apply a light coat to all parts, especially the latch, the pivot post hole, and the pin holes. 12. Bolt and backplate assembly: Apply a light coat to all parts, especially the bolt face, bolt sear, guide rods, recoil springs, bolt rails, cam followers, and cocking lever.		

Intermediate Training Objective 5: Assemble the MK 19.

CONDITIONS: The gunner is given a disassembled, cleaned, inspected, and lubricated MK 19.

STANDARDS: The gunner must assemble the weapon in the proper sequence within four minutes.

PERFORMANCE MEASURES	GO	NO-GO
1. Assemble the sear assembly.		
2. Assemble the charger assemblies (LH, RH).		
3. Assemble the round-positioning block.		
4. Insert the ogive plunger assembly.		
5. Insert the alignment guide assembly.		
6. Attach the vertical cam.		
7. Attach the primary drive lever.		
8. Attach the feed tray.		
9. Attach the feed slide assembly.		
10. Attach the top cover assembly.		
11. Attach and engage the secondary drive lever.		
12. Insert the bolt and backplate assembly.		
13. Complete within four minutes.		

Intermediate Training Objective 6: Perform a function check.

CONDITIONS: The gunner is given an assembled MK 19.

STANDARDS: The gunner must perform a function check on the weapon.

PERFORMANCE MEASURES	GO	NO-GO
1. Close the top cover.		
2. Charge the weapon.		
3. Lock the charging handles forward.		
4. Place the gun on S (SAFE).		
5. Press the trigger.		
6. If the weapon fires, notify the supervisor.		
7. If the weapon does not fire, continue.		
8. Move the safety to F (FIRE).		
9. Press the trigger. Bolt should spring forward.		
10. Open the top cover.		
11. Check the tip of the firing pin. It should be exposed.		
12. Move the feed slide to the left.		
13. Close the top cover.		
14. Move the safety to S (SAFE).		

Intermediate Training Objective 7: Clean, inspect, and lubricate the MK 64, MOD 7, gun cradle.

CONDITIONS: The gunner is given a gun cradle, TM 9-1010-231-13&P, and lubricant.

STANDARDS: The gunner must clean, inspect, and lubricate the gun cradle.

PERFORMANCE MEASURES	GO	NO-GO
1. Clean the gun cradle. a. Brush or wipe away dirt. b. Use a rag to apply cleaning solvent. c. Wipe dry. 2. Inspect the gun cradle for: a. Bent or cracked retaining pins. b. Damaged threads on depression stop. c. Damage to mounting surface on the gun cradle. d. Loss of mounting hardware (nuts, bolts, pins, or chains). e. Depression stop in place (training use only). f. Frozen pintle lock. g. Missing cotter pins. 3. Lubricate the gun cradle. Apply a light coat to all pins and pivot points.		

STATION THREE

Station Three has four intermediate training objectives.

TASK: Load, unload, correct malfunctions, and clear the MK 19.

CONDITIONS: The gunner is given a location in which to work, a complete M3 tripod-mounted MK 19, and 10 rounds of linked 40-mm ammunition (dummy).

STANDARDS: The gunner must load, unload, and clear the MK 19, and take immediate and remedial action according to TM 9-1010-230-10 and this training program.

Intermediate Training Objective 1: Clear the MK 19.

CONDITIONS: The gunner is given an M3 tripod-mounted weapon in a designated location and 10 rounds of linked 40-mm ammunition (dummy).

STANDARDS: The gunner must clear the MK 19.

PERFORMANCE MEASURES	GO	NO-GO
1. Place the gun on S (SAFE).		
2. Lock the bolt to the rear.		
3. Remove the live round or spent case from the bolt face.		
4. Catch the live round as it falls out.		
5. Open the top cover.		
6. Remove the rounds from the feeder.		
7. Move the safety to F (FIRE).		
8. Ride the bolt forward.		
9. Move the safety to S (SAFE).		
10. Move the feed slide assembly to the left.		
11. Close the top cover.		

Intermediate Training Objective 2: Load the MK 19.

CONDITIONS: The gunner is given an M3 tripod-mounted weapon in a designated location and 10 rounds of linked 40-mm ammunition (dummy).

STANDARDS: The gunner must load the weapon.

PERFORMANCE MEASURES	GO	NO-GO
1. Place the weapon on S (SAFE).		
2. Ride the bolt forward.		
3. Return the charger assemblies to the original upright position.		
4. Open the top cover.		
5. Slide the rounds, female link first, through the feed throat.		
6. Insert the first round into the feeder and across the first pawl.		
7. Move the feed slide assembly to the left.		
8. Close the top cover.		
9. Charge the weapon.		
10. Lock the charger handles forward.		
11. Move the safety to F (FIRE).		
12. Press the trigger (bolt goes forward).		
13. Move the safety to S (SAFE).		
14. Charge the weapon (live round on bolt face).		
15. Lock the charger handles forward.		

Intermediate Training Objective 3: Correct malfunctions.

CONDITIONS: The gunner is given an M3 tripod-mounted weapon in a designated location, 10 rounds of linked 40-mm ammunition (dummy), and a simulated stoppage.

STANDARDS: The gunner must correct the malfunction.

PERFORMANCE MEASURES	GO	NO-GO
1. In a training situation, wait ten seconds after any stoppage.		
2. Charge the weapon and try to fire.		
3. If the weapon fails to fire, recharge and try to fire. If the weapon fails to fire (or when told that the gun did not fire) after the second attempt, take remedial action:		
4. Unload and clear the weapon.		
5. Inspect the weapon and ammunition and try to find the cause of the stoppage.		
6. If possible, correct the cause of the stoppage, reload, and try to fire the weapon.		

Intermediate Training Objective 4: Stop a runaway gun.

CONDITIONS: The gunner is given an M3 tripod-mounted weapon in a designated location, ten rounds of linked 40-mm ammunition (dummy), and a simulated runaway gun.

STANDARDS: The gunner must stop the runaway gun.

PERFORMANCE MEASURES	GO	NO-GO
Lower one of the charging handles.		

STATION FOUR

Station Four has four intermediate training objectives.

TASK: Mount and dismount the MK 19 onto one of the following: M3 tripod, M4 pedestal on a 1/4-ton vehicle, HMMWV with weapon platform, or commander's cupola on a M113.

CONDITIONS: The gunner is given a complete MK 19, a complete MK 64, MOD 7, gun cradle, a location in which to work, TM 9-1010-231-13&P, and one or more of the following: M3 tripod, 1/4-ton vehicle with an M4 pedestal, HMMWV with weapon platform, or M113.

STANDARDS: The gunner must mount and dismount the MK 19 on the gun cradle, mount the gun on other mounts as directed. The time limit for moving the MK 19 from one mount to another is four minutes.

TRAINING NOTE: When the MK 19 is mounted on a unit's organic carrier or the M3 tripod, it is called a weapon system. The weapon system is mounted using the three-man concept. The gunner is responsible for ensuring that it is mounted correctly, with the aid of the other two-team members. The steps for securing the MK 19 to the gun cradle are the same. The MK 19 is moved from one mount to another with the MK 64, MOD 7, gun cradle attached.

Intermediate Training Objective 1: Mount the MK 19 on the M3 tripod.

CONDITIONS: The gunner is given a location in which to work, an M3 tripod with T&E mechanism, the gun cradle, and TM 9-1010-231-13&P.

STANDARDS: The gunner must mount the gun cradle on the M3 tripod.

PERFORMANCE MEASURES	GO	NO-GO
MOUNTING:		
1. Set up the M3 tripod.		
2. Secure the MK 64, MOD 7, gun cradle onto the M3 tripod.		
3. Lift the gun into the gun cradle.		
4. Slide the front portion of the receiver into the gun cradle.		
5. Secure the rear of the gun to the gun cradle.		
6. Attach the T&E mechanism to the M3 tripod.		
7. Attach the T&E mechanism to the gun cradle.		
8. Attach the feed throat.		

PERFORMANCE MEASURES	GO	NO-GO
DISMOUNTING:		
9. Remove the T&E mechanism retaining pin.		
10. Insert the gun cradle stow pin (left side).		
11. Unlock the M3 tripod pintle lock release cam.		
12. Lift the MK 19, with the MK 64, MOD 7, gun cradle attached, from the M3 tripod.		

Intermediate Training Objective 2: Mount and dismount the MK 19 from the M4 pedestal.

CONDITIONS: The gunner is given a location in which to work, a vehicle with an M4 pedestal mount, an M3 tripod-mounted MK 19, the T&E mechanism, pintle adapter, ammunition container bracket, and tools.

STANDARDS: The gunner must remove the MK 19 from the M3 tripod, mount it on the M4 pedestal within four minutes, and dismount it from the M4 pedestal within four minutes.

PERFORMANCE MEASURES	GO	NO-GO
MOUNTING:		
1. Remove the T&E mechanism-retaining pin.		
2. Insert the cradle stow pin (left side).		
3. Unlock the M3 tripod pintle lock lever.		
4. Move the MK 19 to the M4 pedestal.		
5. Loosen the locking lever on the M4 pedestal.		
6. Insert the pintle adapter into the top part of the M4 pedestal.		
7. Remove the quick-release pin from the pintle adapter.		
8. Tighten the locking lever on the M4 pedestal.		
9. Lift up the MK 19 and insert the gun cradle pintle into the top of the pintle adapter.		
10. Insert the quick-release pin into the pintle adapter from either side.		
11. Attach the T&E mechanism to the gun mount.		
12. Center the elevation handwheel.		
13. Remove the stow pin.		
14. Attach the T&E mechanism clamp to the M4 pedestal.		
15. Attach the two support clamps to the M4 pedestal.		
16. Attach the ammunition container bracket (right side).		
17. Attach the empty case catch bag.		
DISMOUNTING:		
18. Remove the ammunition container bracket (right side).		
19. Remove the pin securing the T&E mechanism to the gun cradle.		
20. Insert the gun cradle stow pin.		
21. Remove the quick-release pin from the pintle adapter.		

PERFORMANCE MEASURES	GO	NO-GO
22. Lift the MK 19 from the M4 pedestal.		
23. Loosen the locking lever on the M4 pedestal.		
24. Lift the pintle adapter from the M4 pedestal.		

Intermediate Training Objective 3: Mount and dismount the MK 19 from the HMMWV weapon platform.

CONDITIONS: The gunner is given a mounted MK 19, a location in which to work, a HMMWV with weapon platform, pintle adapter assembly, T&E mechanism, and TM 9-1010-231-13&P.

STANDARDS: The gunner must remove the MK 19 from its mount and install it on the HMMWV weapon platform within four minutes.

PERFORMANCE MEASURES	GO	NO-GO
MOUNTING:		
1. Remove the MK 19 from its mount.		
2. Loosen the locking screws on the HMMWV pedestal.		
3. Insert the pintle adapter into the HMMWV pedestal.		
4. Tighten the locking screws.		
5. Remove the quick-release pin from the pintle adapter.		
6. Lift the MK 19, with MK 64, MOD 7, gun cradle attached to the pintle adapter and insert the gun cradle pintle into the pintle adapter.		
7. Insert the pintle adapter quick-release pin.		
8. Attach the train and elevating assembly to the gun cradle.		
9. Center the elevating handwheel.		
10. Remove the gun cradle stow pin.		
11. Attach the train and elevating mechanism to the HMMWV pedestal.		
12. Attach the ammunition container bracket.		
DISMOUNTING:		
13. Remove the ammunition container bracket.		
14. Remove the train and elevating mechanism from the HMMWV pedestal.		
15. Remove the T&E mechanism from the gun cradle.		
16. Insert the gun cradle stow pin.		

PERFORMANCE MEASURES	GO	NO-GO
17. Remove the quick-release pin from the pintle adapter assembly. 18. Lift the MK 19 from the HMMWV pedestal. 19. Remove the pintle adapter from the HMMWV pedestal.		

Intermediate Training Objective 4: Mount and dismount the MK 19 from the commander's cupola of an M113.

CONDITIONS: The gunner is given a mounted MK 19, a location in which to work, an M113 with the commander's cupola, pintle adapter, ammunition bracket, and TM 9-1010-231-13&P.

STANDARDS: The gunner must remove the MK 19 from its mount and install it on the commander's cupola of the M113 within four minutes.

PERFORMANCE MEASURES	GO	NO-GO
MOUNTING:		
1. Insert the pintle adapter into the mount receptacle.		
2. Remove the pintle adapter quick-release pin.		
3. Lift the MK 19 to the pintle adapter and insert the cradle pintle into the pintle adapter.		
4. Insert the pintle adapter quick-release pin.		
5. Attach the ammunition bracket.		
DISMOUNTING:		
6. Insert the gun cradle stow pin.		
7. Remove the ammunition bracket.		
8. Remove the pintle adapter quick-release pin.		
9. Lift the MK 19 from the commander's cupola.		

STATION FIVE

TASK: Identify characteristics of a defensive position for the MK 19.

CONDITIONS: The gunner is given an M3 tripod, a complete MK 19, a location in which to work, a direction of fire, and a partially dug fighting position.

STANDARDS: The gunner must identify the actions needed to complete the fighting position.

PERFORMANCE MEASURES	GO	NO-GO
1. Check assigned direction of fire.		
2. Check gun traverse.		
3. Check overhead cover.		
4. Check fighting position camouflage.		
5. Check for protection against small-arms fire.		
6. Check for protection against indirect fire fragments.		
7. Check for protection from aerial observation.		
8. Check for mutual support with other fighting positions.		
9. Check for a good field of fire.		

STATION SIX

TASK: Prepare a range card for the MK 19.

CONDITIONS: During daylight, in a defensive position, the gunner is given a tripod-mounted MK 19, a principal direction of fire and secondary sector of fire (both with recognizable targets), paper, pencil, and compass.

STANDARDS: The gunner must prepare range cards that include a data and sketch section for each sector of fire, with appropriate sketches and military symbols.

PERFORMANCE MEASURES	GO	NO-GO
1. Orient the range card.		
2. Sketch the terrain that is to the front of the position (prominent and man-made, including terrain features).		
3. Fill in or construct the marginal data section. a. Gun number. b. Unit designation. c. Date. d. Magnetic north arrow.		
4. Sketch machine gun symbol.		
5. Specify gun location in relation to prominent terrain feature or eight-digit grid coordinate. a. Determine azimuth in mils from the terrain feature to the gun. b. Determine distance between the terrain feature and the gun. c. Sketch and identify the terrain feature on the card. d. Connect the feature and gun with a barbed line. e. Write the distance in meters above the line. f. Write the azimuth in mils below the line from the feature to the gun.		
6. Sketch in the TRPs, if any.		
7. Label TRPs, if any.		

STATION SEVEN

TASK: Mount, place into operation, and dismount the AN/TVS-5 NVD from the MK 19.

CONDITIONS: The gunner is given a mounted MK 19 and an AN/TVS-5.

STANDARDS: The gunner must mount the AN/TVS-5 on the MK 19, place the sight into operation, and dismount the AN/TVS-5 from the MK 19.

PERFORMANCE MEASURES	GO	NO-GO
1. Mount the M2 mounting bracket onto the MK 19.		
2. Mount the AN/TVS-5 onto the M2 bracket.		
3. Install the batteries in the device.		
4. Press your eye against the eye guard.		
5. Turn the ON-OFF/TUBE BRIGHTNESS control to ON.		
6. Adjust the ON-OFF/TUBE BRIGHTNESS control, as needed.		
7. Turn the ON-OFF/RETICLE BRIGHTNESS control to ON.		
8. Turn the diopter focus ring until the clearest image of the reticle pattern is obtained.		
9. Turn the objective focus ring until the target in the field of view is sharply defined.		
10. Turn the ON-OFF/RETICLE BRIGHTNESS control to OFF.		
11. Turn the ON-OFF/TUBE BRIGHTNESS control to OFF.		
12. Remove the batteries from the AN/TVS-5.		
13. Remove the AN/TVS-5 from the M2 bracket.		
14. Remove the M2 bracket from the MK 19.		

APPENDIX E

ANNUAL GUNNERY TRAINING PROGRAM

This appendix contains four suggested annual training programs for the MK 19, MOD 3; one for each of the fighting and supporting active and reserve forces. Because of the high cost of M918TP and M430 HEDP rounds, live fire is kept to a minimum. Gunnery training programs have been prepared to adequately train gunners based on the mission profile of the unit. Refer to Chapter 4, Marksmanship, and Appendix I, Gunnery Exercises, for the firing tables that support these programs. A definition of each force, along with a summary of the annual gunnery training program designed for it, is included in this appendix.

E-1. LIVE-FIRE EXERCISES

LFXs allow the gunner to qualify on the MK 19 and to participate in collective gunnery exercises. Gunners practice and qualify using Tables I through IV in Chapter 4. The gun crews practice individual and collective skills using Tables I-1 through I-7 in Appendix I.

a. **Practice and Qualification Table Matrix** (Table E-1)**.** The tables used for gunner practice and qualification are in Chapter 4. There are four scorecards available and they are used based on the type of target (hull or pop-up silhouettes) and whether the practice-qualification is during the day or during limited visibility. Each scorecard has two tables, one for practice and one for qualification. The tables have versions for hull or pop-up silhouette engagements and for the type of NVD used. Sample scorecards are shown in Figures 4-27 through 4-30 (and reproducible forms are provided in the back of this manual). Gunners will only fire one day practice-qualification and one night practice-qualification. Units should select the practice and qualification based on the light conditions, type of targets available, and type of NVDs used. The following table shows which scorecards should be used.

Conditions	Target	Night Vision Devices	Scorecard
Day	Hull	NA	Scorecard I
Day	Pop-Up	NA	Scorecard III
Limited/Night	Hull	**AN/PEQ-2A** mounted on the TWS mounting bracket. **AN/PAS-13** mounted on the TWS mounting bracket. **AN/TVS-5** upgraded with the 3d generation tube mounted on the TWS mounting bracket.	Scorecard II
Limited/Night	Pop-Up or E-Type	All night vision devices.	Scorecard IV
Limited/Night	Any type	No night vision device.	Scorecard IV

Table E-1. Practice and qualification tables.

b. **Gunnery Exercises.** Gunnery exercises allow individual gun crews to practice firing the MK 19 while performing other tasks such as mounting the gun and preparing range cards. They also allow multiple gun crews to coordinate their fires while engaging targets. Table E-1 outlines each of the seven exercises. Based on their unit's training needs and the resources available, commanders must decide on the number and type of gunnery exercises to include in their annual training program.

Table	Exercise	Number of Guns	Number of Rounds
I-1	Instructional Fire Exercise	1	26 rounds
I-2	Dismounted Range Card Exercise	1	42 rounds
I-3	Mounted Combat Exercise	1	26 rounds
I-4	Night Fire Exercise	1	22 rounds
I-5	Linear and Deep Targets	2 to 4	80 rounds per gun
I-6	Linear Targets with Depth	2 to 4	40 rounds per gun
I-7	Area Targets	4 to 6	80 rounds per gun

Table E-2. Gunnery exercises.

E-2. FIGHTING FORCES ACTIVE COMPONENT

These forces consist of infantry, military police, scout, and engineer units of the active Army. Gunners assigned to these units are required to qualify semiannually. They are also required to participate in squad/platoon-level live fires quarterly and combined arms live fire exercises (CALFEX) annually. Assistant gunners are also required to qualify or fire familiarization exercises annually.

a. *Every quarter* the unit conducts preliminary marksmanship instruction and crew drills. The Gunnery Skills Test is also conducted.

b. During the *first quarter*, semiannual gunner practice qualification and qualification exercises are performed (see Table E-1 for the correct tables to use). Assistant gunner familiarization (with a zeroed gun) using Tables I and III are also conducted (see table E-1). Unit LFXs are performed (see Tables I-5 through I-7).

c. During the *second quarter,* unit LFXs are performed (see Tables I-1 through I-4, and I-5 through I-7).

d. During the *third quarter,* semiannual gunner practice qualification and qualification exercises are performed (see Table E-1 for the correct tables to use). Unit LFXs are also performed (see Tables I-5 through I-7).

e. During the *fourth quarter,* conduct LFXs or CALFEX using exercises from Tables I-5 through I-7.

E-3. SUPPORTING FORCES ACTIVE COMPONENT

These forces consist of all other units in the active component not identified as fighting force units. These guns are used primarily in defense of battalion, brigade, and division rear operations. Gunners from these forces participate in instruction fire and qualification annually.

a. *Every quarter* the unit conducts preliminary marksmanship instruction and crew drills. The Gunnery Skills Test is also conducted.

b. During the *first quarter,* there is no additional training.

c. During the *second quarter,* perform LFXs (see Tables I-5 through I-7).

d. There is no additional training during the *third quarter.*

e. During the *fourth quarter,* annual day gunner practice qualification and qualification exercises are performed. Night practice qualification and qualification is conducted once every two years (see Table E-1 for the correct tables to use). Once every two years, assistant gunner familiarization (with a zeroed gun) using Tables I and III is also conducted (see table E-1).

E-4. FIGHTING FORCES RESERVE COMPONENTS

These forces consist of military police, engineer, and infantry units of the reserve components. Gunners for these forces must maintain a level of proficiency higher than that of the active support forces. At the same time, limited time and range resources must be considered. Gunners of these forces conduct sustainment on the trainer at least quarterly. Live fires consist of annual qualification and a unit level LFX. These forces are allocated 192 or 118 rounds each year for live fire and qualification.

a. *Every quarter* the unit conducts preliminary marksmanship instruction and crew drills. The Gunnery Skills Test is also conducted.

b. There is no additional training during the *first quarter.*

c. There is no additional training during the *second quarter.*

d. During the *third quarter,* annual gunner practice qualification and qualification exercises are performed. See Table E-1 for the correct tables to use. Once every two years, the assistant gunner fires familiarization (with a zeroed gun) using Tables I and III (see table E-1). Unit LFXs are performed (see Tables I-5 through I-7).

e. There is no additional training during the *fourth quarter only.*

E-5. SUPPORTING FORCES RESERVE COMPONENT

These forces consist of all reserve units not specifically outlined under the paragraph on fighting forces. They are allocated 78 rounds for each year to qualify.

a. *Every quarter* the unit conducts preliminary marksmanship instruction and crew drills. The Gunnery Skills Test is also conducted.

b. There is no additional training during the *first quarter.*

c. There is no additional training during the *second quarter.*

d. During the *third quarter,* gunner practice qualification and qualification exercises are performed once every two years. See Table E-1 for the correct tables to use. Once every two years, the assistant gunner fires familiarization (with a zeroed gun) using Tables I and III (see table E-1). Unit LFXs are performed (see Table I-7).

e. There is no additional training during the *fourth quarter.*

APPENDIX F

FIRING TABLE

The table in this appendix provides information about the weapon elevation, time of flight, and other factors of MK 19 projectiles.

NOTE: 40-mm TP, XM918 is reduced time-of-flight experimental ammunition.

ELEVATION (MILS)					FLIGHT TIME (SECONDS)				REMAINING VELOCITY			
									M430		M385	
RANGE (M)	M430	M385	M918	XM918*	M430	M385	M918	XM918*	FPS	MPS	FPS	MPS
0	0.0	0.0	0.0	0.0	0.00	0.00	0.00	0.00	790	240.8	790	240.8
100	9.0	9.0	8.9	4.6	0.43	0.43	0.43	0.31	747	227.7	744	226.8
200	18.6	18.7	18.5	9.5	0.88	0.88	0.88	0.63	706	215.2	701	213.7
300	29.1	29.3	28.8	14.9	1.36	1.37	1.35	0.97	667	203.3	660	201.2
400	40.4	40.8	39.9	20.6	1.87	1.88	1.85	1.33	630	192.0	622	189.6
500	52.6	53.4	51.8	26.7	2.40	2.43	2.38	1.70	596	181.7	586	178.6
600	65.9	66.9	64.7	33.3	2.97	3.07	2.93	2.10	564	171.9	552	168.2
700	80.3	81.8	78.5	40.4	3.58	3.63	3.52	2.52	533	162.5	520	158.5
800	96.0	98.2	93.5	48.1	4.22	4.29	4.14	2.97	505	153.9	491	149.7
900	113.1	116.1	109.8	56.3	4.90	4.99	4.80	3.43	478	145.7	463	141.1
1,000	131.9	135.8	127.5	65.2	5.63	5.74	5.50	3.93	453	138.1	438	133.5
1,100	152.5	157.6	146.8	74.9	6.41	6.55	6.24	4.45	430	131.1	414	126.2
1,200	175.2	181.9	167.8	85.3	7.24	7.42	7.03	5.00	409	124.7	393	119.8
1,300	200.4	209.0	191.0	96.6	8.14	8.37	7.88	5.58	389	118.6	373	113.7
1,400	228.6	239.6	216.6	108.9	9.11	9.40	8.79	6.19	372	113.4	356	108.5
1,500	260.3	274.6	245.0	122.2	10.1	10.53	9.77	6.84	356	108.5	341	103.9
1,600	296.7	315.3	277.1	136.7	7	10.53	10.84	7.53	343	104.5	329	100.3
1,700	339.2	364.2	313.7	152.6	11.3	11.80	12.01	8.27	333	101.5	320	97.5
1,800	390.8	426.4	356.5	169.9	4	13.25	13.34	9.05	325	99.1	314	95.7
1,900	458.3	518.4	408.5	189.0	12.6	15.01	14.88	9.88	322	98.1	314	95.7
2,000	570.1	X	477.2	210.0	5	17.44	16.82	10.77	324	98.8	X	X
2,100	x	X	757.3	233.4	14.1	X	23.66	11.73	X	X	X	X
2,200	X	X	X	259.6	7	X	X	12.77	X	X	X	X
2,300	X	X	X	289.3	16.0	X	X	13.91	X	X	X	X
2,400	X	X	X	323.4	6	X	X	15.18	X	X	X	X
2,500	X	X	X	363.8	18.9	X	X	16.62	X	X	X	X
2,600	X	X	X	413.8	7	X	X	18.32	X	X	X	X
2,700	X	X	X	482.4	X	X	X	20.54	X	X	X	X
2,800	X	X	X	X	X	X	X	X	X	X	X	X
					X							
*REDUCED TIME-OF-FLIGHT EXPERIMENTAL AMMUNITION.												

Table F-1. Comparison of direct-fire aiming data.

	REMAINING VELOCITY				ANGLE OF FALL (DEGREES)				MAXIMUM ORDINATE (FEET)			
	M918		XM918*									
RANGE (M)	FPS	MPS	FPS	MPS	M430	M385	M918	XM918*	M430	M385	M918	XM918*
0	790	240.8	1100	335.3	0.0	0.0	0.0	0.0	0.0	0.0	0.0	0.0
100	750	228.6	1044	318.2	0.5	0.5	0.5	0.3	0.7	0.7	0.7	0.4
200	712	217.0	991	302.1	1.1	1.2	1.1	0.6	3.1	3.1	3.1	1.6
300	675	205.7	940	286.5	1.8	1.9	1.8	0.9	7.4	7.5	7.3	3.8
400	641	195.4	892	271.9	2.6	2.7	2.6	1.3	14.0	14.2	13.8	7.1
500	608	185.3	847	258.2	3.6	3.7	3.5	1.8	23.3	23.7	22.8	11.7
600	578	176.2	804	245.1	4.7	4.8	4.5	2.3	35.7	36.5	34.7	17.9
700	549	167.3	763	232.6	5.9	6.1	5.6	2.9	51.7	53.1	50.1	25.7
800	521	158.8	724	220.7	7.3	7.6	6.9	3.6	72.0	74.4	69.4	35.6
900	496	151.2	688	209.7	8.9	9.3	8.4	4.3	97.4	101.0	93.4	47.7
1,000	471	143.6	653	199.0	10.7	11.3	10.1	5.2	128.7	134.1	122.7	62.5
1,100	449	136.9	620	189.9	12.8	13.6	12.0	6.1	167.0	174.8	158.2	80.3
1,200	428	130.5	590	179.8	15.2	16.2	14.1	7.2	213.6	224.9	201.2	101.6
1,300	409	124.7	561	171.0	17.9	19.1	16.5	8.4	270.3	286.3	252.9	126.8
1,400	391	119.2	534	162.8	21.0	22.5	19.2	9.8	339.3	361.8	315.0	156.6
1,500	376	114.6	508	154.8	24.5	26.4	22.3	11.4	423.6	455.3	390.0	191.6
1,600	362	110.3	484	147.5	28.4	30.9	25.8	13.1	527.6	527.9	480.8	232.7
1,700	350	106.7	462	140.8	32.9	36.1	29.7	15.0	658.4	724.6	592.2	280.8
1,800	341	103.9	442	134.7	38.2	42.2	34.1	17.2	828.4	931.6	731.6	337.0
1,900	334	101.8	423	128.9	44.4	50.0	39.2	19.6	1,065.6	1,259.3	912.3	402.8
2,000	332	101.2	406	123.7	53.2	X	45.3	22.4	1,486.5	X	1,166.1	479.9
2,100	350	106.7	392	119.5	X	X	63.6	25.3	X	X	2,298.7	570.7
2,200	X	X	379	115.5	X	X	X	28.6	X	X	X	678.0
2,300	X	X	368	112.2	X	X	X	32.3	X	X	X	806.2
2,400	X	X	359	109.4	X	X	X	36.3	X	X	X	961.6
2,500	X	X	354	107.9	X	X	X	40.8	X	X	X	1,154.6
2,600	X	X	351	107.0	X	X	X	45.8	X	X	X	1,405.6
2,700	X	X	352	107.3	X	X	X	52.0	X	X	X	1,767.3
2,800	X	X	X	X	X	X	X	X	X	X	X	X
	X	X			X	X	X		X	X	X	

*REDUCED TIME-OF-FLIGHT EXPERIMENTAL AMMUNITION.

Table F-1. Comparison of direct-fire aiming data (continued).

APPENDIX G

AIMING DEVICES

This appendix provides information on the employment of the AN/PEQ-2A target pointer illuminator aiming light (TPIAL), AN/TVS-5, and the AN/PAS-13B (V) 3 heavy weapon thermal sight (HWTS). These devices have the advantage of greater magnification and resolution for the employment of the MK 19 during periods of degraded visibility. These devices, when properly boresighted and employed under proper conditions, allow infantrymen to engage targets at the weapon's maximum effective ranges during periods of degraded visibility.

G-1. AN/PEQ-2A DESCRIPTION AND OPERATION

The TPIAL projects an infrared laser beam that cannot be seen with the eye but can be seen with NVD. It is also capable of projecting a much wider infrared illuminating beam from an integral illuminator. The TPIAL works with night vision goggles and mounts on various weapons with mounting brackets and adapters (Figure G-1 and Table G-1). Leaders can also use the AN/PEQ-2A in the hand-held mode to illuminate and designate targets.

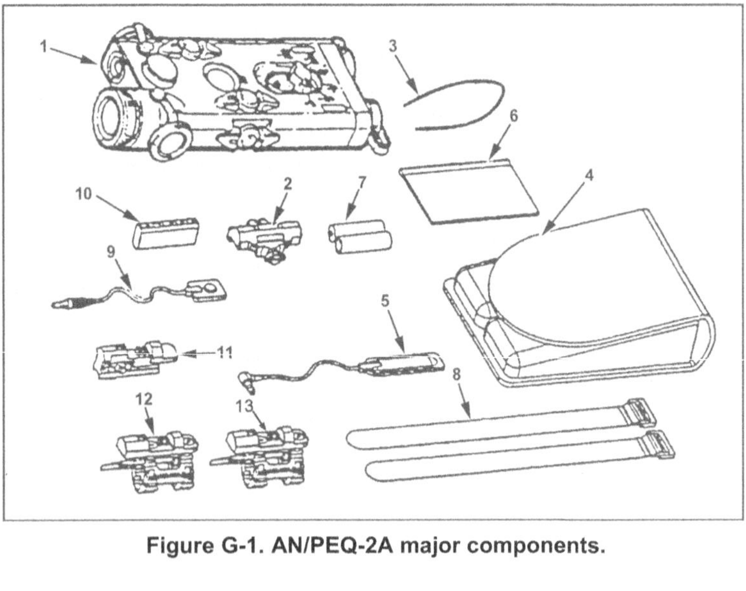

Figure G-1. AN/PEQ-2A major components.

ITEM	DESCRIPTION
1	TPIAL Assembly with Safety Block
2	Rail Grabber Mounting Bracket
3	Neck Cord
4	Carrying Bag
5	Cable Switch, 20 inch Remote, Button
6	Operator's Manual
7	Batteries, 2 AA
8	Straps, Retention
9	Cable Switch, 12 inch, Membrane
10	Bracket Adapter
11	Training Extender (Army only)
12	M4/M16A2 Bracket Assembly (Army only)
13	M16A2 Bracket Assembly (Marine only)

Table G-1. Names of the AN/PEQ-2A major components.

a. **Controls and Indicators.** The AN/PEQ-2A has controls and indicators that allow the user to operate the device and to select its different modes.

(1) ***Battery Installation.*** To install batteries in this device, first unscrew the battery caps and install two AA batteries. Orient the batteries as indicated by the markings on the AN/PEQ-2A body (Figure G-2).

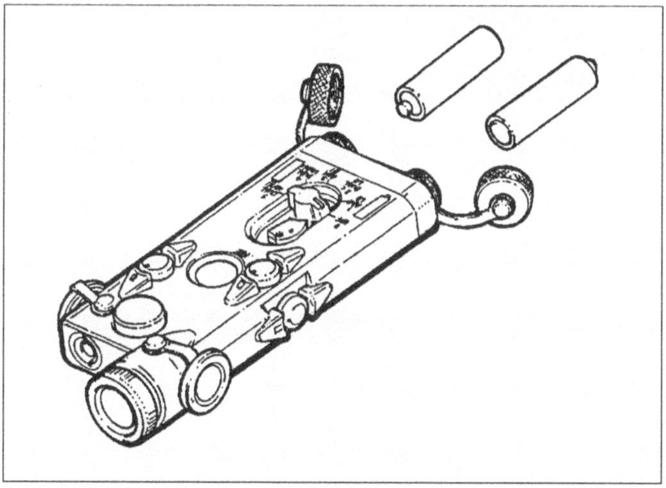

Figure G-2. AN/PEQ-2A battery installation.

(2) ***Safety Block Installation.*** The safety block installed in the training mode (blue side up) prevents the operator from accessing the non-eye safe modes (AIM HI, DUAL

LO/HI, DUAL HI/HI) (Figure G-3). A .050 hex-head Allen Wrench is needed to unscrew the block from the body and re-install it in the tactical mode (black side up).

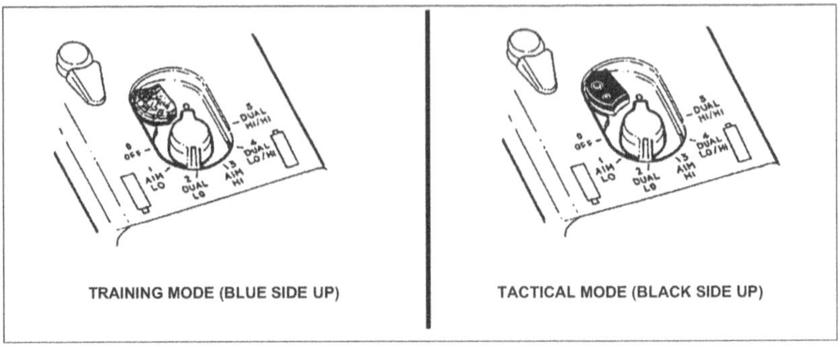

Figure G-3. Safety block installation.

(3) ***Mode Selector.*** The mode selector is used to set the mode in which the AN/PEQ-2A will operate when the cable switch button or push button is depressed. The mode selector has six positions (Table G-2).

KNOB POSITION		OPERATION
0	OFF	The AN/PEQ-2A will not operate.
1	AIM LO	The aiming beam operates at low power.
2	DUAL LO	The aiming beam and the illuminating beam operate at low power.
3	AIM HI	The aiming beam operates at high power.
4	DUAL LO/HI	The aiming beam operates at low power and the illuminating beam operates at full power.
5	DUAL HI/HI	The aiming beam operates at high power and the illuminating beam operates at full power.

Table G-2. Mode selector position.

(4) ***Button Switch.*** The button switch is used when the AN/PEQ-2A is hand held. Pressing the button switch operates the AN/PEQ-2A in the operational mode set by the

mode selector switch (Figure G-4). When the button is released, the AN/PEQ-2A turns off.

(a) A green LED is incorporated into the body of the AN/PEQ-2A to indicate that the unit is ON. Whenever the AN/PEQ-2A is activated, the green LED will light and stay lit until the unit is turned OFF.

(b) If continuous operation of the AN/PEQ-2A is desired, pressing the button switch twice in rapid succession will latch the unit ON. The unit will remain ON until the button switch is pressed the third time.

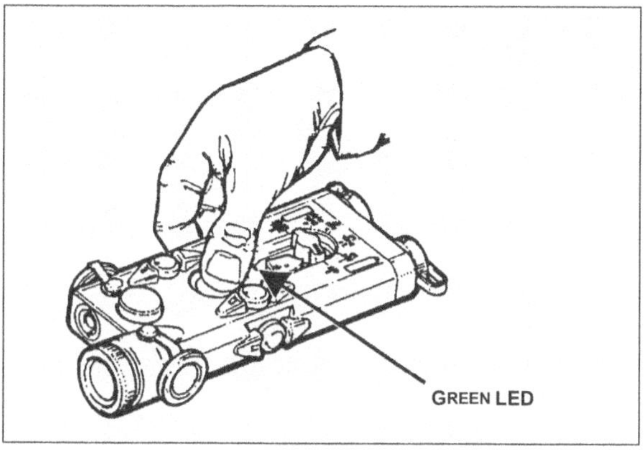

Figure G-4. Operation of the button switch.

(5) ***Cable Switch.*** The cable switch is used when the AN/PEQ-2A is mounted on a weapon (Figure G-5). The cable switch plugs into the back of the AN/PEQ-2A assembly. Pressing the button or pad at the end of the cable switch causes the unit to turn on in the operational mode selected by the mode select switch. When released, the AN/PEQ-2A turns off.

(a) If continuous operation of the AN/PEQ-2A is desired, pressing the cable switch twice in rapid succession will latch the unit ON. The unit will remain on until the push button is pressed a third time.

(b) When the cable switch plug is installed in the AN/PEQ-2A, it automatically locks into place. To remove the switch, pull back on the plug sleeve and pull the plug out.

CAUTION

Do not try to remove the plug by pulling on the cable.

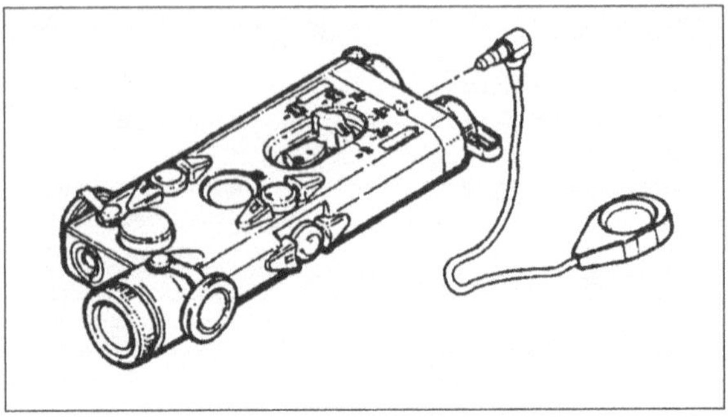

Figure G-5. Installation of the cable switch.

(6) ***Focus Knob.*** The focus knob is used to vary the spread of the illumination beam based on the range and size of the area to be illuminated (Figure G-6).

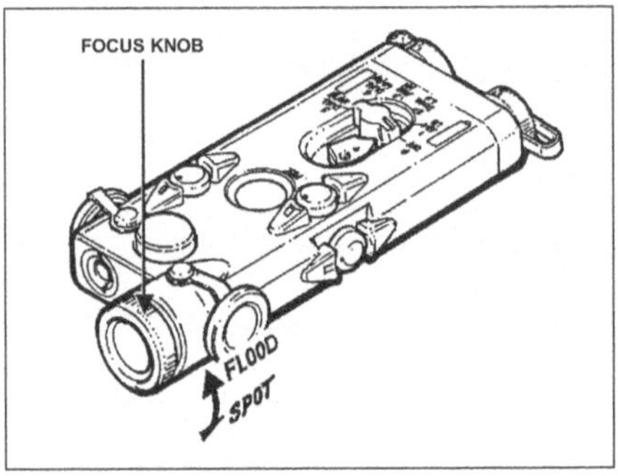

Figure G-6. Using the focus knob.

(7) ***Lens Caps.*** Several lens caps can be used with the AN/PEQ-2A for different purposes (Figure G-7).

(a) The *black lens cap* blocks the AN/PEQ-2A illuminator or aiming laser beam should the device be activated. To use the black lens cap, pull it from its stored location on the side of the unit and stretch it over the front of the focus knob or aiming beam so that it fits snugly in place.

(b) The *diffuser lens cap* enables the illuminator or aiming laser to emit in a 45-degree cone (10 feet at 10 feet). To use the diffuser lens cap, pull it over the front of the focus knob or aiming beam so that it fits snugly in place.

(c) The *neutral density lens cap* enables the AN/PEQ-2A illuminator or aiming laser to be operated in low power. To use the neutral density lens cap, pull it from its stored location on the side of the unit and stretch it over the front of the focus knob or aiming beam so that it fits snugly in place.

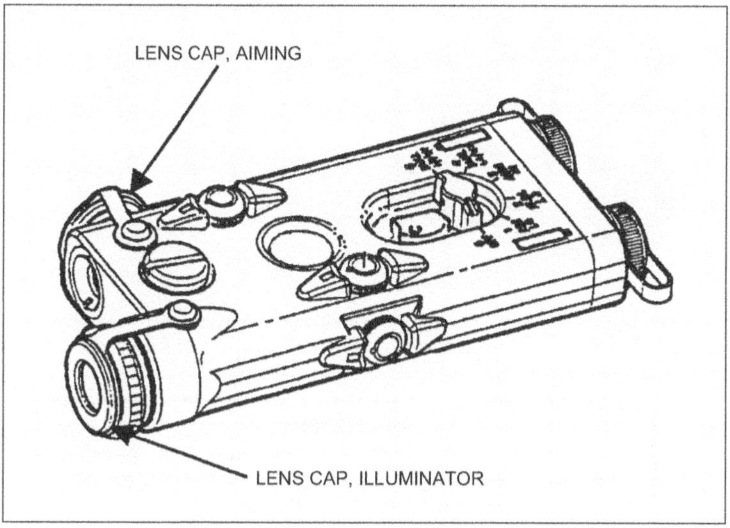

Figure G-7. Installing the lens caps.

(8) ***Boresight Adjusters.*** The AN/PEQ-2A is equipped with boresight adjusters for zeroing the aiming beam and illumination beam to the weapon (Figure G-8). The AN/PEQ-2A adjusters move the beam in true horizontal and vertical directions with the top adjusters used for elevation and the side adjusters for windage. When zeroing the AN/PEQ-2A, it is best to zero the aiming beam to the weapon and then align the illumination beam to the aiming beam.

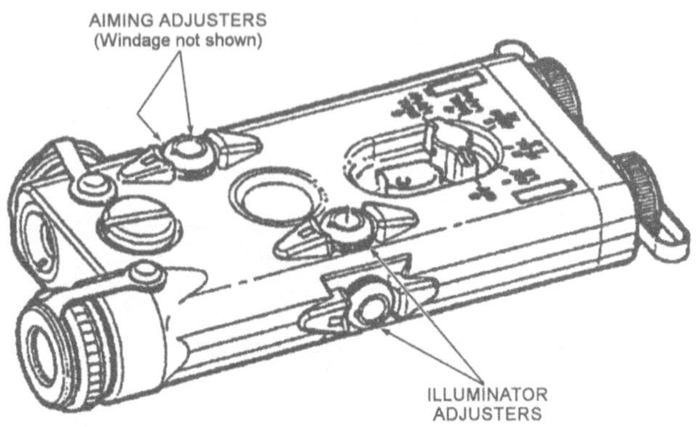

Figure G-8. Boresight adjusters for both aiming and illumination beams.

> **WARNING**
> **Eye damage can occur if the laser is handled carelessly. The danger area is 15 meters in the training mode and 220 meters in the tactical mode.**

b. **Mounting and Dismounting Procedures.** The AN/PEQ-2A can be mounted on various weapons or used in the hand-held mode to illuminate and designate targets.

(1) ***Mounting Procedures.*** Ensure that the mounting bracket is installed before proceeding.

(a) Secure the bracket adapter to the underside of the AN/PEQ-2A.

(b) Place the bracket adapter into the MK 19 bracket mounting groove, located in the center of the rail grabber.

(c) Tighten the bracket's lever screw into the bracket adapter hole.

(d) Plug the cable into the AN/PEQ-2A and run the cable to the front of the MK 19 bracket. The "cable switch, 12-inch, membrane" or the "cable switch, 20-inch, button" may be used for remote access.

(e) Once the cable is secure at the front of the MK 19 bracket (ensure the cable will not become damaged by the movement of the quadrant portion of the bracket) run it between the MK 19 and the bracket to the handles on the backplate assembly. If using the button cable, attach Velcro to the top of the night handle and secure excess cable to the handle. If using the membrane pad, run it the same way as with the button cable and attach the membrane to the inside of the right handle.

(2) ***Dismounting Procedures.*** Dismount the AN/PEQ-2A when the mission is complete, when it is needed on another weapon, or when it will be used in its hand-held mode.

(a) Untie the cable and disconnect the remote button from the backplate handles.

(b) Unplug the cable from the AN/PEQ-2A and place the cable back in the carrying case.

(c) Turn the AN/PEQ-2A mounting screw counterclockwise and remove the AN/PEQ-2A from the bracket adapter.

(d) Unscrew the bracket adapter from the MK 19 bracket rail and place the bracket adapter and the AN/PEQ-2A in its carrying case.

c. **Boresighting Procedures**. Boresighting aligns the sighting system to the bore of the weapon. The AN/PEM-1 Borelight is used for this procedure.

(1) Place the MK 19 in the ready to fire position 10 meters from the bore light offset zero target. Use the 10 meter distance gauge in the borelight kit to measure the distance. Ensure the target and weapon are level and stable (the bore sight zero will be off if not) before making any adjustments to the laser.

(2) Set the range on the MK 19 bracket to 500 meters.

(3) Mount the bore light and turn it on. While holding the MK 19 steady, zero the bore light. Follow the instructions in the AN/PEM-1 borelight operations manual for further details. The 40-mm MK 19 mandrel interface (NSN 3460-01-502-0575) is used with the 5.56-mm mandrel interface and has to be ordered separately.

(4) Make adjustments with the T&E mechanism until the bore light is on the bore light aiming point.

(5) Adjust the windage and elevation of the aiming laser until the laser is on the laser aiming point.

(6) Re-check the bore light aiming point and then the laser aiming point again.

(7) To zero the IR light source narrow the beam as much as possible and adjust its windage and elevation until the beam is 4 centimeters horizontally to the left of the laser aiming point while the aiming laser is still on the laser aiming point. The AN/PEQ-2A is now boresighted.

d. **Zeroing Procedures.** Zeroing the weapon aligns the line of sight of the AN/PEQ-2A with the trajectory of the round so that they intersect at 500 meters.

(1) Set the MK 19 mounting bracket to 500 meters by loosening the friction knob on the left side of the mount. Press and hold the friction knob in (towards the mount), then slide the mount assembly to the 500-meter increment. Release and tighten the friction knob.

(2) Identify a 500-meter target downrange and fire one round; note the impact of the round in relation to the target.

(3) Adjust the aiming light on the AN/PEQ-2A to the impact of the round by adjusting the elevation and windage adjusters.

(4) Manipulate the T&E mechanism so that the aiming light is center mass on the target. Repeat until the point of impact is the same as the point of aim.

(5) Adjust the illuminating beam's elevation and windage adjusters so that the aiming light is in the center of the illuminating beam.

G-2. THE AN/PAS-13B (V) 3 HEAVY WEAPON THERMAL SIGHT

The AN/PAS-13B (V) 3 HWTS is a silent, lightweight, compact, and durable battery-powered infrared imaging sensor that operates with low battery consumption (Figure G-9).

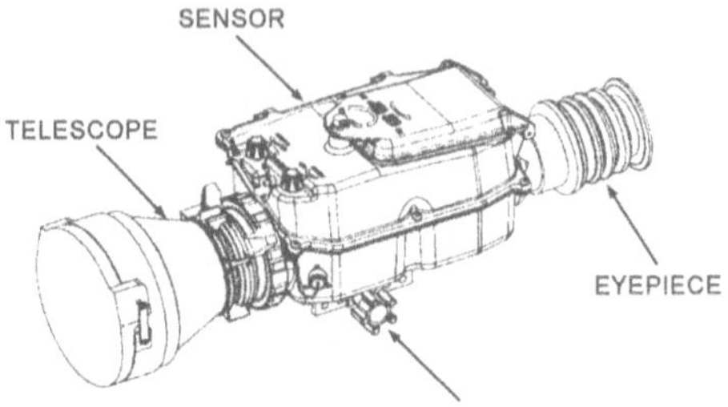

Figure G-9. AN/PAS-13B (V) 3 heavy weapon thermal sight.

a. **Components.** The HWTS is capable of target acquisition under conditions of limited visibility such as darkness, smoke, fog, dust, and haze. The HWTS operates effectively at night and can also be used in the daytime. Infrared light is received through the telescope, detected by an IR sensor, converted to digital data, processed, and then displayed for the user. Besides the carrying cases, the HWTS is composed of four major components: the telescope, the basic sensor, the eyepiece, and the mount.

(1) ***Telescope.*** The telescope receives IR light emitting from an intended target and its surroundings. The telescope magnifies and projects the IR light onto the basic sensor's scanner.

(2) ***Basic Sensor.*** The scanner reflects the IR light received from the telescope onto the detective assembly. The detective assembly senses the IR light and converts it to video. The sensor's electronics condition the video for display on the LED array. The LED array illuminates the IR image along with the reticle. The light from the LED array is reflected off the scanner to form an image at the eyepiece.

(3) ***Eyepiece.*** The eyepiece displays the thermal image and all system indicators on the cathode ray tube.

(4) ***Mount.*** A universal attachment interface between the HWTS and the MK 19 bracket (figure only shows the torque limiting knob).

b. **Modes of Operation.** The HWTS has three modes of operation: STANDBY, ON, and EMERGENCY.

(1) ***STANDBY Mode.*** When the system is first turned on, the HWTS begins a cool down period of approximately two minutes. After the cool down period, the HWTS

enters the STANDBY mode. During the STANDBY mode, power is not applied to the scanner or display in order to extend the life of the battery.

(2) ***ON Mode.*** When the HWTS is in the STANDBY mode and pressure is applied to the eyecup, the HWTS switches to the ON mode, and a switch engages to provide power to the scanner and display. After a three-second delay, the system is fully operational.

(3) ***EMERGENCY Mode.*** When switched to the EMERGENCY mode, the HWTS continuously applies power to the entire system. This allows the operator to bypass the three-second delay experienced when switching from the STANDBY to the ON mode. Since power is applied to the entire system while in the emergency mode, battery life is greatly reduced.

c. **Controls and Indicators.** Controls allow the gunner to configure the sight to the situation and individual preferences while the indicators display the current configuration (Figure G-10 and Table G-3).

(1) The CONTRAST CONTROL adjusts the contrast of the thermal image displayed on the raster. It has an automatic and a manual mode.

CAUTION

Ensure the brightness control switch is depressed before turning it on or off.

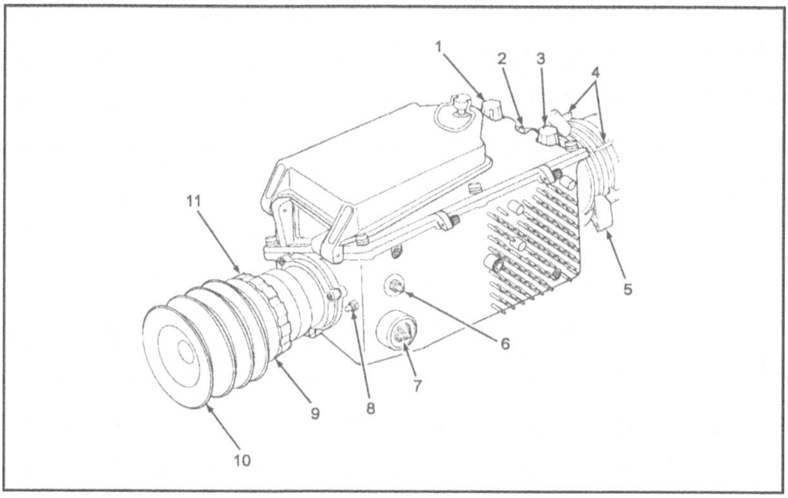

Figure G-10. Location of the HWTS controls and indicators.

ITEM	DESCRIPTION
1	Contrast Control
2	Emergency Switch Control
3	Brightness Control
4	Focus Ring
5	Field of View Ring
6	Zoom/Reticle Select Switch
7	Reticle Adjust Switch
8	Black/White Polarity Switch
9	Diopter Holding Device
10	Eyecup
11	Diopter Focus Ring

Table G-3. HWTS controls and Indicators.

(2) The EMERGENCY SWITCH CONTROL places the HTWS in emergency mode. The eyecup must be depressed in order to for the CRT to illuminate.

(3) The BRIGHTNESS CONTROL is a rotary switch with an off detent position (turned fully counterclockwise). The purpose is to turn the system on or off and adjust the brightness of the eyepiece display.

(4) The FOCUS RING adjusts the telescope focus from 20 meters to infinity. It requires a manual adjustment and affects both the wide and narrow fields of view.

(5) The FIELD OF VIEW RING (FOV) is located on the telescope. It has a wide and a narrow field of view. The wide FOV is for using low magnification during target detection, and the narrow FOV is for using high magnification during recognition and engagement.

(6) The RETICLE SELECT SWITCH selects one of the available reticles depending on the type of thermal sight (medium or heavy) and the weapon. It must be held for two seconds to enable reticle changes. After two seconds, release the switch to cycle to the next reticle. This control is disabled after ten seconds of inactivity.

(7) The RETICLE ADJUST SWITCH adjusts the reticle aiming features in azimuth and elevation. It is used during zeroing, and it must be held for two seconds to allow changes to be made. After two seconds, each press moves the reticle aiming features one increment. This control is also disabled after ten seconds of inactivity.

(8) The BLACK/WHITE POLARITY SWITCH selects the polarity of the thermal image displayed on the raster. The initial setting is "white hot." The polarity switch affects the appearance of the target.

(9) The DIOPTER HOLDING DEVICE allows the diopter to be adjusted when pulled back.

(10) The EYECUP controls the STANDBY/ON operating mode. When forward pressure is applied to the eyecup, the system is in the ON mode. When forward pressure is removed for more than 30 seconds, the system returns to the STANDBY mode.

(11) The DIOPTER FOCUS RING adjusts the focus of the raster and indicators to the operator's eye. It ranges from +2 to -6 diopters.

d. **Status Indicators and Raster** (Figure G-11). The status indicators and raster are visible when the user presses his eye against the eyecup. The indicators display the system status and configuration. The raster displays the thermal image, the selected reticle, and azimuth/elevation zeroing adjustment indicators.

(1) The status indicators are:

- NOT COOL (1): Displayed when the system is not cool enough for proper operation.
- WHT HOT/BLK HOT (2): Indicates the target polarity.
- EMER (3): Displayed when system is in emergency mode.
- LOW (4): Displayed when battery power is low (approximately fifteen minutes of useful power left).
- FIELD OF VIEW (7): Indicates when the wide field of view (WFOV) or the narrow field of view (NFOV) is in use
- ZOOM (8): Displayed when the zoom mode is selected
- RET ADJ (9): Indicates the reticle adjustment mode is selected
- RET SEL (10): Indicates the reticle select mode is selected

(2) The DISPLAY RASTER displays the thermal image with a superimposed reticle and indicators:

- RETICLE SELECT (5): Identifies the selected reticle/weapon.
- ELEVATION INDICATOR (6): Indicates the elevation zeroing adjustment of the reticle.
- AZIMUTH INDICATOR (11): Indicates the elevation zeroing adjustment of the reticle.
- RANGE SCALE (12): The range scale allows the operator to estimate the range to a target based on a known height or width. The numbers on the scale are the range in hundreds of meters (example: = 400 meters). The vertical lines on the scale reflect the height of a five-foot man at that range. The distance between two consecutive lines reflects the width of a 10-foot tank at the range specified by the vertical line on the left.
- CROSSHAIR/AIMING POINT (13): The vertical line at the aiming point reflects the height of a five-foot man at 800 meters. The horizontal line of the cross-hair reflects the width of a 10-foot tank at 800 meters.

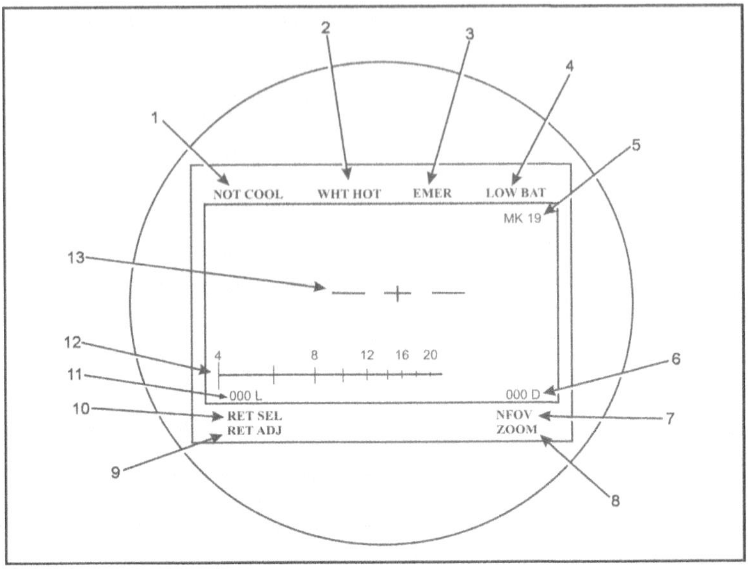

Figure G-11. Status indicators and raster.

e. **Mounting and Dismounting Procedures.** The HWTS is mounted on the MK 19 with a mounting bracket (Figure G-12). See TM 11-5855-312-10 for details.

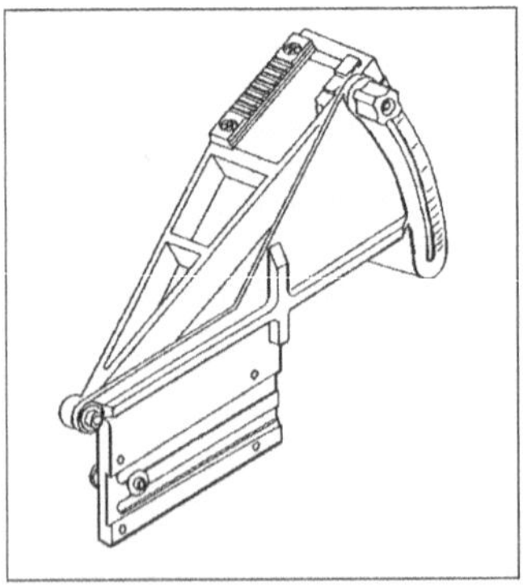

Figure G-12. MK 19 bracket.

(1) ***Mounting Procedures.*** Mounting the HWTS to the MK 19 is a two-step process (Figure G-13):

(a) Mount the bracket to the MK 19 by sliding the bracket into the dovetail slot until the locking pin engages the mount.

(b) Install the HWTS onto the bracket mount.

(c) Shake the HWTS to make sure the sight is secured properly.

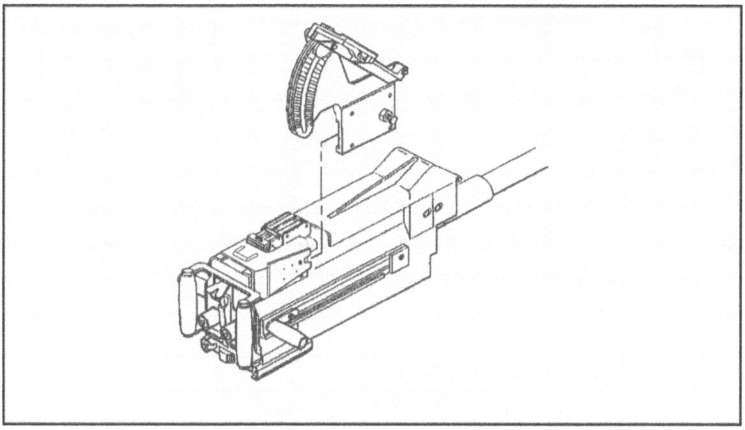

Figure G-13. Mounting the HWTS on MK 19.

(2) ***Dismounting Procedures.*** To dismount the HWTS:

(a) Take the HWTS off the mounting bracket rail and place the HWTS back in its carrying case.

(b) Pull out the locking pin and slide the bracket off the mount.

G-3. AN/TVS-5

The AN/TVS-5 is a portable, battery-operated electro-optical instrument used for observation and aimed fire of weapons at night (Figure G-14 and Table G-4). It amplifies reflected light such as moonlight, starlight, and sky glow so that the viewed scene becomes clearly visible to the operator. The sight does not emit visible or infrared light (except from the eyepiece) that can be detected by the enemy. By using this device, the gunner can observe the area and detect and engage any suitable target.

WARNING

Ensure the weapon is not loaded and is on S (SAFE) before installing the HWTS on the weapon. A loaded weapon may accidentally discharge causing severe injury or death.

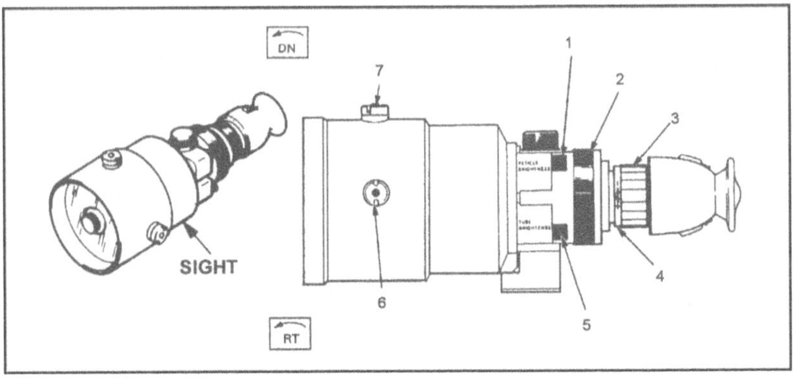

Figure G-14. Location of the AN/TVS-5 components.

ITEM	DESCRIPTION
1	ON-OFF/ Reticle Brightness
2	Objective Focus Ring
3	Diopter Focus Ring
4	Diopter Indicator
5	ON-OFF/ Tube Brightness
6	Reticle Azimuth Adjustment Acutator
7	Reticle Elevation Adjustment Acutator

Table G-4. AN/TVS-5 component names.

a. **Controls and Indicators.**Controls allow the gunner to configure the sight to the situation and individual preferences while indicators display the current configuration.

(1) The ON-OFF/TUBE BRIGHTNESS control applies power to the sight and controls the brightness of the image intensifier tube. It also enables the ON-OFF RETICLE BRIGHTNESS control to function.

(2) The ON-OFF RETICLE BRIGHTNESS control applies power to the reticle and controls the brightness of the reticle.

(3) The OBJECTIVE FOCUS RING adjusts the range focus from 25 meters to infinity.

(4) The DIOPTER FOCUS RING adjusts the focus of the eyepiece.

(5) The DIOPTER INDICATOR indicates the direction of rotation of the DIOPTER FOCUS RING for + and diopters.

(6) The RETICLE ELEVATION ADJUSTMENT ACTUATOR controls the reticle adjustment in the up and down direction. Each click of adjustment moves the strike of the round 1.0 inch at 100 meters.

(7) The RETICLE AZIMUTH ADJUSTMENT ACTUATOR controls reticle adjustment right or left. Each click or adjustment moves the strike of the round 1.0 inch at 100 meters.

b. **AN/TVS-5 Sight Reticle.** The MK 19 and the M2 machine guns share the same AN/TVS-5 sight reticle (Figure G-15).

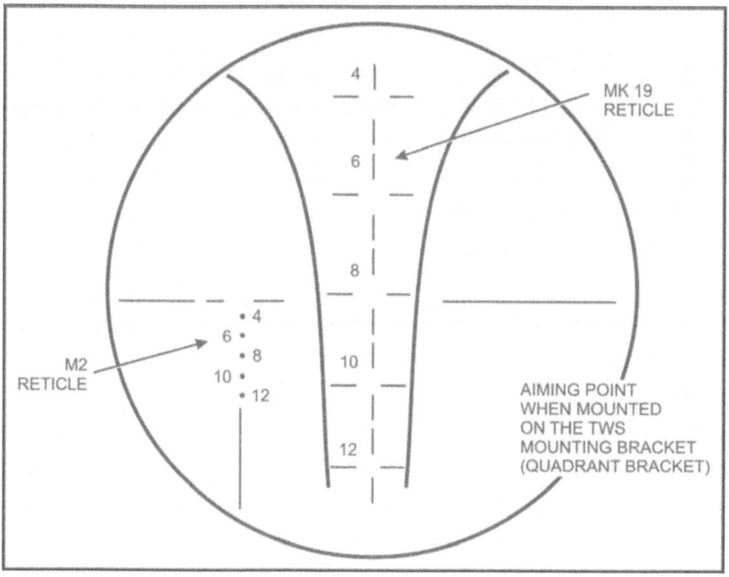

Figure G-15. AN/TVS-5 reticle for the MK 19 and M2 machine guns.

WARNING

1. **Eye guard:**
   - **To avoid recoil injury, attach the eye guard before using the sight.**
   - **When activated by pressure on the eye guard, the sight emits illumination the enemy can detect. Therefore, avoid touching the eye guard except when you wish to use the sight, and then press your eye area against it to activate it.**
2. **Batteries:**
   - **Dispose of batteries as directed. The contents of the mercury batteries cause extreme irritation to the eyes and to oral and nasal passages.**
   - **To prevent explosion, avoid disposing of batteries by burning.**
   - **Avoid short circulating batteries.**
   - **Avoid recharging batteries.**
   - **Remove batteries before storing sight.**
   - **Always replace both mercury batteries at the same time.**

c. **Mounting Procedures**. To mount the AN/TVS-5:

(1) Align the scribe line on the sight with the scribe line on the bracket.

(2) Place the sight in the grove of the bracket and tighten the lever screw clockwise.

(3) Secure the lever screw with lacing wire or tape to ensure the sight does not vibrate loose.

(4) Seat the device. After firing the initial burst, retighten the lever screw to ensure the sight is securely mounted. (If you are unable to fire at this time, lightly shake the sight to ensure the sight is mounted correctly.)

d. **Dismounting Procedures.** To dismount the AN/TVS-5:

(1) Remove the lacing wire or tape from the lever screw.

(2) Loosen the lever screw until the sight is free to be lifted off the bracket.

(3) Remove the batteries and place the sight in its carrying case.

e. **Boresighting.** The AN/TVS-5 cannot be boresighted when used with the MK 19. The sight cannot focus on a target closer than 25 meters, and the offset is too great at that distance to boresight accurately.

f. **Zeroing.** To zero the AN/TVS-5 to the MK 19, follow the procedures in TM 11-5855-214-10.

APPENDIX H

CREW DRILLS

The purpose of a crew drill is to build precision, speed, skill, and teamwork in checking equipment, placing the MK 19 into and out of action, and moving the MK 19 to a new position. The main goal is achieved by following prescribed procedures. Precision, speed, skill, and teamwork grow with practice. During the crew drill, all oral and visual signals are repeated. Once crewmembers can perform their duties well, they rotate to allow each member of the crew to learn the duties of others. The ground-mounted MK 19 has three crewmembers: the gunner, the assistant gunner, and the ammunition bearer. The vehicle-mounted MK 19 also has three crewmembers: the squad leader, the gunner, and the driver. Normally the crew is not required to carry the MK 19 and ammunition over long distances. A vehicle usually carries the crew, ammunition, and equipment to the firing position. The terrain and the enemy situation dictate how close the vehicle may be brought to the firing position.

H-1. CREW EQUIPMENT AND DRILL AREA

Dismounted crew drills are conducted with a gunner, assistant gunner, and ammunition bearer. For drills involving a vehicle, the squad leader, gunner, and driver are crewmembers. In addition to individual weapons and equipment, each crewmember for dismounted drills carries specific equipment used with the tripod-mounted MK 19 (Table H-1).

CREW MEMBER	SUGGESTED MINIMUM EQUIPMENT
Gunner	MK 19, binoculars, and compass
Assistant Gunner	M3 tripod, MK 64 gun cradle, with T&E mechanism attached
Ammunition Bearer	One box of ammunition

Table H-1. Equipment.

A crew drill is conducted in an area large enough to allow crewmembers to move freely with equipment and without interfering with the movements of other crewmembers.

WARNING

Each soldier must wear all the protective equipment required by range safety regulations. This may include helmet, earplugs, and protective vest.

H-2. FORMATION FOR CREW DRILL

To form the crew for crew drill, the leader commands FORM FOR CREW DRILL. Upon hearing the leader's command, the crew repeats the command and forms into a column, five paces apart, in the following order: assistant gunner, gunner, and ammunition bearer (Figure H-1).

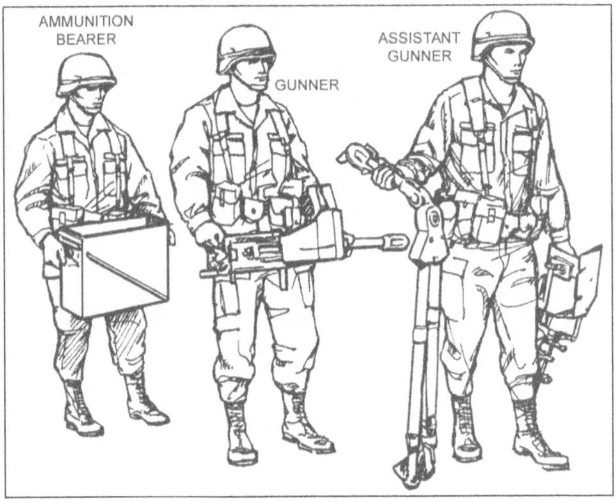

Figure H-1. Three-man MK 19 crew.

Each crewmember carries the following equipment in the specified manner:

- The assistant gunner carries the tripod, with the front leg positioned at a 60-degree angle and the trail legs collapsed, in his right hand; he carries the gun cradle, with the T&E mechanism attached, in his left hand.
- The gunner carries the MK 19, binoculars, and compass.
- The ammunition bearer carries one can of ammunition.

H-3. INSPECTION OF EQUIPMENT

The leader commands CHECK EQUIPMENT. All crewmembers repeat the command and inspect the components. The equipment is inspected and verified by the following crewmembers.

a. The assistant gunner inspects the M3 tripod, gun cradle, and T&E mechanism and ensures:

(1) The indexing levers and clamps on the front and trail legs work, and that the legs are in the "short (low)" position.

(2) The front leg is placed at a 60-degree angle and the trail legs are closely folded and the front leg clamp is tight.

(3) The sleeve lock latch and pintle lock release cam work.

(4) The pintle bushing is clean and has no burrs.

(5) The cradle pintle is clean, and all pins and bolts are seated.

(6) The T&E mechanism is centered.

b. The gunner inspects the MK 19 and ensures:

(1) The weapon is clean.

(2) The bolt is forward, the top cover is closed, and the weapon is on S (SAFE).

(3) The charger handles are forward and locked in the "up" position.

(4) The binocular lenses are clean and the focus mechanism works.

(5) The compass magnetic arrow and bezel ring work freely.

c. The ammunition bearer inspects the ammunition and ensures:

(1) The correct type of ammunition is on hand.

(2) The ammunition can is dent-free.

(3) All ogives are tight and the ammunition is clean.

d. The crewmembers report when the inspections are finished in the following manner:

(1) The ammunition bearer reports "Ammunition correct," or any defects.

(2) The gunner reports "Gun and ammunition correct," or any defects.

(3) The assistant gunner reports, "All correct," or any defects.

H-4. PLACING THE MK 19 INTO ACTION

The leader commands and signals GUN TO BE MOUNTED, and points to the position where the gun is to be mounted; FRONT, and points in the direction of fire; ACTION, and indicates with his fist the MK 19's direction in the chosen position. Crewmembers take the following steps to place the gun into operation.

a. The assistant gunner:

(1) Drags the tripod to the firing position by its front leg with his right hand, and carries the gun cradle with the T&E mechanism attached in his left hand.

(2) When he arrives at the firing position, aligns the front tripod leg for direction and, with one quick motion, snaps the trail legs apart. This action makes the sleeve lock-latch snap in place (Figure H-2).

(3) Stamps each leg into the ground and secures the gun cradle to the tripod by using the pintle lock release cam (Figure H-3).

(4) Secures the T&E mechanism to the tripod, by locking the traversing slide lock lever (Figure H-4).

(5) Helps the gunner mount and load the MK 19.

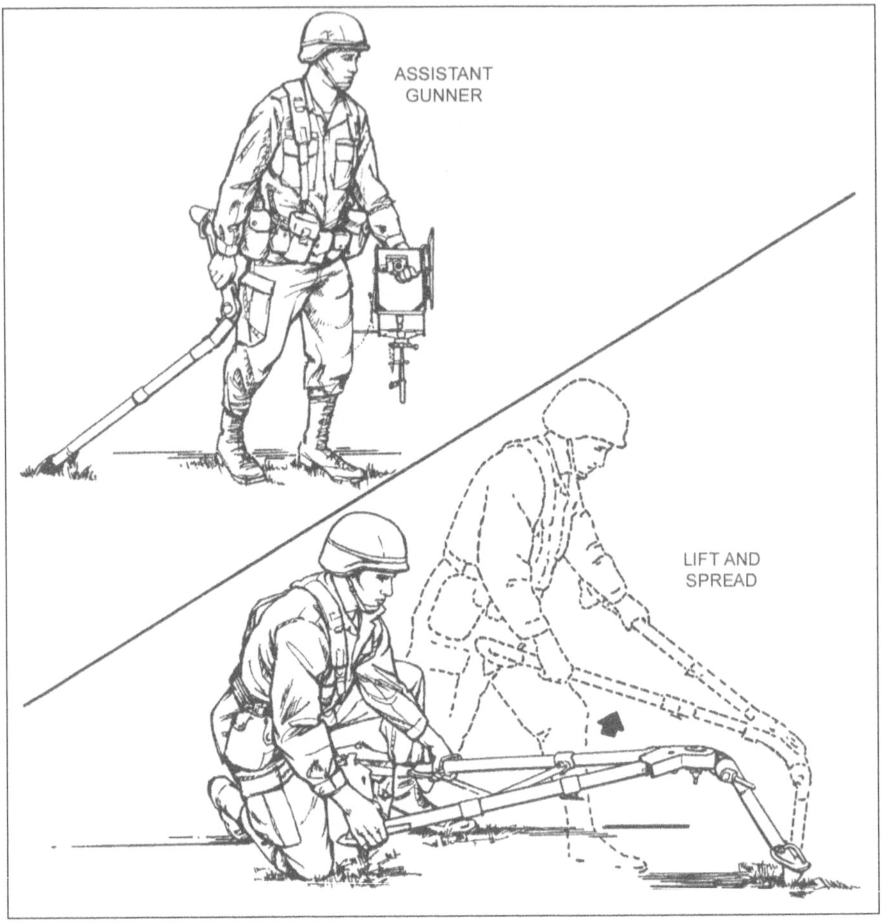

Figure H-2. Positioning the tripod.

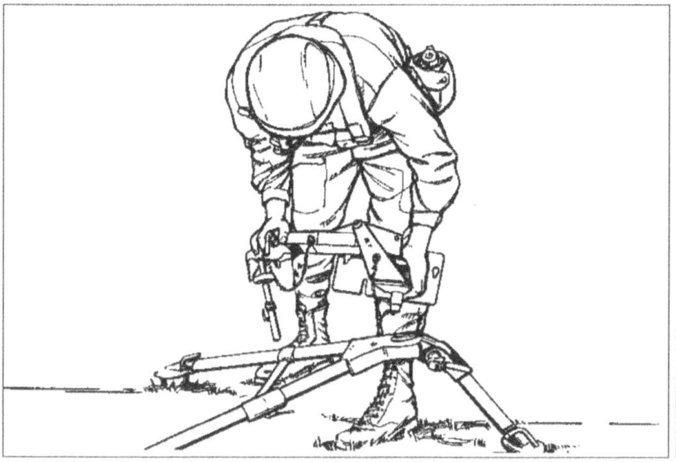

Figure H-3. Mounting the MK 19 mount on the tripod.

Figure H-4. Securing the T&E mechanism.

b. The gunner:

(1) Carries the MK 19 receiver to the tripod, with the barrel in his left hand, and the control grip in his right hand (Figure H-5).

(2) At the tripod, the assistant gunner grasps the barrel while the gunner holds the grips.

(3) The gunner and assistant gunner align the grooves on the receiver with the lugs in the gun cradle, and slide the receiver forward (Figure H-6).

(4) Secures the rear of the weapon by inserting the retaining pin through the cradle and sear assembly (Figure H-7).

Figure H-5. Gunner moving into position.

Figure H-6. Placing the MK 19 in the cradle.

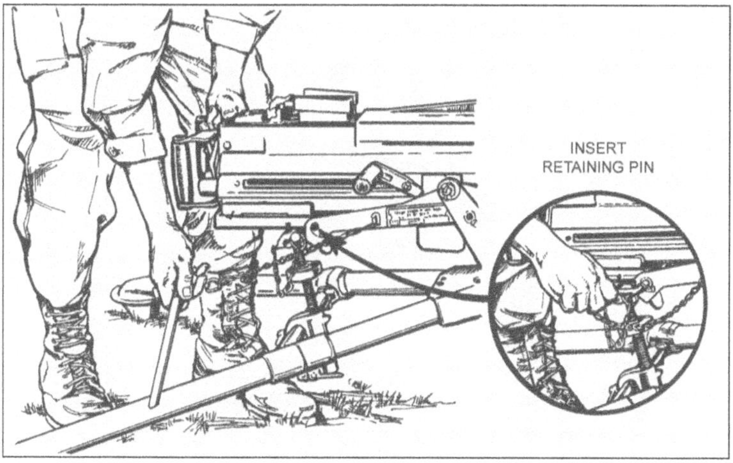

Figure H-7. Securing the MK 19 to the cradle.

c. The ammunition bearer:

(1) Carries the ammunition can to the MK 19 position (Figure H-8).

(2) Takes the lid off and ensures the ogive ends of the rounds are pointed downrange (Figure H-9).

(3) Gives the can to the assistant gunner.

(4) Returns to the ammunition point and stands by to do whatever is needed to complete the mission.

Figure H-8. Ammunition bearer with ammunition can.

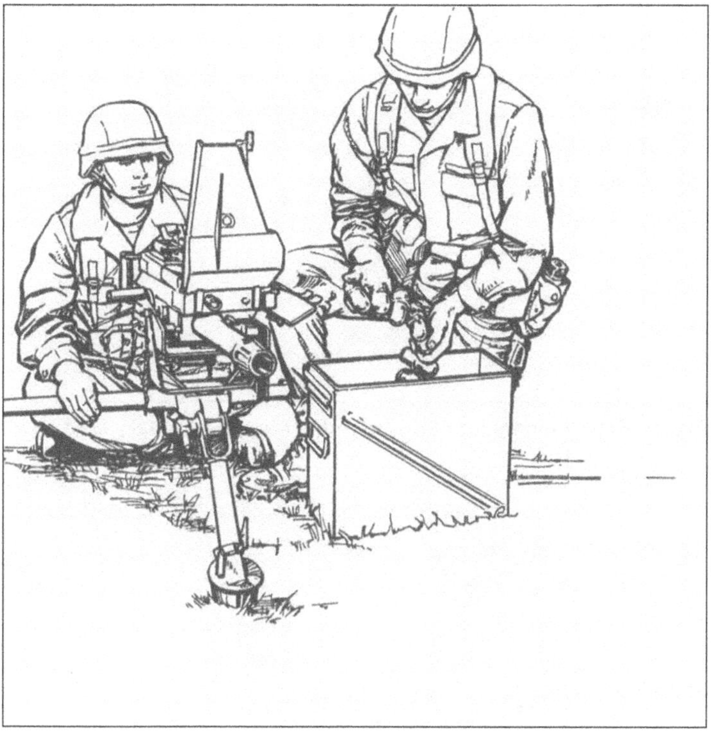

Figure H-9. Alignment of ammunition belt.

d. The gunner opens the top cover. The assistant gunner inserts the first round into the feeder (female link first), pushes the first round across the primary feed pawl (Figure H-10), and moves the feed slide assembly to the left by pushing the secondary drive lever to the right (Figure H-11).

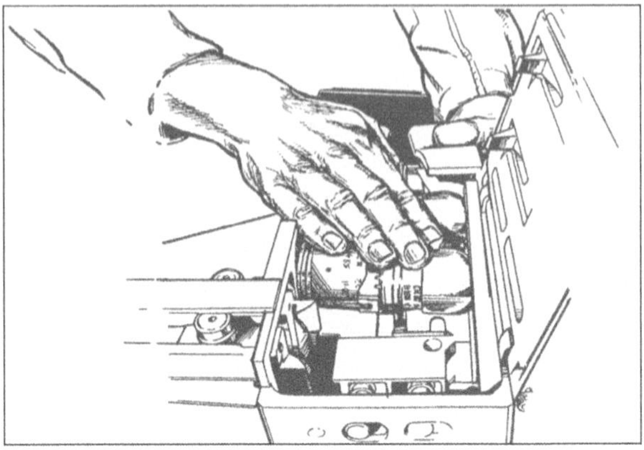

Figure H-10. Placement of the round across the primary feed pawl.

Figure H-11. Feed slide assembly moved to the left.

e. The gunner closes the top cover and charges the weapon (Figure H-12). The gunner presses the trigger, which places the lead round in the bolt extractor. He charges the weapon a second time, forcing the lead round from the bolt extractors into the bolt fingers. The weapon is ready to fire. When the assistant gunner raises the rear sight, the gunner calls "Up" (Figure H-13).

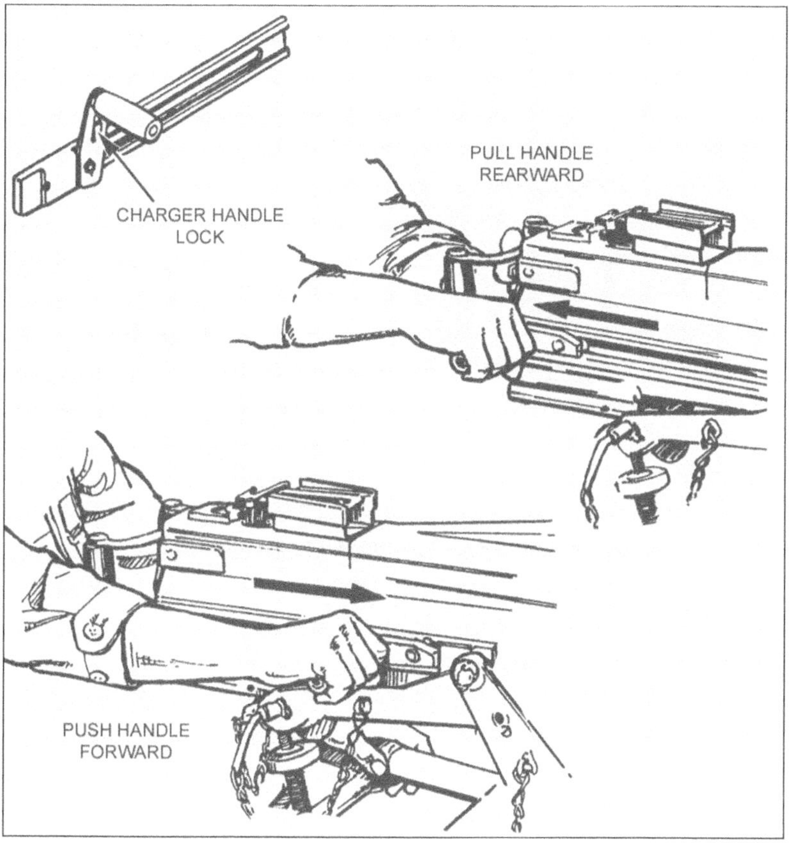

Figure H-12. Gunner charging the MK 19.

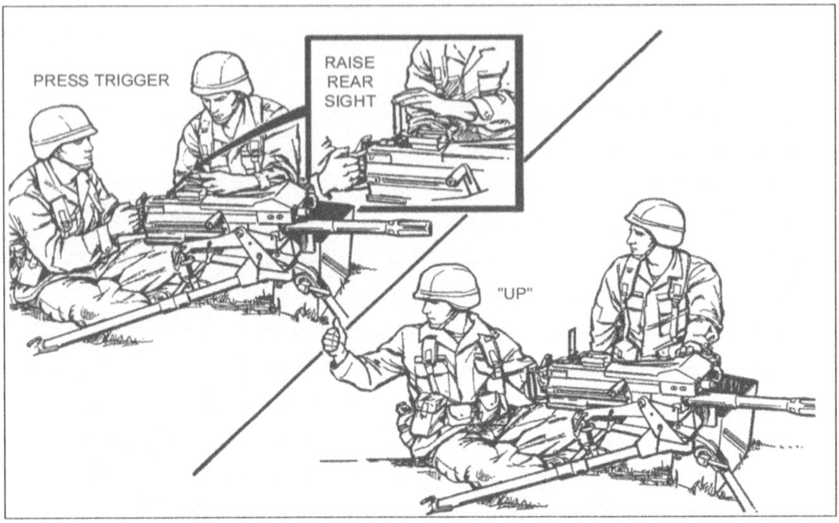

Figure H-13. Raising the rear sight.

H-5. TAKING THE MK 19 OUT OF ACTION

To take the gun out of action, the leader commands OUT OF ACTION. Each crewmember accomplishes the following in sequence.

a. The gunner raises the top cover of the MK 19. The assistant gunner releases the ammunition belt from the feed tray and places it in the ammunition can. The gunner takes the round or spent cartridge case from the bolt fingers, places it in the ammunition can, and closes the top cover.

b. The assistant gunner releases the MK 19 from the gun cradle. The gunner lifts the MK 19 off the mount and moves to a position 10 paces behind the tripod. He places the MK 19 on the ground in front of him with the flash suppressor to the left and with the sights on the top.

c. Once the gunner has moved away from the tripod, the assistant gunner:

(1) Releases the T&E mechanism and takes off the gun cradle/ T&E mechanism.

(2) Releases the tripod sleeve lock latch.

(3) Grasps the front leg and rotates the tripod to a vertical position.

(4) Releases the sleeve latch, and collapses the trail legs.

(5) Loosens the front leg clamp, folds down the front leg to a 60-degree angle, and tightens the clamp.

(6) Returns, with the tripod and gun cradle, to a position five paces in front of the gunner.

d. The ammunition bearer comes forward and gets the ammunition can(s) or any other loose equipment, and returns to a position five paces behind the gunner.

e. The gunner announces "Up."

H-6. TRANSPORT OF THE TRIPOD-MOUNTED MK 19

To move the MK 19 and equipment a short distance, the gunner commands SECURE EQUIPMENT, FOLLOW ME. The MK 19 can be moved using one of three methods:

a. **Dragging.** The gunner grasps the front leg and pulls the MK 19 to the new position, and the assistant gunner secures any loose equipment.

b. **Two-man Carry.** The gunner stands behind the tripod with a trail leg in each hand. The assistant gunner grasps the front leg and avoids touching the barrel if it is hot. The gunner and assistant gunner carry the mounted MK 19. The assistant gunner returns to secure any loose equipment.

c. **Three-man Carry.** The gunner and assistant gunner each holds a trail leg just above the traverse bar. The ammunition bearer grasps the front leg and avoids touching the barrel if it is hot. The crew moves to the new position. The ammunition bearer and the assistant gunner go back and get the ammunition or loose equipment, and return to the new position.

H-7. DISMOUNT THE MK 19

To dismount the MK 19 from the vehicle-mounted or stowed position, each member of the crew performs the following steps in order within four minutes.

a. To dismount the MK 19 from the mounted position the crew performs the following actions.

(1) The squad leader:

(a) Dismounts the vehicle with binoculars and compass.

(b) Obtains the M3 tripod from the stow position.

(c) Takes the M64 gun cradle and T&E mechanism from the gunner and moves to the ground firing position.

(d) At the firing position, the gunner aligns the front tripod leg for direction and snaps the trail legs apart.

(e) Stamps the tripod legs into the ground.

(f) Zero's the T & E mechanism, if it is not zeroed.

(g) Secures the gun cradle to the M3 tripod and the T&E mechanism to the traverse bar and gun cradle.

(h) Assists the gunner in mounting the MK 19.

(2) The gunner:

(a) Raises the top cover, releases the ammunition belt from the feed tray, and places it in the ammunition can.

(b) Takes the round or spent cartridge case from the bolt fingers, places it in the ammunition can, and closes the top cover.

(c) Lifts the MK 19 from the gun cradle after the driver releases it, and hands it to the driver.

(d) Releases the T&E mechanism from the pedestal with the train lock handle and removes the empty case catch bag assembly.

(e) Removes the gun cradle with the T&E mechanism attached and hands it to the squad leader.

(f) Takes the MK 19 from the driver and moves to the firing position.

(g) With the squad leader's assistance, the gunner secures the MK 19 to the gun cradle.

(3) The driver:

(a) Dismounts the vehicle and assists the gunner in clearing the weapon.

(b) Removes the ammunition can and releases the MK 19 and takes it from the gunner.

(c) Gives the MK 19 to the gunner when the gunner is ready.

(d) Stows the empty case catch bag assembly and takes the ammunition can to the firing position, while also removing the lid and ensuring the ogive ends of the rounds are pointed downrange.

(e) Returns to the vehicle and monitors the radio and/or takes more ammunition to the firing point, and maintains rear security.

(4) The squad leader and the gunner load the weapon (as in a crew drill), and place the MK 19 into action.

b. To dismount the MK 19 from the stowed position the crew performs the following actions.

(1) The squad leader:

(a) Secures the binoculars and compass.

(b) Obtains the M3 tripod, the M64 gun cradle, and the T&E mechanism from their stowed position.

(c) Moves to the ground firing position where he aligns the front tripod leg for direction and snaps the trail legs apart.

(d) Stamps the tripod legs into the ground.

(e) If required, zeros' the T&E mechanism to the traverse bar and gun cradle.

(f) Assists the gunner in mounting the MK 19.

(2) The gunner:

(a) Dismounts the vehicle and obtains the MK 19 from its stowed position.

(b) Moves to the firing position with the MK 19.

(c) Secures the MK 19 into the gun cradle with the squad leader's assistance.

(3) The driver:

(a) Dismounts the vehicle and moves to the firing position with a can of ammunition.

(b) Assists in obtaining both the M3 tripod and the MK 19 from the stowed position.

(c) Removes the lid from the ammunition can and ensures the ogive ends of the rounds are pointed downrange.

(d) Returns to the vehicle, monitors the radio and/or takes more ammunition to the firing position, and maintains rear security.

(4) The squad leader and the gunner load the weapon as in the crew drill and place the MK 19 into action.

H-8. REMOUNT THE MK 19

The vehicle crew must be able to mount the MK 19 from the ground-mount to a vehicle-mount or stowed position within four minutes.

a. To mount the MK 19 from the ground to the vehicle the crew performs the following actions.

(1) The squad leader assists the gunner in clearing the weapon by releasing the ammunition belt from the feed tray, and placing it in the ammunition can. He then:

(a) Removes the ammunition can and places it on the ground.

(b) Releases the MK 19 from the gun cradle.

(c) Releases the T&E mechanism from the tripod traverse rod and takes off the gun cradle and T&E mechanism, after the gunner removes the MK 19.

(d) Grasps the front leg of the tripod and rotates the tripod to vertical position, releases the sleeve latch, and collapses the trail legs.

(e) Loosens the front leg clamp, folds down the front leg, and tightens the clamp.

(f) Returns to the vehicle, gives the gun cradle/T&E mechanism to the gunner, and stows the M3 tripod.

(g) Stows the binoculars and compass in the vehicle and resumes his duties.

(2) The gunner raises the top cover of the MK 19 and takes the round or spent cartridge case from the bolt fingers, and places it in the ammunition can, and closes the top cover. He then:

(a) Lifts the MK 19 off the gun cradle when the squad leader releases it and moves to the vehicle.

(b) Gives the MK 19 to the driver and mounts the vehicle.

(c) Takes the gun cradle and T&E mechanism from the squad leader and mounts the gun cradle to the pedestal.

(d) Zeros' the T&E mechanism if needed and reattaches it to the gun cradle.

(e) Attaches the T&E mechanism to the pedestal and clamps it with the train lock handle.

(f) Takes the MK 19 from the driver and installs it on the gun cradle.

(g) Attaches the travel lock assembly if movement is imminent.

(h) Takes the empty case catch bag assembly from the driver and attaches it to the gun cradle.

(3) The driver receives the MK 19 from the gunner and gives it to him when he is ready. He then:

(a) Assists the gunner in installing the MK 19 in the gun cradle.

(b) Moves to the firing position, retrieves the ammunition can and any loose equipment, and returns to the vehicle.

(c) Places the ammunition can on top of the vehicle, with the ogive ends of the rounds pointed down range.

(d) Hands the empty case catch bag assembly to the gunner.

(4) The squad leader and the gunner load the weapon (as in the crew drill) and place the MK 19 into action.

b. To dismount the MK 19 from the ground mount to a stowed position, the crew performs the following actions.

(1) The squad leader:

(a) Assists the gunner in clearing the weapon by releasing the ammunition belt from the feed tray, and places it in the ammunition can.

(b) Removes the ammunition can and places it on the ground. He then releases the MK 19 from the gun cradle.

(c) Releases the T&E mechanism from the tripod traverse rod and takes off the gun cradle/T&E mechanism after the gunner removes the MK 19.

(d) Grasps the front leg of the tripod and rotates the tripod to a vertical position, releases the sleeve latch, collapses the trail legs, loosens the front leg clamp, folds down the front leg, and tightens the clamp.

(e) Returns to the vehicle and stows the M3 tripod, T&E mechanism, binoculars, and compass.

(f) Resumes his duties.

(2) The gunner:

(a) Raises the top cover of the MK 19 and takes the round or spent cartridge case from the bolt fingers.

(b) Places the round in the ammunition can, and closes the top cover.

(c) Lifts the MK 19 off the gun cradle when the squad leader releases it and moves to the vehicle.

(d) Stows the MK 19 in its stowage position.

(3) The driver:

(a) Moves to the firing position.

(b) Collects the ammunition can and any loose equipment.

(c) Stows the ammunition can and assists the squad leader and gunner in stowing any of their equipment.

H-9. CONVERT THE M966 HMMWV TOW MISSILE CARRIER FROM MK 19 MODE TO TOW 2 MODE

To convert the M966 weapon system from the MK 19 mode to the TOW 2 mode, the crew performs the following actions.

a. The squad leader:

(1) Places the traversing unit in the traveling position.

(2) Obtains the nightsight and collimation cases from the stowed position and places them on top of the vehicle near the gunner.

(3) Stows the gun cradle/ammo bracket, T&E mechanism, and empty case catch bag assembly.

(4) Unstraps the missile guidance system (MGS), unlatches its cover, and stows the MGS in the commander's footwell.

(5) Connects the J-1 cable and the battery power conditioner cable to the MGS and MGS battery power conditioner.

(6) Connects the nightsight power conditioner cable to the TOW vehicle power cable (TVPC), uncoils the cable and hands it up to the gunner.

(7) Ensures that the power cable is connected to the TVPC, and that the TVPC switch is on.

(8) Assists the gunner in boresighting the system.

b. The gunner :

(1) Raises the top cover and releases the ammunition belt from the feed tray and places it in the ammunition can.

(2) He takes the round or spent cartridge case from the bolt fingers, places it in the ammunition can, and closes the top cover.

(3) The gunner releases the MK 19 from the gun cradle, lifts the MK 19 out of the gun cradle, and gives it to the driver.

(4) The gunner releases the T&E mechanism from the pedestal with the train lock handle.

(5) Removes the empty case catch bag assembly and gives it to the squad leader. The gunner moves the ammo can and gives it to the driver.

(6) Removes the gun cradle with the T&E mechanism attached and hands it to the squad leader.

(7) Opens the pedestal mount cover and secures it. Lifts the traversing unit with adapter from the stowage base and secures the traversing unit to the pedestal mount with the coupling clamp.

(8) Receives the launch tube from the driver and secures it to the traversing unit.

(9) Receives the optical sight from the driver and installs it on the traversing unit.

(10) Opens the nightsight case, removes nightsight device and installs it on the optical sight and hands the case to the driver.

(11) Takes the TVPC cable from the squad leader and installs the nightsight device.

(12) Opens the collimation case, removes the collimator, attaches to nightsight, and connects collimator power cable.

(13) Assists the squad leader in boresighting.

(14) Collimates sights, removes collimator and power cable, stores collimator and cable in case, and hands case to the driver.

(15) Receives the missile from the driver and loads it in the system.

c. The driver:

(1) Dismounts the vehicle and enters the right rear passenger door and then unlocks and opens the cargo hatch.

(2) Stows the MK 19 when received from the gunner.

(3) Stows the ammunition can when received from the gunner.

(4) Places the optical sight on top of the vehicle for the gunner.

(5) Unstraps a missile, inspects the missile, and gives the missile to the gunner.

(6) Ensures every item in the cargo hatch is secure, closes the cargo hatch, exits the vehicle through the right rear passenger door and takes his place as driver.

H-10. CONVERT THE M966 HMMWV TOW MISSILE CARRIER FROM TOW 2 MODE TO MK 19 MODE

This exercise requires the crew to accomplish its task within three minutes. To convert the M966 weapon system from the TOW 2 mode to the MK 19 mode and put the TOW 2 into its stowed position, the crew performs the following actions.

a. The squad leader:

(1) Disconnects the J-1 cable from the MGS, coils the cable and stows in the traversing unit.

(2) Disconnects the battery power conditioner cable from the MGS battery power cable, and winds it on the stowage handle.

(3) Disconnects the MGS battery cable, recovers the MGS cover, stows the cable in the MGS, replaces the cover on the MGS, and straps the MGS down.

(4) Disconnects the nightsight power cable from the TVPC, and winds and stows the cable.

(5) Receives the nightsight case from the gunner and stows it in the vehicle.

(6) Gives the gunner the gun cradle/ammo bracket, T&E mechanism, and empty case catch bag assembly.

b. The gunner:

(1) Arms the missile (if armed), unlocks the bridge clamp, removes the missile and hands it to the driver.

(2) Disconnects the TVPC power cable from the night sight and drops down the hatch.

(3) Disconnects the nightsight and places it in the case, and hands the case to the squad leader.

(4) Disconnects the optical sight and hands it to the driver.

(5) Disconnects the launch tube, and hands it to driver.

(6) Unclasps the traversing unit, lifts the traversing unit out of the pedestal mount, and takes it through the gunners hatch.

(7) Secures it to the TU stowage base, and closes the pedestal mount access cover.

(8) Takes the gun cradle/ammo bracket and T&E mechanism from the squad leader.

(9) Installs the cradle/ammo bracket on the heavy interdiction missile system (HIMS) mount.

(10) Ensures the T&E mechanism is zeroed, installs the T&E mechanism to the gun cradle and the pedestal mount on the HIMS.

(11) Takes the MK 19 from the driver and installs it on the gun cradle.

(12) He installs the empty case catch bag assembly on the gun cradle.

(13) Receives ammunition can from the driver, while ensuring that the ogive ends are pointing down range, and loads the MK 19.

c. The driver:

(1) Dismounts the vehicle and enters the right rear passenger door, and unlocks and opens the cargo hatch.

(2) Takes the missile from the gunner and stows it in the missile rack.

(3) Unstraps the nightsight case and gives it to the gunner.

(4) Takes the optical sight from the gunner and stows it in the vehicle.

(5) Takes the launch tube from the gunner and stows it in the vehicle.

(6) Unstraps the MK 19 from its stowage, and gives it to the gunner.

(7) Unstraps one can of ammunition and places it on top of the vehicle near the gunner.

(8) Replaces the forward handling ring and dust cover on the missile.

(9) Ensures that every item in the cargo hatch is secure and closes the cargo hatch.

(10) Exits the vehicle through the right rear passenger door and takes his place as the driver.

H-11. ENGAGE TARGETS WITH THE MK 19

The MK 19 squad must be able to successfully engage a target while occupying a firing position. This crew drill is based on the procedures found in Section III of Chapter 5, Application of Fire. The purpose of this drill is to exercise proper fire techniques to successfully engage a target while not hitting friendly troops or attempting to hit a target out of range. The crew occupies a dismounted position when the squad leader announces, "Fire mission."

a. The squad leader then performs the following tasks:

(1) Announces direction using one of the following methods: orally (example:"Right front," or "Left flank"), by a reference point (example:"TRP 1," "Lone tree."), fires tracer from a weapon or another MK 19, or by laying the gun on the target himself.

(2) Announces target description "Troops," "Tank," or if more than one type of target, designated target, "Lead truck," "Right building, far end."

(3) Announces target range, for example, "Four five zero," "One one hundred."

(4) Takes the following corrective actions, as required. If the gunner announces, "Cannot identify," the squad leader attempts to identify and gives further instructions. If the gunner announces "Lost," the squad leader gives additional target location. If the gunner announces "Friendly, the squad leader gives new target description or gives end of mission.

(5) Announces the method of engagement by announcing the class and rate of fire.

(6) Announces method of control.

(7) Issues subsequent fire commands as needed.

(8) Terminates the alert: CEASE FIRE, END OF MISSION.

b. The gunner performs the following tasks:

(1) Attempts to acquire target. If he cannot, the gunner announces "Cannot identify," "Lost," or "Friendly."

(2) When identified, announces "Identified," or "Up."

(3) Makes an adjustment to the lay of the gun as the squad leader gives new data.

c. The driver performs the following tasks:

(1) Provides security for the vehicle or position.

(2) Prepares to get more ammunition to the gun as required.

(3) Prepares to move if the squad leader announces "Cease fire," or "End of mission."

APPENDIX I

GUNNERY EXERCISES

There are seven exercises used to train MK 19 gunners and crews. The first four exercises are individual gunnery firing exercises and allow the gun crew to practice firing prior to the collective gunnery exercises. Commanders dictate the level of MOPP used during MK 19 gunnery.

These seven tables are used for single and multiple gun-crew practice and can be conducted on any MK 19 range. Each table has columns for the task, condition, standard, ammunition required, time required, and a column to record a satisfactory or unsatisfactory (SAT Y/N) score for the task.

I-1. INDIVIDUAL GUNNERY

These first four tables are used for gunnery practice. Each single gun exercise requires the gun crew to mount and dismount the MK 19, and engage point and area targets at various ranges. The two dismounted day exercises require the gun crew to prepare a range card and engage targets using data from the range card (Tables I-1 and Table I-2). The third exercise (Table I-3) is a mounted combat exercise and the fourth exercise (Table I-4) is a night fire exercise.

a. **Instructional Fire Exercise** (Table I-1). The instructional firing exercise gives the crew practice firing on a range.

ACTION	CONDITIONS	STANDARDS	AMMO	TIME	SAT Y/N
1. Mount the MK 19 on the M3 tripod.	Given a MK 19, M3 tripod, T&E mechanism, MK 64 gun cradles, and a firing position.	Gun crew must be able to mount the MK 19 in its tripod mode within 1 min.	0	1 min	
2. Prepare a range card.	Given a MK 19 mounted on a M3 tripod, and a firing position with a designated sector of fire and targets.	Gunner must prepare a completed range card within 15 min.	0	15 min	
3. Zero the MK 19 while using a 400-meter BMP target.	Given a tripod-mounted MK 19, 400-meter BMP target, selected firing position, and 4 rds of 40-mm TP.	Gunner must impact on the target with at least 2 of the 4 rds on the BMP target at 400 meters.	4 rds	N/A	
4. Engage a BMP target at 1,100 meters while using range card data (point target).	Given a tripod-mounted MK 19, completed range card, BMP target at 1,100 meters, selected firing position, and 8 rds of 40-mm TP.	Gunner must impact on the 1,100 meter BMP target with at least 2 out of 8 rounds.	8 rds	N/A	
5. Engage a BMP target at 1,500 meters while using range card data.	Given a tripod-mounted MK 19, completed range card, BMP target at 1,500 meters, selected firing position, and 8 rds of 40-mm TP.	Gunner must impact on the 1,500 meter BMP target area with at least 1 out of 8 rds.	8 rds	N/A	
6. Engage dismounted troops at 600 meters while using range card data (area target).	Given a tripod-mounted MK 19, completed range card, dismounted troop targets at 600 meters, selected firing position, and 6 rds of 40-mm TP.	Gunner must impact at least 4 out of 6 rds within 5 meters of a silhouette at a range of 600 meters.	6 rds	N/A	
7. Dismount the MK 19.	Given a MK 19, M3 tripod, and a selected firing position.	Gun crew must dismount the MK 19 from the M3 tripod within 1 min.	0	1 min	
		Total:	**26 rds**		

Exercise Scoring

Fully Trained (T) = Satisfactory on 7 of 7 tasks
Need practice (P) = Satisfactory on 5 or 6 tasks
Untrained (U) = Unsatisfactory on 4 or fewer tasks

Table I-1. Instructional fire exercise.

During scoring procedures, the grader positions himself so that he can observe both the gunner and the target. Once the exercise commences, he:

- Times the tasks having a time standard.
- Observes and informs the gunner of the strike of the rounds.
- Records the results of each task in the right hand column with a "Y" for a satisfactory completion of the task or an "N" for an unsatisfactory completion.
- Sums the results and assigns an overall score in accordance with the scores in the bottom box of the figure above.

b. **Dismounted Range Card Exercise** (Table I-2). The dismounted range card exercise is held after instructional fire and gives the gunner and crew a chance to test their skills against a time-and-hit exercise. It is held as often as the commander feels necessary to maintain crew skills.

(1) Time the gunner and crew while they place the tripod-mounted MK 19 into operation in a selected location.

(2) Time the gunner while he completes a range card with the ranges of all targets within his sector of fire.

(3) Have the gunner suppress or kill single or multiple targets in his sector of fire while using his range card data.

(4) Time the gunner and crew while they take the MK 19 out of operation.

ACTION	CONDITIONS	STANDARDS	AMMO	TIME	SAT Y/N
1. Mount the MK 19 on an M3 tripod.	Given a MK 19, M3 tripod, M64 gun mount, T&E mechanism, and a selected firing position.	Gun crew must mount the MK 19 on the M3 tripod within 1 min.	0	1 min	
2. Prepare a range card.	Given a MK 19 mounted on a M3 tripod, a firing position with an assigned sector of fire, and designated targets.	Gunner must prepare a completed range card with all targets in his sector of fire within 15 min.	0	15 min	
3. Zero the MK 19 while mounted on the M3 tripod at a 400-meter BMP.	Given a tripod-mounted MK 19, selected firing position, 400-meter BMP target, and 4 rds of 40-mm TP.	Gunner must impact at least 2 out of 4 rds on the 400-meter BMP target.	4 rds	N/A	
4. Engage a 600-meter and tripod-mounted BMP target while using range card data (point target).	Given a tripod-mounted MK 19, selected firing position, 600-meter BMP target, and 4 rds of 40-mm TP.	Gunner must impact at least 3 out of 4 rds on the 600-meter BMP target within 1 min.	4rds	1 min	

Table I-2. Dismounted range card exercise.

ACTION	CONDITIONS	STANDARDS	AMMO	TIME	SAT Y/N
5. Engage a 1,100-meter BMP target while using range card data.	Given a tripod-mounted MK 19, selected firing position, BMP target at 1,100 meters, completed range card, and 8 rds of 40-mm TP.	Gunner must impact on BMP target at 1,100 meters with at least 2 out of 8 rds within 2 min.	8 rds	2 min	
6. Engage a 600-meter (area) troop target while using the range card data.	Given a tripod-mounted MK 19, 6 rds of 40-mm TP, selected firing position, and a troop target at 600 meters.	Gunner must impact at least 4 out of 6 rds within 5 meters of a troop silhouette at a range of 600.	6 rds	1.5 min	
7. Engage a BMP target at 1,500 meters while using the range card data.	Given a tripod-mounted MK 19, 8 rds of 40-mm TP, selected firing position, BMP target at 1,500 meters, and a completed range card.	Gunner must impact on the 1,500 meter BMP target with at least 4 out of 8 rds within 2 min.	8 rds	2 min	
8. Engage a BMP target at 1,100 meters and dismounted troops at 600 meters while using range card data.	Given a tripod-mounted MK 19, 12 rds of 40-mm TP, BMP target at 1,100 meters, dismounted troop target at 600 meters, and a completed range card.	Gunner must impact at least 2 rds on the BMP target at 1,100 meters, then shift to a dismounted troop target and place at least 2 rds within 5 meters of a troop silhouette at a range of 600 meters.	12 rds	3 min	
9. Dismount the MK 19 from the M3 tripod.	Given a mounted MK 19.	Gun crew must take the MK 19 out of operation from the tripod mount within 1 min.	0	1 min	
		Total:	**42 rds**		

Exercise Scoring

Fully Trained (T) = Satisfactory on 9 of 9 tasks
Need Practice (P) = Satisfactory on 7 or 8 tasks
Untrained (U) = Unsatisfactory on 6 or fewer tasks

Table I-2. Dismounted range card exercise (continued).

During scoring procedures, the grader positions himself so that he can observe both the gunner and the target. Once the exercise commences, he:

- Times the tasks having a time standard.
- Observes and informs the gunner of the strike of the rounds.
- Records the results of each task in the right hand column with a "Y" for a satisfactory completion of the task or an "N" for an unsatisfactory completion.
- Sums the results and assigns an overall score in accordance with the scores in the bottom box of the figure above.

c. **Mounted Combat Exercise.** The mounted combat exercise gives the gunner and the crew practice fighting from a vehicle. It is held as often as the commander feels necessary to maintain crew skills (Table I-3).

(1) Time the gun crew while they place the MK 19 into operation on its organic carrier. Carriers may be HMMWVs, 2 1/2-ton trucks, M151-series utility trucks, or M113-series APCs.

(2) Using the time-and-hit standards, give the gunner and crew a series of single and multiple targets to suppress or kill in their sector of fire.

(3) Upon completion of mounted firing, time the gunner and crew while they take the MK 19 out of operation.

ACTION	CONDITIONS	STANDARDS	AMMO	TIME	SAT Y/N
1. Mount the MK 19 on its organic carrier.	Given a MK 19, organic carrier, and a selected firing position.	Gun crew must mount the MK 19 on its organic carrier within 2 min.	0	2 min	
2. Engage a BMP target at 400 meters while mounted.	Given a MK 19, organic carrier, 4 rds of 40-mm TP, and a selected firing position.	Gunner must impact on the BMP target with at least 3 out of 4 rds at 400 meters within 1 min.	4 rds	1 min	
3. Engage a BMP target at 800 meters while mounted.	Given a MK 19, organic carrier, 8 rds of 40-mm TP, and a selected firing position.	Gunner must impact a BMP target at 800 meters with at least 3 out of 8 rds within 1.5 min.	8 rds	1.5 min	
4. Engage a 600-meter (area) target.	Given a MK 19, organic carrier, 6 rds of 40-mm TP, and a selected firing position.	Gunner must impact at least 4 out of 6 rds within 5 meters of a troop silhouette at a range of 600 meters.	6 rds	1.5 min	
5. Engage a BMP target at 1,100 meters.	Given a MK 19, organic carrier, 8 rds of 40-mm TP, and a selected firing position.	Gunner must impact the BMP target area at 1,100 meters with at least 2 out of 8 rds within 2 min.	8 rds	2 min	
6. Dismount the MK 19 from its organic carrier.	Given a MK 19 and an organic carrier.	Gun crew must dismount the MK 19 from its organic carrier within 2 min.	0	2 min	
			Total: 26 rds		

Exercise Scoring

Fully Trained (T) = Satisfactory on 6 of 6 tasks
Need Practice (P) = Satisfactory on 5 tasks
Untrained (U) = Unsatisfactory on 4 or fewer tasks

Table I-3. Mounted combat exercise.

During scoring procedures, the grader positions himself so that he can observe both the gunner and the target. Once the exercise commences, he:

- Times the tasks having a time standard.
- Observes and informs the gunner of the strike of the rounds.
- Records the results of each task in the right hand column with a "Y" for satisfactory completion of the task or an "N" for an unsatisfactory completion.
- Sums the results and assigns an overall score in accordance with the scores in the bottom box of the figure above.

d. **Night Fire Exercise** (Table I-4). This LFX develops and improves MK 19 gunnery skills at night or during degraded conditions. This exercise is conducted as often as the commander feels it is necessary to maintain crew skills.

ACTION	CONDITIONS	STANDARDS	AMMO	TIME	SAT Y/N
1. Mount the AN/TVS-5 to the MK 19.	At night, given an AN/TVS-5 complete, a MK 19, M3 tripod, and a T&E mechanism.	Gunner must mount the AN/TVS-5 on the MK 19 within 2 min.	0	2 min	
2. Engage a 400-meter BMP target using a zeroed MK 19 with an AN/TVS-5 mounted for observation.	At night, given an AN/TVS-5 complete, a MK 19, an M3 tripod, 6 rds of 40-mm TP, and a 400-meter BMP target, during the hours of daylight.	Gunner must impact 3 out of 6 rds on a BMP target at 400 meters within 1 min.	6 rds	1 min	
3. Engage a BMP target at 600 meters.	At night, given a MK 19, 8 rds of 40-mm TP, and a target at a distance of 600 meters.	Gunner must impact on a BMP target with at least 4 out of 8 rds at 600 meters within 2 min.	8 rds	2 min	
4. Engage a dismounted troop target at 800 meters.	At night, given a MK 19, 8 rds of 40-mm TP, and a target at a distance of 800 meters.	Gunner must impact within 5 meters of a silhouette target with at least 4 out of 8 rds at 800 meters within 2 min.	8 rds	2 min	
5. Dismount AN/TVS-5 from the MK 19.	At night, given a MK 19 and an AN/TVS-5.	Gun crew must take the AN/TVS-5 out of operation within 2 min.	0	2 min	
			Total: 22 rds		

Exercise Scoring
Fully Trained (T) = Satisfactory on 5 of 5 tasks **Need Practice (P)** = Satisfactory on 4 tasks **Untrained (U)** = Unsatisfactory on 3 or fewer tasks

Table I-4. Night fire exercise.

During scoring procedures, the grader positions himself so that he can observe both the gunner and the target. Once the exercise commences, he:

- Times the tasks having a time standard.
- Observes and informs the gunner of the strike of the rounds.
- Records the results of each task in the right hand column with a "Y" for a satisfactory completion of the task or a "N" for an unsatisfactory completion.

- Sums the results and assigns an overall score in accordance with the scores in the bottom box of the figure above.

I-2. COLLECTIVE GUNNERY

These three exercises test the application of fire from two or more guns against linear, deep, linear with depth, and area targets. They test collective individual skills, fire control, leader skills, adjustment of fire, methods of target engagement, and the control of one or more fire units. Tables I-5 through I-7 should be used with unit LFXs. They can be held on any range or piece of terrain that supports the MK 19. Targets may be made up of 55-gallon drums, hulks of old vehicles, silhouettes, chalk, panel markers, or engineer tape.

ACTION	CONDITIONS	STANDARDS	AMMO	TIME	SAT Y/N
1. Engage a linear target with a pair of MK 19s (two guns). (see also Task 2)	Given a linear target 100 meters wide from 800 to 1,200 meters, two MK 19s on tripods with T&E mechanisms, during the hours of daylight or under degraded conditions.	Gun crews must cover 90 percent of the linear target with effective fire while performing the following: 1. The leader lays the guns on his respective flanks. 2. The leader issues the fire command for engaging a linear target. 3. Gunners engage using traversing fire. 4. Observers ensure that fires remain on the target and adjust appropriately.	40 rds	2 min	
2. Engage a linear target with two pairs of MK 19s (four guns).	Given a linear target 200 meters wide from 800 to 1,200 meters, four MK 19s on tripods with T&E mechanisms, during the hours of daylight or under degraded conditions.	Same as Task 1, except the leader may have to subdivide the target.	80 rds	3 min	

Table I-5. Linear and deep targets.

ACTION	CONDITIONS	STANDARDS	AMMO	TIME	SAT Y/N
3. Engage a deep target with a pair of MK 19s (two guns).	Given a deep target 100 meters long with midrange to the target from 800 to 1,200 meters, two MK 19s on tripods with T&E mechanisms, during the hours of daylight or under degraded conditions.	Gun crews must cover 90 percent of the deep target with effective fire while performing the following: 1. The leader lays the guns on his respective ends of the target. 2. The leader issues the fire command for engaging a deep target. 3. Gunners engage using searching fire. 4. Observers ensure that fire remains on the target and adjust appropriately.	40 rds	2 min	
4. Engage a deep target with two pairs of MK 19s (four guns).	Given a deep target 200 meters long with midrange to the target from 800 to 1,200 meters, four MK 19s on tripods with T&E mechanisms, during the hours of daylight or under degraded conditions.	Gun crews should cover 90 percent of the deep target with effective fire while performing the same steps in task 3, except the leader may have to subdivide the target.	80 rds	3 min	
			Total: 240 rds		

Table I-5. Linear and deep targets (continued).

ACTION	CONDITIONS	STANDARDS	AMMO	TIME	SAT Y/N
1. Engage a linear target with depth using a pair of MK 19s (two guns).	Given a target 100 meters wide and 50 meters long (or 50 meters wide and 100 meters long), from 800 to 1,200 meters, two MK 19s with tripods and T&E mechanisms, during daylight or under degraded conditions.	Gun crews must cover 80 percent of the linear with depth target with effective fire while performing the following: 1. The leader lays the guns on the gun crews' respective flanks (ends) of the target. 2. The leader issues the fire command for engaging a linear target with depth. 3. Gunners engage using traversing and searching fire. 4. Observers ensure that fire remains on the target and adjust appropriately.	40 rds for each target.	3 min	
2. Engage a linear target with depth using two pairs of MK 19s (four guns).	Given a target 200 meters wide and 50 meters long (or 50 meters wide and 200 meters long), from 800 to 1,200 meters, four MK 19s with tripods and T&E mechanisms, during daylight or under degraded conditions.	Same as Task 1, except the leader may want to subdivide the target.	80 rds for each target.	4 min	
			Total: 120 rds		

Table I-6. Linear targets with depth.

ACTION	CONDITIONS	STANDARDS	AMMO	TIME	TPU
1. Engage an area target with two pairs of MK 19s (four guns).	Given an area target from 1,000 to 1,400 meters (midrange) extending over 200 meters in length and width, four MK 19s on tripods with T & E mechanisms, during the hours of daylight or under degraded conditions.	The combined gun crews must cover 70 percent of the area target with suppressive fire while performing the following: 1. The leader determines direction and elevation to the area target (midrange). 2. The leader divides the area target and ensures that each gun is laid on its portion, and assigns TRPs. 3. Each gun registers on its respective TRP. 4. The leader issues the fire command for an area target. 5. All MK 19s engage; gunners use traverse and search manipulations to adjust and distribute fires. 6. Observers ensure the fires remain on target and adjust appropriately.	160 rds	5 min	
2. Engage an area target with three pairs of MK 19s (six guns).	Given an area target from 1,000 to 1,400 meters (midrange), extending over 300 meters in length and 200 meters in width, six MK 19s on tripods with T&E mechanisms, during the hours of daylight or under degraded conditions.	The combined gun crews must cover 70 percent of the area target with suppressive fire while performing the following: (Performance measures are the same as for Task 1.)	240 rds	8 min	
			Total: 400 rds		

Table I-7. Area targets.

GLOSSARY

ACRONYMS AND ABBREVIATIONS

AAR	after-action review
AE	angle of elevation
AI	assistant instructor
AIT	advanced individual training
APC	armored personnel carrier
AS	angle of sight
BMP	fighting vehicle manufactured by the former Soviet Union
BOD	bore obstruction detector
BRDM	Soviet Union vehicle
C4	composition four
CALFEX	combined arms live-fire exercise
CLP	cleaner, lubricant, preservative
cm	centimeter
DA	Department of the Army
deg	degrees
DODIC	Department of Defense Identification Code
EA	engagement area
elev	elevation
EOD	explosive ordinance disposal
FM	field manual
FOV	field of view
fps	feet per second
ft	feet
FTX	field training exercise
GMD	grease, molybdenum disulfide
HE	high explosive
HEDP	high-explosive dual-purpose
HIMS	heavy interdiction missile system
HMMWV	high-mobility multipurpose wheeled vehicle
HWTS	heavy weapon thermal sight
IAW	in accordance with
IR	infrared
JP-4	jet propulsion fuel

LAW	lubricant, automatic weapons, arctic
LBE	load-bearing equipment
LED	light emitting diode
LFX	live-fire exercise
LH	left hand
LSA	lubricant oil, semifluid, automatic weapon
LSAT	lubricant oil, semifluid, automatic weapons, temperate
MACS	multipurpose arcade combat simulator
MGS	missile guidance system
min	minute
mm	millimeter
MOD	model
MOPP	mission-oriented protective posture
MPMG	multipurpose machine gun
MPRC	multipurpose range complex
MTP	mission training plan
MWTS	medium weapon thermal sight
NBC	nuclear, biological, chemical
NCO	noncommissioned officer
NCOES	noncommissioned officers education system
NFOV	narrow field of view
No.	number
NVD	night vision device
PEWS	platoon early warning system
PIBD	point initiating, base detonating
PMCS	preventive maintenance checks and services
QE	quadrant elevation
RBC	rifle bore cleaner
rds	rounds
RH	right hand
RP	reference point
RPG	rocket propelled grenade
sec	seconds
SOP	standing operating procedure
STX	situational training exercise
T&E	traverse and elevating
TESS	tactical engagement simulator system

TM	technical manual
TOE	table of organization and equipment
TP	training practice
TPIAL	target pointer illuminator aiming light
TPT	training practice trainer
TRADOC	U.S. Army Training and Doctrine Command
TRP	target reference point
TVPC	TOW vehicle power cable
TWS	thermal weapon sight
VDT	video disk trainer
W	width
WFOV	wide field of view

REFERENCES

SOURCES USED

These are the sources quoted or paraphrased in this publication.

DA Cir 350-38	Standards in Weapons Training. 3 July 1997.
DA Form 5517-R	Standard Range Cards. February 1986.
DA PAM 385-63	Range Safety. 19 May 2003.
FM 3-22.68	Crew-Served Machine Guns, 5.56-mm and 7.62-mm. 31 January 2003.
FM 3-25.26	Map Reading and Land Navigation. 20 July 2001.
FM 25-2	Unit Training Management. 10 September 1984.
FM 25-3	Training in Units. 12 December 1984.
TC 25-8/ with change	Training Ranges. 16 November 1992.

DOCUMENTS NEEDED

These documents must be available to the intended users of this publication.

* DA Form 5517-R	Standard Range Cards (LRA). February 1986.
DA Form 7518-R	MK 19 Day Practice and Qualification (Hull Targets).
DA Form 7519-R	MK 19 Night Practice and Qualification with Hull Targets and Night Vision Devices.
DA Form 7520-R	MK 19 Day Practice and Qualification (Pop-up Targets).
DA Form 7521-R	MK 19 Night Practice and Qualification with Pop-up Targets or without Night Vision Devices.
* TC 23-13	Crew-Served Weapon Night Vision Sight. 20 January 1967.

* This source was also used to develop this production.

* TM 9-1005-245-13&P/ with 5 changes	Operator's, Unit, and Direct Support Maintenance Manual with Repair Parts and Special Tools List for Machine Gun Mounts and Combinations for Tactical/Armored Vehicles and Ground Mounting M122A1 Machine Gun Tripod Mount, M3 Tripod Mount, M66 Machine Gun Mount Ring, M240E1 Flexible Machine Gun Mount, M142 Machine Gun Mount, .50 Caliber Machine Gun Mount. 22 March 2002.
* TM 9-1010-230-10	Operator's Manual for Machine Gun, 40-mm, MK 19, MOD 3. 30 May 2001.
* TM 9-1010-231-13&P	Operator's, Organizational and Intermediate Maintenance with Repair Parts List and Components List Mount, Machine Gun MK 64: MOD 4, 5, 6, 7, 8 used with Machine Gun, MK 19 MOD 3, 40-mm Machine Gun, M2, Caliber .50 Machine Gun, M60 Machine Gun, 7.62-mm Machine Gun, M60E3, 7.62-mm. 31 October 2000.
* TM 11-5855-214-10	Operator's Manual for Night Vision Sight, Crew-Served Weapon, AN/TVS-5. 15 February 1989.
* TM 11-5855-302-12&P	Operator's and Unit Maintenance Manual, Sight, Thermal, AN/PAS-13 (V2), AN/PAS-13 (V3). 1 January 1997.
* TM 11-5855-308-12&P	Operator's and Unit Maintenance Manual Target Pointer Illuminator/Aiming Light (TPIAL) AN/PEQ-2A. 15 May 2000.
* TM 11-5855-312-10	Operator's Manual, Sight, Thermal AN/PAS- 13B (V2), AN/PAS-13B (V3). 5 January 2001.
* TM 11-5860-201-10	Operator's Manual for Laser Infrared Observation Set, AN/GVS-5. 2 February 1982.

READINGS RECOMMENDED

These readings contain relevant supplemental information.

DoD Reg 5100-76-M	Physical Security of Sensitive Conventional Arms, Ammunition, and Explosives. August 2000.
FM 3-5/ with change	NBC Decontamination. 31 January 2002.
FM 3-21.71	Mechanized Infantry Platoon and Squad (Bradley). 20 August 2002.

* This source was also used to develop this production.

FM 5-250	Explosives and Demolitions. 30 July 1998.

FM 7-8/ with change	Infantry Rifle Platoon and Squad. 22 April 1992.
FM 7-20/ with change	The Infantry Battalion. 29 December 2000.
FM 31-70	Basic Cold Weather Manual. 12 April 1968.

INTERNET WEBSITES

Reimer Digital Library, http://www.adtdl.army.mil.

USAMC Logistics Support Activity, http://www.logsa.army.mil/.

Army Publishing Directorate, http://www.usapa.army.mil.

INDEX

MK 19, 40-mm GRENADE MACHINE GUN, MOD 3 FIRING TABLE I DAY PRACTICE AND QUALIFICATION WITH HULL TARGETS SCORECARD

For use of this form, see FM 3-22.27; the proponent agency is TRADOC.

PRIVACY ACT STATEMENT

AUTHORITY: 10 USC 3012(g)/Executive order 9397
PRINCIPAL PURPOSE: To aid individual training on targets at various ranges.
ROUTINE USES: To evaluate individual proficiency.
DISCLOSURE: Voluntary. However, mass rating and recording require some tracking method.

1a. LAST NAME	1b. FIRST NAME	1c. MI	2. RANK	3. UNIT

TABLE I (A). DISMOUNTED AND MOUNTED DAY PRACTICE

4. RANGE	5. LANE	6. GRADER	7. DATE

TASK	RANGE (Meters)	AMMO	TIME (Minutes)	ENGAGEMENT STANDARDS	GO	NO GO
ZERO	400	4	NA	2 ROUNDS HIT		
2	1,100 (+/- 200)	8	2.5	2 ROUNDS HIT		
3	1,500 (+/- 200)	10	3.5	2 ROUNDS HIT		
4	600 (+/- 100)	6	2	2 ROUNDS HIT		
5	800 (+/- 100)	6	2	2 ROUNDS HIT		
6	400	4	1.5	2 ROUNDS HIT		
MULTIPLE TARGETS						
7	1,100 (+/- 200)	10	4	1 ROUND HIT		
8	600 (+/- 100)			1 ROUND HIT		
9	800 (+/- 100)	14	4.5	1 ROUND HIT		
10	1,500 (+/- 200)			1 ROUND HIT		
				TOTALS		

8. COMMENTS

9. NUMBER OF ENGAGEMENT MET *(Choose One)*

☐ 10 - EXPERT ☐ 8-7 - MARKSMAN
☐ 9 - SHARPSHOOTER ☐ 6 AND BELOW - UNQUALIFIED

10. GUNNER'S SIGNATURE	11. GRADER'S SIGNATURE

TABLE I (B). DISMOUNTED AND MOUNTED DAY QUALIFICATION

12. RANGE	13. LANE	14. GRADER	15. DATE

TASK	RANGE (Meters)	AMMO	TIME (Minutes)	ENGAGEMENT STANDARDS	GO	NO GO
ZERO	400	4	NA	2 ROUNDS HIT		
2	1,100 (+/- 200)	8	2	2 ROUNDS HIT		
3	1,500 (+/- 200)	10	3	2 ROUNDS HIT		
4	600 (+/- 100)	6	1.5	2 ROUNDS HIT		
5	800 (+/- 100)	6	1.5	2 ROUNDS HIT		
6	400	4	1	2 ROUNDS HIT		
MULTIPLE TARGETS						
7	1,100 (+/- 200)	10	3.5	1 ROUND HIT		
8	600 (+/- 100)			1 ROUND HIT		
9	800 (+/- 100)	14	4	1 ROUND HIT		
10	1,500 (+/- 200)			1 ROUND HIT		
				TOTALS		

16. COMMENTS

17. NUMBER OF ENGAGEMENT MET *(Choose One)*

☐ 10 - EXPERT ☐ 8-7 - MARKSMAN
☐ 9 - SHARPSHOOTER ☐ 6 AND BELOW - UNQUALIFIED

18. GUNNER'S SIGNATURE	19. GRADER'S SIGNATURE

DA FORM 7518-R, AUG 2003 APD V1.00

MK 19, 40-mm GRENADE MACHINE GUN, MOD 3 FIRING TABLE II
NIGHT PRACTICE AND QUALIFICATION WITH HULL TARGETS SCORECARD

For use of this form, see FM 3-22.27; the proponent agency is TRADOC.

PRIVACY ACT STATEMENT

AUTHORITY: 10 USC 3012(g)/Executive order 9397
PRINCIPAL PURPOSE: To aid individual training on targets at various ranges.
ROUTINE USES: To evaluate individual proficiency.
DISCLOSURE: Voluntary. However, mass rating and recording require some tracking method.

NOTE: Use these tables if the targets being used are hulls, and any of the following applies:
AN/PEQ-2A mounted on the TWS mounting bracket.
AN/PAS-13 mounted on the TWS mounting bracket.
AN/TVS-5 with the 3d generation tube mounted on the TWS mounting bracket.

1a. LAST NAME	1b. FIRST NAME	1c. MI	2. RANK	3. UNIT

TABLE II (A). DISMOUNTED AND MOUNTED NIGHT PRACTICE

4. RANGE	5. LANE	6. GRADER	7. DATE

TASK	RANGE (Meters)	AMMO	TIME (Minutes)	ENGAGEMENT STANDARDS	GO	NO GO
ZERO	400	4	NA	2 ROUNDS HIT		
2	1,100 (+/- 200)	8	2.5	2 ROUNDS HIT		
3	1,500 (+/- 200)	10	3.5	2 ROUNDS HIT		
4	600 (+/- 100)	6	2	2 ROUNDS HIT		
5	800 (+/- 100)	6	2	2 ROUNDS HIT		
6	400	4	1.5	2 ROUNDS HIT		
MULTIPLE TARGETS						
7	1,100 (+/- 200)	10	4	1 ROUND HIT		
8	600 (+/- 100)			1 ROUND HIT		
9	800 (+/- 100)	14	4.5	1 ROUND HIT		
10	1,500 (+/- 200)			1 ROUND HIT		
				TOTALS		

8. TYPE DEVICE *(Choose One)*
☐ AN/PEQ-2A ☐ AN/PAS-13 ☐ AN/TVS-5

9. NUMBER OF ENGAGEMENT MET *(Choose One)*
☐ 10 - EXPERT ☐ 8-7 - MARKSMAN
☐ 9 - SHARPSHOOTER ☐ 6 AND BELOW - UNQUALIFIED

10. GUNNER'S SIGNATURE	11. GRADER'S SIGNATURE

TABLE II (B). DISMOUNTED AND MOUNTED NIGHT QUALIFICATION

12. RANGE	13. LANE	14. GRADER	15. DATE

TASK	RANGE (Meters)	AMMO	TIME (Minutes)	ENGAGEMENT STANDARDS	GO	NO GO
ZERO	400	4	NA	2 ROUNDS HIT		
2	1,100 (+/- 200)	8	2	2 ROUNDS HIT		
3	1,500 (+/- 200)	10	3	2 ROUNDS HIT		
4	600 (+/- 100)	6	1.5	2 ROUNDS HIT		
5	800 (+/- 100)	6	1.5	2 ROUNDS HIT		
6	400	4	1	2 ROUNDS HIT		
MULTIPLE TARGETS						
7	1,100 (+/- 200)	10	3.5	1 ROUND HIT		
8	600 (+/- 100)			1 ROUND HIT		
9	800 (+/- 100)	14	4	1 ROUND HIT		
10	1,500 (+/- 200)			1 ROUND HIT		
				TOTALS		

16. TYPE DEVICE *(Choose One)*
☐ AN/PEQ-2A ☐ AN/PAS-13 ☐ AN/TVS-5

17. NUMBER OF ENGAGEMENT MET *(Choose One)*
☐ 10 - EXPERT ☐ 8-7 - MARKSMAN
☐ 9 - SHARPSHOOTER ☐ 6 AND BELOW - UNQUALIFIED

18. GUNNER'S SIGNATURE	19. GRADER'S SIGNATURE

MK 19, 40-mm GRENADE MACHINE GUN, MOD 3 FIRING TABLE III
DAY PRACTICE AND QUALIFICATION WITH POP-UP TARGETS SCORECARD

For use of this form, see FM 3-22.27; the proponent agency is TRADOC.

PRIVACY ACT STATEMENT

AUTHORITY: 10 USC 3012(g)/Executive order 9397
PRINCIPAL PURPOSE: To aid individual training on targets at various ranges.
ROUTINE USES: To evaluate individual proficiency.
DISCLOSURE: Voluntary. However, mass rating and recording require some tracking method.

1a. LAST NAME	1b. FIRST NAME	1c. MI	2. RANK	3. UNIT

TABLE III (A). DISMOUNTED AND MOUNTED DAY PRACTICE

4. RANGE	5. LANE	6. GRADER	7. DATE

TASK	RANGE (Meters)	AMMO	TIME (Minutes)	ENGAGEMENT STANDARDS	GO	NO GO
ZERO	400	4	NA	2 ROUNDS HIT		
2	600 (+/- 100)	6	2	1 ROUND HIT		
3	800 (+/- 100)	8	2.5	1 ROUND HIT		
4	400	4	1.5	1 ROUND HIT		
MULTIPLE TARGETS						
5	800 (+/- 100)	12	4	1 ROUND HIT		
6	400			1 ROUND HIT		
7	400	10	3	1 ROUND HIT		
8	600 (+/- 100)			1 ROUND HIT		
				TOTALS		

8. COMMENTS

9. NUMBER OF ENGAGEMENT MET *(Choose One)*

☐ 8 - EXPERT ☐ 6 - MARKSMAN
☐ 7 - SHARPSHOOTER ☐ 5 AND BELOW - UNQUALIFIED

10. GUNNER'S SIGNATURE

11. GRADER'S SIGNATURE

TABLE III (B). DISMOUNTED AND MOUNTED DAY QUALIFICATION

12. RANGE	13. LANE	14. GRADER	15. DATE

TASK	RANGE (Meters)	AMMO	TIME (Minutes)	ENGAGEMENT STANDARDS	GO	NO GO
ZERO	400	4	NA	2 ROUNDS HIT		
2	600 (+/- 100)	6	1.5	1 ROUND HIT		
3	800 (+/- 100)	8	2	1 ROUND HIT		
4	400	4	1	1 ROUND HIT		
MULTIPLE TARGETS						
5	800 (+/- 100)	12	3.5	1 ROUND HIT		
6	400			1 ROUND HIT		
7	400	10	2.5	1 ROUND HIT		
8	600 (+/- 100)			1 ROUND HIT		
				TOTALS		

16. COMMENTS

17. NUMBER OF ENGAGEMENT MET *(Choose One)*

☐ 8 - EXPERT ☐ 6 - MARKSMAN
☐ 7 - SHARPSHOOTER ☐ 5 AND BELOW - UNQUALIFIED

18. GUNNER'S SIGNATURE

19. GRADER'S SIGNATURE

MK 19, 40-mm GRENADE MACHINE GUN, MOD 3 FIRING TABLE IV
NIGHT PRACTICE AND QUALIFICATION WITH POP-UP TARGETS SCORECARD

For use of this form, see FM 3-22.27; the proponent agency is TRADOC.

PRIVACY ACT STATEMENT

AUTHORITY: 10 USC 3012(g)/Executive order 9397
PRINCIPAL PURPOSE: To aid individual training on targets at various ranges.
ROUTINE USES: To evaluate individual proficiency.
DISCLOSURE: Voluntary. However, mass rating and recording require some tracking method.

NOTE: Use this table if you do not have a MK 19/sight combination that applies to Table II, and you are using hull targets OR if targets are pop-up/E type silhouettes.

1a. LAST NAME	1b. FIRST NAME	1c. MI	2. RANK	3. UNIT

TABLE IV (A). DISMOUNTED AND MOUNTED NIGHT PRACTICE

4. RANGE	5. LANE	6. GRADER	7. DATE

TASK	RANGE (Meters)	AMMO	TIME (Minutes)	ENGAGEMENT STANDARDS	GO	NO GO
ZERO	400	4	NA	2 ROUNDS HIT		
2	600 (+/- 100)	6	2	2 ROUNDS HIT		
3	800 (+/- 100)	6	2	2 ROUNDS HIT		
4	400	4	1.5	2 ROUNDS HIT		
MULTIPLE TARGETS						
5	800 (+/- 100)	10	3.5	1 ROUND HIT		
6	400			1 ROUND HIT		
7	400	14	2.5	1 ROUND HIT		
8	600 (+/- 100)			1 ROUND HIT		
				TOTALS		

8. COMMENTS

9. NUMBER OF ENGAGEMENT MET *(Choose One)*

- ☐ 8 - EXPERT
- ☐ 6 - MARKSMAN
- ☐ 7 - SHARPSHOOTER
- ☐ 5 AND BELOW - UNQUALIFIED

10. GUNNER'S SIGNATURE	11. GRADER'S SIGNATURE

TABLE IV (B). DISMOUNTED AND MOUNTED NIGHT QUALIFICATION

12. RANGE	13. LANE	14. GRADER	15. DATE

TASK	RANGE (Meters)	AMMO	TIME (Minutes)	ENGAGEMENT STANDARDS	GO	NO GO
ZERO	400	4	NA	2 ROUNDS HIT		
2	600 (+/- 100)	6	1.5	2 ROUNDS HIT		
3	800 (+/- 100)	6	1.5	2 ROUNDS HIT		
4	400	4	1	2 ROUNDS HIT		
MULTIPLE TARGETS						
5	800 (+/- 100)	10	3	1 ROUND HIT		
6	400			1 ROUND HIT		
7	400	14	2	1 ROUND HIT		
8	600 (+/- 100)			1 ROUND HIT		
				TOTALS		

16. COMMENTS

17. NUMBER OF ENGAGEMENT MET *(Choose One)*

- ☐ 8 - EXPERT
- ☐ 6 - MARKSMAN
- ☐ 7 - SHARPSHOOTER
- ☐ 5 AND BELOW - UNQUALIFIED

18. GUNNER'S SIGNATURE	19. GRADER'S SIGNATURE

Made in United States
Cleveland, OH
05 November 2026

47672471R00327